Studies in Computational Intelligence 416

For further volumes:
http://www.springer.com/series/7092

Aleksander Byrski, Zuzana Oplatková,
Marco Carvalho, and Marek Kisiel-Dorohinicki (Eds.)

Advances in Intelligent Modelling and Simulation

Simulation Tools and Applications

Editors
Aleksander Byrski
AGH University of Science and Technology
Kraków
Poland

Marco Carvalho
Florida Institute of Technology
Melbourne, FL
USA

Zuzana Oplatková
Faculty of Applied Informatics
Tomas Bata University in Zlín
Zlín
Czech Republic

Marek Kisiel-Dorohinicki
AGH University of Science and Technology
Kraków
Poland

ISSN 1860-949X
e-ISSN 1860-9503
ISBN 978-3-642-44743-3
ISBN 978-3-642-28888-3 (eBook)
DOI 10.1007/978-3-642-28888-3
Springer Heidelberg New York Dordrecht London

Softcover reprint of the hardcover 1st edition 2012

Printed on acid-free paper

Springer is part of Springer Science+Business Media (www.springer.com)

Preface

The human capacity to abstract complex systems and phenomena into simplified models has played a critical role in the rapid evolution of our modern industrial processes and scientific research. As a science and an art, Modelling and Simulation have been one of the core enablers of this remarkable human trace, and have become a topic of great importance for researchers and practitioners.

In the last several years, the increasing availability of massive computational resources, and interconnectivity has helped fuel tremendous advances in the field, collapsing previous barriers and redefining new horizons for its theories, capabilities and applications.

The acceptance of current modelling and simulation practices is directly related to the growing quality and capability of modern simulation frameworks, tools and applications. As more advanced and capable tools become available, new applications and solutions will emerge to join the growing list of success stories.

This book was created to compile some of the most recent concepts, advances, challenges and ideas associated with Intelligent Modelling and Simulation frameworks, tools and applications. Our goal was to build from key insights and discussions taken place at that meeting to create a volume containing the state of the art in the area of Intelligent Simulation.

After an extensive review of the selected current publications and ongoing research efforts in the field, the editors of this book have identified a set of representative topics to compose the volume. The focus was on applied modelling and simulation tools and techniques, as well as applications and case studies.

The book covers a diverse set of topics on simulation tools and applications. The first chapter discusses the important aspects of a human interaction and the correct interpretation of results during simulations. The second chapter gets to the heart of the analysis of entrepreneurship by means of agent-based modelling and simulations. The following three chapters bring together the central theme of simulation frameworks, first describing an agent-based simulation framework, then a simulator for electrical machines, and finally an airborne network emulation environment.

The two subsequent chapters discuss power distribution networks from different points of view—anticipation and optimization of multi-echelon inventory policy.

After that, the book includes also a group of chapters discussing the mathematical modelling supported by verification simulations, and a set of chapters with models synthesised by means of artificial intelligence tools and complex automata framework. Lastly, the book includes a chapter introducing the use of graph-grammar model for generation of three-dimensional computational meshes and a chapter focused on the experimental and computational results regarding simulation of aero engine vortexes.

We are grateful to all contributors of this book, for their willingness to work on this project. We thank the authors for their interesting proposals of the book chapters, their time and efforts that helped in making this volume a valuable reference to researchers and practitioners in the field, as well as an inspiration to those interested in the area of Intelligent Modelling and Simulation.

We would like also to express our thanks to reviewers, who have helped us to ensure high quality of this volume. We gratefully acknowledge their time, assistance and valuable remarks and comments.

Our special thanks go to Prof. Janusz Kacprzyk, editor-in-chief of the Springer's Studies in Computational Intelligence Series, to Dr. Thomas Ditzinger, Senior Editor and to Mr. Holger Schäpe, Editorial Assistant of Springer Verlag in Heidelberg, for their assistance and excellent cooperative collaboration in this book project.

Kraków, Zlín and Melbourne,
January 2012

Aleksander Byrski
Zuzana Oplatková
Marco Carvalho
Marek Kisiel-Dorohinicki

Contents

List of Contributors

Marco Arguedas
Institute for Human and Machine Cognition,
15 SE Osceola Ave, Ocala, FL 34471, USA;
e-mail: marguedas@ihmc.us

İnci Batmaz
Department of Statistics,
Middle East Technical University,
06531 Ankara, Turkey;
e-mail: ibatmaz@metu.edu.tr

Aleksander Byrski
AGH University of Science and Technology,
Al. Mickiewicza 30, 30-059 Kraków, Poland;
e-mail: olekb@agh.edu.pl

Marco Carvalho
Florida Institute of Technology,
150 W. University Blvd. Melbourne, FL 32901, USA,
and
Institute for Human and Machine Cognition,
15 SE Osceola Ave, Ocala, FL 34471, USA;
e-mail: mcarvalho@fit.edu

Virgil Chindriş
Department of Electrical Machines and Drives,
Technical University of Cluj,
RO-400114 Cluj, 28, Memorandumului, Romania;
e-mail: virgil.chindris@mae.utcluj.ro

Wojciech Czech
AGH University of Science and Technology,
Al. Mickiewicza 30, 30-059 Kraków, Poland;
e-mail: czech@agh.edu.pl

Witold Dzwinel
AGH University of Science and Technology,
Al. Mickiewicza 30, 30-059 Kraków, Poland;
e-mail: dzwinel@agh.edu.pl

Łukasz Faber
AGH University of Science and Technology,
Al. Mickiewicza 30, 30-059 Kraków, Poland;
e-mail: faber@student.agh.edu.pl

Adrián Granados
Institute for Human and Machine Cognition,
15 SE Osceola Ave, Ocala, FL 34471, USA;
e-mail: agranados@ihmc.us

Daniel Hague
U.S. Air Force Research Laboratory,
26 Electronic Parkway, Rome NY, USA;
e-mail: daniel.hague@rl.af.mil

Wei Hua Ho
Department of Mechanical and Industrial Engineering,
University of South Africa,
Private Bag X6, Florida 1710, South Africa;
e-mail: howh@unisa.ac.za

Martin Ihrig
Snider Entrepreneurial Research Center,
The Wharton School, University of Pennsylvania,
418 Vance Hall, 3733 Spruce Street, Philadelphia, PA 19104, USA;
e-mail: ihrig@wharton.upenn.edu

Roman Jašek
Tomas Bata University in Zlín,
Faculty of Applied Informatics,
Nam. T.G. Masaryka 5555, 760 01 Zlín, Czech Republic;
e-mail: jasek@fai.utb.cz

Elçin Kartal-Koç
Department of Statistics,
Middle East Technical University,
06531 Ankara, Turkey;
e-mail: kartalelcin@gmail.com

Elif Kayış
Middle East Technical University,
Institute of Applied Mathematics, 06531, Ankara, Turkey;
e-mail: elif.kayis@metu.edu.tr

Marek Kisiel-Dorohinicki
AGH University of Science and Technology,
Al. Mickiewicza 30, 30-059 Kraków, Poland;
e-mail: doroh@agh.edu.pl

Katja Klingebiel
Chair of Factory Organisation,
Technical University of Dortmund,
Leonhard-Euler-Str. 5, 44227 Dortmund, Germany;
e-mail: katja.klingebiel@tu-dortmund.de

Cong Li
Department of Supply Chain Engineering,
Fraunhofer Institute for Material Flow and Logistics,
Joseph-von-Fraunhofer StraSSe 2-4, 44227 Dortmund, Germany;
e-mail: cong.li@iml.fraunhofer.de

Michael Muccio
U.S. Air Force Research Laboratory,
26 Electronic Parkway, Rome NY, USA;
e-mail: michael.muccio@rl.af.mil

Pavel Nahodil
Department of Cybernetics, Faculty of Electrical Engineering,
Czech Technical University in Prague,
Technická 2, 166 27 Prague, Czech Republic;
e-mail: nahodil@fel.cvut.cz

Gaby Neumann
Technical University of Applied Sciences,
Bahnhofstraße, 15745 Wildau, Germany;
e-mail: gaby.neumann@th-wildau.de

Zuzana Oplatková
Tomas Bata University in Zlín,
Faculty of Applied Informatics,
Nam. T.G. Masaryka 5555, 760 01 Zlín, Czech Republic;
e-mail: oplatkova@fai.utb.cz

Anna Paszyńska
Jagiellonian University,
ul. Reymonta 4, 30-059 Kraków, Poland;
e-mail: anna.paszynska@uj.edu.pl

Maciej Paszyński
AGH University of Science and Technology,
Al. Mickiewicza 30, 30-059 Kraków, Poland;
e-mail: paszynsk@agh.edu.pl

Carlos Pérez
Institute for Human and Machine Cognition,
15 SE Osceola Ave, Ocala, FL 34471, USA;
e-mail: cperez@ihmc.us

Kamil Piętak
AGH University of Science and Technology,
Al. Mickiewicza 30, 30-059 Kraków, Poland;
e-mail: kpietak@agh.edu.pl

Brendon Poland
U.S. Air Force Research Laboratory,
26 Electronic Parkway, Rome NY, USA;
e-mail: brendon.poland@rl.af.mil

Vilda Purutçuoğlu
Middle East Technical University,
Department of Statistics, 06531, Ankara, Turkey;
e-mail: vpurutcu@metu.edu.tr

Mircea Ruba
Department of Electrical Machines and Drives,
Technical University of Cluj,
RO-400114 Cluj, 28, Memorandumului, Romania;
e-mail: mircea.ruba@mae.utcluj.ro

Robert Schaefer
AGH University of Science and Technology,
Al. Mickiewicza 30, 30-059 Kraków, Poland;
e-mail: schaefer@agh.edu.pl

Roman Šenkeřík
Tomas Bata University in Zlín,
Faculty of Applied Informatics,
Nam. T.G. Masaryka 5555, 760 01 Zlín, Czech Republic;
e-mail: senkerik@fai.utb.cz

Joseph Suprenant
U.S. Air Force Research Laboratory,
26 Electronic Parkway, Rome NY, USA;
e-mail: joseph.suprenant@rl.af.mil

Loránd Szabó
Department of Electrical Machines and Drives,
Technical University of Cluj,
RO-400114 Cluj, 28, Memorandumului, Romania;
e-mail: Lorand.Szabo@mae.utcluj.ro

Rareş Terec
Department of Electrical Machines and Drives,
Technical University of Cluj,
RO-400114 Cluj, 28, Memorandumului, Romania;
e-mail: rares.terec@mae.utcluj.ro

Juri Tolujew
Fraunhofer Institute for Factory Operation and Automation IFF,
Postfach 1453, 39004 Magdeburg, Germany;
e-mail: juri.tolujew@iff.fraunhofer.de

Jaroslav Vitků
Department of Cybernetics, Faculty of Electrical Engineering,
Czech Technical University in Prague,
Technická 2, 166 27 Prague, Czech Republic;
e-mail: vitkujar@fel.cvut.cz

Gerhard-Wilhelm Weber
Middle East Technical University,
Institute of Applied Mathematics, 06531, Ankara, Turkey;
e-mail: gweber@metu.edu.tr

Ivan Zelinka
Technical University of Ostrava,
Faculty of Electrical Engineering and Computer Science,
17. listopadu 15, 708 33 Ostrava-Poruba, Czech Republic;
e-mail: ivan.zelinka@vsb.cz

Requirements and Solutions for a More Knowledgeable User-Model Dialogue in Applied Simulation

Gaby Neumann and Juri Tolujew

Abstract. Generally, it is pretty clear and widely accepted that the human actor plays a significant role in any simulation project—although in recent years some authors proclaimed a revival of ideas for a human-free simulation at least related to distinct parts of a simulation study. Therefore, the paper aims to provide an overview on needs and challenges in model-user interaction. Furtheron, approaches, methods and tools are presented that support the user in bringing in his/her knowledge in all phases of a simulation project from model building via understanding a model and using it for experimentation to correctly interpreting simulation outcome. In addition to this, concepts are introduced and demonstrated for more flexible, but generic tools supporting comfortable simulation output analysis in a context beyond pre-defined simulation goals. With this the paper wants to contribute to bringing simulation tools and algorithms on one side and the simulation user—novice and expert—on the other closer together in order to achieve a true dialogue and exchange within a discrete event simulation supportive infrastructure.

1 Introduction and Motivation

One of today's challenges in applied simulation consists in seeing it in the context of human-centered processes. Here, simulation needs to be understood in its entire characterisation as complex problem-solving, collaboration, knowledge-generation and learning process at the same time. This view is in line with literature characterising modelling and simulation in general as both, knowledge-processing activity

Gaby Neumann
Technical University of Applied Sciences, Bahnhofstraße, 15745 Wildau, Germany
e-mail: gaby.neumann@th-wildau.de

Juri Tolujew
Fraunhofer Institute for Factory Operation and Automation IFF,
Postfach 1453, 39004 Magdeburg, Germany
e-mail: juri.tolujew@iff.fraunhofer.de

A. Byrski et al. (Eds.): Advances in Intelligent Modelling and Simulation, SCI 416, pp. 1–26.
springerlink.com

and goal-directed knowledge-generation activity [18]. Based upon this, advanced methodologists and technologists were expected to be allowed to integrate simulation with several other knowledge techniques. But looking at today's situation in simulation projects it still has to be considered that a sound application of a knowledge management perspective to modelling and simulation is still missing. Instead, the term "knowledge-based simulation" is typically used for applying AI approaches to automatically create simulation models from expert knowledge as it can be found in [8].

Research focuses, for example, on developing efficient and robust models and formats to capture, represent and organize the knowledge for developing conceptual simulation models which can be generalized and interfaced with different applications and implementation tools [30]. Other work aims to develop concepts for modelling human decisions, e.g., in manufacturing systems [31]. Furthermore, ongoing research is targeted towards modelling and simulation of human behaviour to support workplace design by use of digital human models [23]. Latest research works for advanced modelling of human behaviour by incorporating the emotional level in terms of fear, aggressiveness, fatigue and stress into simulation models in order to receive more realistic human action patterns in particularly challenging situations [7].

In all of these approaches expert knowledge is modeled or represented by logical rules even though they might be fuzzy, i.e. in a formalized way. In contrast to this, the fact that non-formalized expert knowledge also finds its way into the simulation model on one hand or is created throughout the simulation lifecycle and needs to be externalized on the other is not in the focus of research in this field. That is why, information about decisions taken when building the model or running experiments quite often stay in the heads of the people involved in the project. The same applies to really new knowledge about the particular application or even about the simulation methodology that is gained in the course of a simulation project. Typically this experience-based knowledge is even forgotten before being reused. Furthermore, the simulation model itself also forms a kind of dynamic repository containing knowledge about parameters, causal relations and decision rules gathered through purposeful experiments. This knowledge is being somewhat hidden in the model as long as not being discovered, understood and interpreted by another person.

Against this background, research on implementing a knowledge management perspective in simulation projects should address the following questions:

- Which information and knowledge is needed by whom at what stage of a simulation project?
- Which knowledge and information is provided by whom in which step of a simulation project?
- Which knowledge is generated with whom in which step of the simulation project?
- Which knowledge is "stored" in the conceptual and simulation models, evolves from simulation experiments, and is "hidden" in the input/output data of simulation runs?

- How simulation knowledge with the different stakeholders or repositories can be accessed, extracted, externalized and distributed, shared, applied?

In order to contribute to this research the paper particularly investigates further the needs and approaches for a true dialogue between the simulation user on one hand and the simulation tool and algorithm on the other. For this the impact of the user on the simulation project is being demonstrated before looking into simulation projects from a knowledge management perspective. Here, logistics simulation projects are used as example. Based upon this, a general framework for user-model interaction is elaborated that particularly focuses on activities for model-based experimentation and simulation output interpretation. Finally, research findings are summarized, conclusions derived from them are presented, and needs for further research are identified.

2 Knowledge and Knowledge Management in Simulation Projects

Knowledge is generally defined as reasoning about information and data to actively enable performance, problem-solving, decision making, learning, and teaching [4]. In simulation as in any other kind of problem-solving this knowledge is to be related to both the subject of the simulation study and the procedure of the simulation project. In general, simulation knowledge can be described as entirety of specific or generalized theoretical or experienced knowledge about the simulation problem and its solution (subject-related knowledge), but also about the procedure and organisation of the simulation project (procedure-related knowledge) that either explicitly or implicitly exists or is created in the course of the simulation project [21, pp. 435–468]. Thus, simulation knowledge combines aspects from a variety of subjects and simulation projects require a respective collection of interdisciplinary expertise.

Therefore, simulation projects are typically organized in the form of a collaborative service involving both, simulation experts and domain experts with individual knowledge to be of use at certain stages of the project [16]:

- *Simulation experts* bring in knowledge and prior project experience concerning their typical area of simulation application, underlying theoretical and applied simulation methodology and tools, and procedures, methods and tools for simulation project management. Consequently, they are primarily responsible for model building and implementation steps.
- *Domain experts* provide knowledge on the domain the problem under investigation is settled down in and especially problem-specific knowledge. Therefore, they are mainly involved in problem description, identification of input data and evaluation/interpretation of results.

The simulation expert typically thinks in rather abstract knowledge categories derived from generalising experiences from prior simulation projects or matching modelling philosophy and syntax of the simulation tool used for model implementation and experimentation. This way of thinking and the terminology used to express it can then be found in the simulation model. In contrast to this the domain expert thinks more in concrete terms of the particular (simulation) problem and its context which not necessarily finds its direct expression in the simulation model.

Consequently, a major task in any simulation project consists in bringing both worlds together and creating a sound basis of approved mutual understanding. In the *problem specification and analysis stage* of the simulation project the domain expert has to explain in detail what the simulation model should reproduce, whilst the simulation expert inevitably has to work on his or her own understanding of the situation and problem right from the beginning by using techniques for systematic analysis and abstraction. To generalize, multifaceted expert background knowledge forms the basis for any serious simulation project (Fig. 1). Project-specific input information usually come with the tender specification or are to be identified and generated in the problem analysis phase of the simulation, i.e. during problem definition and data collection. Here, it needs to be decided (and brought in) what is to be taken into consideration for model building and which information is required for the investigation. In the end, the outcome of this phase already forms first results in the simulation project and at the same time the framework for all further steps. Proceeding within this framework again and again requires more input and background information based upon the knowledge and experience of the users, i.e. the simulation expert and the domain expert, but also creates knowledge with them and within the products of the simulation project.

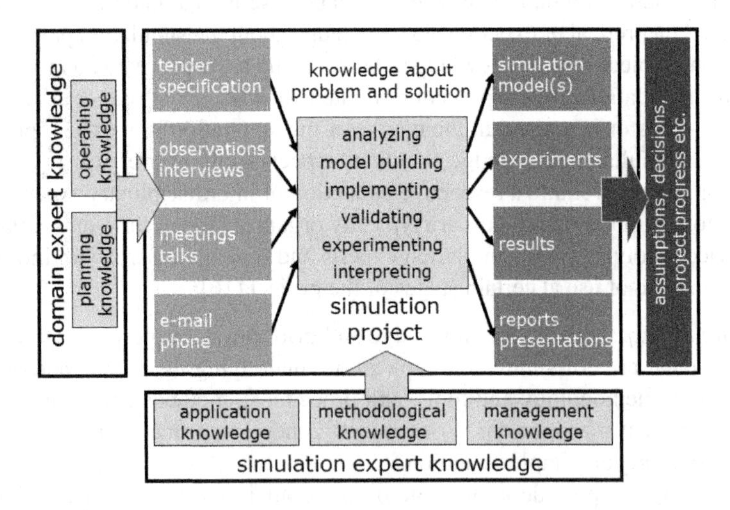

Fig. 1 Sources and evolution in simulation knowledge

Consequently, *model building, implementation and validation processes* should be seen as other important phases of collecting, evaluating and structuring information. In the course of a simulation project a series of models is being built to represent the object of investigation in different ways [26]: The conceptual model informally (textually or graphically) specifies the simulation problem, its objects and their behaviour, data structures etc. The formalized model applies a particular modelling paradigm for formal specification of simulation objects (i.e. data types, functions). Finally, the (executable) computer model allows dynamic experimentation for problem solving. Each of these models forms a specific milestone in the simulation lifecycle. It does not represent the problem and its components only, but also what currently is known about its solution. The simulation model resulting from those efforts is developed, modified, used, evaluated and extended within an ongoing process. It therefore must not be seen as a tool necessary to achieve certain objectives of experimentation and cognition only. In the end, the knowledge stored in the simulation model can be considered proven, independently of whether it was developed by the domain expert him- or herself or by the simulation expert [14]. It refers to modelling settings and assumptions like system boundaries and interaction with its environment, system elements and their links, their level of detail in representation, input data, probability distributions etc. And it also covers model implementation decisions, such as the simulation tool to be used, model components to be used, modelling tricks to be applied etc. Therefore, this knowledge specifies HOW a simulation model looks like in greater detail and relates it to WHY it looks like this. It typically comes with the model building or implementing person and indirectly also with the developer of simulation software.

The "knowledgeable" simulation model is used for systematically "creating" even further knowledge through simulation based on a systematic design of experiments (including a meaningful definition of parameters and strategies) and an intelligent interpretation of results. In principle, simulation runs provide results which are admittedly subject to randomness, but their statistical reliability can be increased to the desired extent by an appropriately large number of replications (samples). With this, guaranteed (abstract) knowledge about quantitative dependencies, times etc. can be gained that has to be related to the real application. In the end deepened, readily generalized knowledge about dynamic behaviour, relations and dependencies is produced—if correctly interpreting simulation output. For this it is necessary to understand what the objectives, parameters and procedures of a certain series of experiments were and to relate them to the results and findings.

Here, the circle of knowledge providing and knowledge accessing is closed; as elaborated it finds its expression in the products of a simulation project, i.e. simulation model(s), experiments, results and eventually project report or presentation (Fig. 1). At the same time really new knowledge is created accordingly with both the domain expert and the simulation expert. The domain expert learns about methodology, steps and tools of simulation; the simulation expert gains enhanced knowledge and experience in the application area and its specific problems from the joint developmental process [16]. The simulation model then mediates between the two types of experts.

With regard to its accessibility, knowledge stored in the simulation model can be described as being of both explicit and implicit natures. According to [17] *implicit or tacit knowledge* is that kind of knowledge (i) that a person carries in his or her mind often not even being aware of it, (ii) that the person cannot express and (iii) that is therefore not directly externalisable. It is typically based on experiences and expressed in the form of action patterns and intuitive decisions. As a consequence the simulation model as result of applying this implicit knowledge somehow carries at least parts of it without allowing direct access. Indirect access might eventually be possible via analysing a person's model building and implementation behaviour and interpreting observations. This way of accessing simulation knowledge then is more a kind of knowledge explication strategy helping a person to recognize and discover own decisions and actions him/herself. But this is difficult and very challenging if possible at all without direct involvement of the primary knowledge stakeholder. In contrast to this explicit knowledge is that kind of knowledge that (i) exists independently of a person e. g. in the form of any document, (ii) has been or can be articulated, codified, stored and accessed by other persons, and (iii) might be transferred into rules and algorithms. Therefore, this kind of knowledge is (quite) easily accessible for being reused in another simulation project or shared with other persons, no matter if they are experts or novices in simulation or experts from the application area. If not articulated yet, this knowledge is still carried by the person, but could be externalized (and therefore accessed) by use of the right means.

Corresponding to the bi-directional flows of knowledge—from the user into the model and vice versa—this kind of purposeful knowledge creation and application requires continuous interaction between simulation user (no matter what kind of expert) and simulation model (no matter at which stage of model development). But to what extent the simulation user and/or simulation model design does influence simulation outcome?

3 Case Study on the Human Impact in Simulation Projects

In order to investigate this and demonstrate the role of the user within a simulation project, the same problem description and input data might be given to different types of users for building the simulation model, running experiments and deriving simulation results. These users might vary in their domain-specific background (e.g. logistics or computing expert, engineer or management person, simulation service provider) and experience (e.g. novice or expert), but they all should run the full simulation project from problem analysis to interpretation of simulation outcome. Here, it should be left up to the users which analysis or modelling tools or simulation package they might use. Based upon the finalized project comparative analysis can be run in order to understand what findings and recommendation from the simulation have in common and which differences are caused by the individual approaches.

A first example of such a comparative case has been run and analysed by [15]. Here, the same simulation problem from the logistics application area was given to two individually working groups of students from different domains and different German universities. The first group was composed of computing students from the University of Hamburg, whereas the second group was formed by logistics students from the University of Magdeburg. All students had already a certain simulation background corresponding to their educational profile. The simulation problem they had to deal with aimed to identify the performance limit of two different designs and two varying operational scenarios for automated stacking at high-performance container terminals (Fig. 2). Both groups of students were to run a complete simulation project from problem analysis to output interpretation.

	Saanen/ Valkengoed (2005)	Computing students	Logistics students
Twin RMGs	49/53*	45 (d) 47.9 (n)	55.3 (d) 47.6 (n)
Cross-over RMGs	49/53/45*	49.4 (d) 68 (n)	50.6 (d) 61.5 (n)

d/n = day-time/night-time scenario

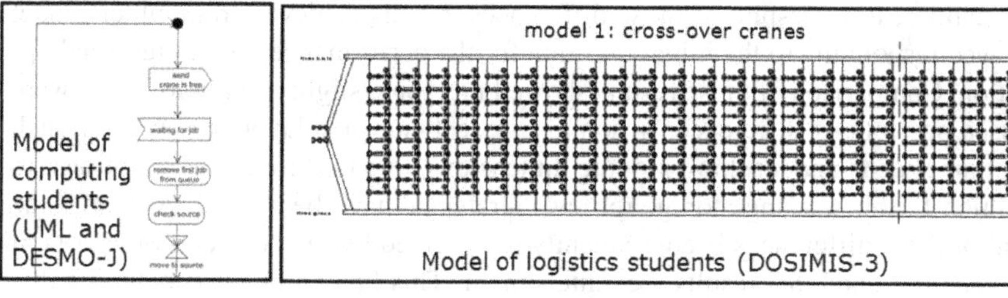

Fig. 2 The impact of the user: a case study

The main differences between the approaches in both cases consist in the preparation of model implementation and the simulation models themselves. The logistics students (so to say as domain experts) started with a detailed analytical investigation of the system and process to gain a clear understanding of the situation and more detailed specification of the problem. From this they derived a conceptual model which they had in their minds when starting to implement the model using the DOSIMIS-3 simulation package (for further information on this tool see [21, pp. 435-468]. At no stage of model development they documented it in any formalized way. For model implementation they used predefined building blocks which already represented a particular amount of functionality and logic. Due to this, they had to deal with certain limitations (or at least special challenges) in modelling. Therefore they spent many thoughts in advance on what really is needed to be represented in the simulation model at what level of detail. This led to a number of simplifications, for example in the representation of the stacking cranes' movements on the basis of a detailed understanding on the material flow backgrounds of the simulation problem.

In contrast to this the computing students focused on a very detailed representation of the stacking cranes' movements including precise tracking of the cranes' positions while moving, but they set aside representation of the storage places and individual containers. This also required detailed modelling of the cranes' management and control system and algorithms, whereas neither warehouse nor stock management was needed. Although these students had not to deal with the large number of storage locations, detailed representation of the stacking cranes required a lot of extra efforts, especially because of the complex and complicated control algorithms. Here, neither limitations nor specific restrictions have been set by the simulation tool used (Java-based programming within the DESMO-J framework as described in [21]). Therefore, the students were not forced to re-think about what really would be required to be represented in the model.

As comparison with a similar study from literature [24] shows both student projects produced valid and usable simulation models. However, efforts for model implementation, for modification eventually necessary in the course of experimentation and for visualisation of results were quite different. Although the students were not highly experienced simulation experts yet which of course had an influence on the effectiveness of the model building process, those different modes of approaching and solving a simulation problem can be found in more professional simulations, too. Despite of those differences, results achieved from either model allowed responding to the initial question for the performance limit in terminal operation. Comparison of those results showed just some slight deviations. They were caused possibly by differences in some basic technical and layout parameters, such as crane speeds and stacking module dimensions, due to different assumptions. In addition to this varying storage/retrieval strategies have been used. Although the total of these differences is consequently represented by the deviation of results, it also can be stated that results are quite similar. This finally allows concluding that despite of different modelling approaches comparable simulation results of similar quality could be achieved.

The case study gave proof of the fact that different persons of different backgrounds might produce different but in the same way correct and usable simulation models of the same problem and situation just because of their individual knowledge and experiences. From this it can be concluded that the individual background of the user significantly impacts several aspects of a simulation (Fig. 3):

- Different modelling approaches and the use of different simulation tools result in very different ways of achieving the (same) intended outcome.
- Simulation models are quite as individual as, for example, resulting trace files or collected statistical results are—especially what concerns level-of-detail.

That is why the objectives of a simulation and the questions to be answered by experiments should already be taken into consideration when designing the conceptual model. Specific opportunities and features offered by the selected simulation tool (and the user's capability in working with it) then influence transformation of the conceptual model into the computer model, when it comes to model implementation. As demonstrated the final appearance of a simulation model strongly depends on the modelling experience and philosophy of the modelling person. Based upon this model and following the user's experimentation strategy simulation output is being produced. Its interpretation and understanding again requires direct involvement of and input by the user.

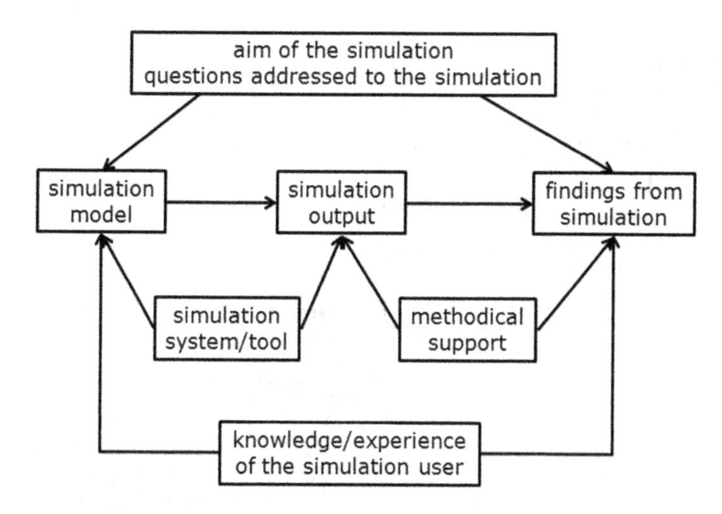

Fig. 3 Impact of the simulation user on the outcome of a simulation project

Consequently, it is worth to go into more detail when discussing about how knowledge-based support can be provided to simulation projects. For this, needs and challenges in model-user interaction are identified first, before approaches for mediating between user and model in a simulation project are presented.

4 Needs and Challenges in Model-User Interaction

According to previous discussions key success factor of any simulation project is an ongoing interaction of the simulation user (simulation expert or domain expert) with the simulation model. The kind of and need for interaction varies with the evolution of the simulation model (Fig. 4).

4.1 Model-User Interaction for Model-Building

In the beginning of the simulation project it is crucial to transfer the user's intentions with regard to the simulation and understanding of what needs to be modelled at what level-of-detail into the textual or graphical representation of the conceptual model. Due to its informal representation, appearance of the conceptual model typically depends on the user's individual preferences (and what the domain area is used to). It is required to be suitable for discussions between domain expert and simulation expert which also form the basis for its validation. The conceptual model itself is a means for mediating between the different backgrounds and experiences the individual experts involved might have. In its final version it can therefore be characterized as jointly agreed, explicated and therefore easy-accessing knowledge on model boundaries, components, level-of-detail and objectives of the simulation.

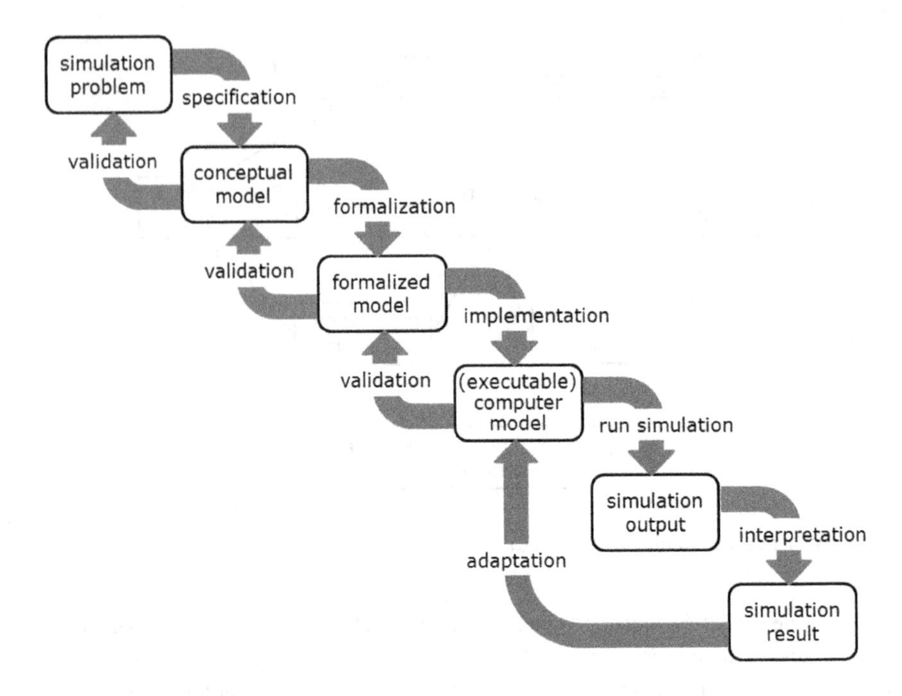

Fig. 4 Model-user interaction throughout a simulation project

Translation of the conceptual model into a *formalised model* is a typical task for the simulation expert. As s/he applies a certain modelling paradigm the resulting model might not directly be understandable to the domain expert anymore. Even further, it might already represent some tacit knowledge from both sources the simulation expert and the formalisation method/tool. Consequently, model validation requires particular expertise with the user on one hand and particular accessibility to the knowledge stored in the model on the other.

The *computer model* results from implementing the formalised model using a certain simulation tool. Nowadays, simulation packages that provide components for model design and support for gathering simulation output data are often the tool of choice. As those tools are typically designed to match modelling and analysis requirements of a specific application area, a wide range of knowledge implicitly comes with them. Consequently this knowledge is to be found in the (executable) simulation model as well. Although having been externalized once when included in the simulation package it now appears as being of tacit nature within the simulation model again. This is because the developer of the simulation package is (usually) a person different from the one (e.g., the simulation expert) who uses it for implementing the simulation model. That is why expert knowledge for model validation but also for experimentation is required again. Furthermore, support for accessing the knowledge stored in the structure, parameters and logic of the simulation model is needed.

Especially the latter is a challenge as the knowledge flow into the simulation model is usually not very well documented. To be used when the results of the simulation project are put into practice, it needs to be explained in such a way as to be accessible to the domain expert in the subject-specific terminology and to be applicable without any loss of information or misrepresentation. Otherwise the technical or organisational solution in the real world cannot be expected to work in the way demonstrated by the respective simulation model. To this day the translation of simulation knowledge into the domain expert's terminology or even directly into the source code of, for example, a control system or a control station remains an unsolved problem. Thus one continues to find a break and discontinuity in applying simulation results. Knowledge important for the realisation of simulated functionality is lost and needs to be re-developed by renewed implementation and testing.

4.2 Model-User Interaction for Deriving Simulation Results

In the further course of a simulation project simulation users, i.e. domain expert and simulation expert, use to face the ever challenging task to interpret numerous and diverse data in a way being correct with respect to the underlying subject of the simulation study and directly meeting its context. These data are usually produced and more or less clearly presented by the simulation tool in the form of trace files, condensed statistics and performance measures derived from them, graphical representations or animation. Problems mainly consist in:

- Clearly specifying questions the domain experts needs to get answered;
- Purposefully choosing measures and selecting data enabling the simulation expert to reply to the domain expert's questions;
- Processing and interpreting data and measures according to the application area and simulation problem.

To overcome these problems and give support in defining simulation goals and understanding simulation results, methods and tools are required that are easy to use and able to mediate between knowledge and understanding of the domain expert (the expert who plans or operates the process and system to be simulated) and the simulation expert (the expert from the point of view of data and their representation inside computers). Within this context, it is worth thinking in more detail about what a domain expert might look for when analysing the outcome of simulation experiments [21, pp.435-468]:

- *Typical events*. The domain expert specifically looks for moments at which a defined situation occurs. This kind of query can be related, for example, to the point in time at which the first or last or a specific object enters or leaves the system in total or an element in particular.
- *Typical phases*. The domain expert is especially interested in periods characterized by a particular situation. In this case s/he asks for the duration of the warm-up period, for the period of time the system, an element or object is in a particular state, or how long a particular change of state takes.

- *Statements*. The domain expert looks for the global characteristics of processes, system dynamics or object flows such as process type (e.g. steady-state, seasonal changes, terminating/non-terminating), performance parameters of resources (e.g., throughput, utilisation, availability), parameters of object flows (e.g. mix of sorts, inter-arrival times, processing times).

Those rather generic information are related to the particular purpose of a simulation run or series of runs, i.e. either model validation or experimentation. Whereas validation of the computer model still belongs to the model building phase of the simulation project, experimentation uses the valid simulation model as tool for different types of investigations:

- In a *what-if* analysis the user discovers how a system reacts on changing conditions or performance requirements, i.e. system loads. During experimentation a particular type of changes is introduced to the model in a systematic way in order to understand sensitivity of a certain parameter, design or strategy.
- A *what-to-do-to-achieve* study aims to answer questions like how to set system parameters or how to improve process control in order to reach a certain behaviour or performance level. Experimentation might be multidimensional including different types of changes to the model; it is strongly oriented towards identifying modification strategies for reaching a particular performance objective or target behaviour.
- *Performance optimisation* experiments serve to solve a particular target function such as minimising job orders' time in system or stock level, maximising service level or resources' utilisation, etc. Here, the limits of typical performance characteristics are to be identified with the respective limit value itself forming the goal of the investigation.

No matter which type of investigation is on the agenda the user always needs to interact with the model in order to implement the intended experimentation strategy and to gain simulation results (Fig. 4):

- *Interaction prior to the simulation run* (or a batch of simulation runs) might consist in
  - adjusting the structure of the simulation model,
  - purposefully changing one or more model or simulation parameters,
  - inserting observers for measuring and gathering system/process states and characteristics, or even simply
  - starting the simulation in order to produce and collect simulation output data that are expected to be of use for the investigation.
- *Post-run interaction* focuses on accessing and dealing with simulation output data in the form of
  - dynamic visualisation (i.e. watching animations) or
  - statistical analysis (i.e. checking original or condensed data, viewing diagrams or other types of graphical representation)

  in order to achieve findings with regard to the focus and aim of the investigation.

Consequently, the entire interaction cycle can be characterized as a *user-model dialogue*: In the model building phase there is a sequence of interrelated activities for design (model creation) and for checking (model validation). In the experimentation phase any pre-run interaction with the model corresponds to the concept of asking questions; post-run interaction is adequate to the concept of responding to questions. Pre-condition for a successful user-model dialogue is true understanding in both directions. The simulation model needs to "understand" what the user is interested in and looking for. This requires the ability to ask the right questions from the user. Those questions might either be very specific and clearly matching "the language of the model" (i.e. directly addressing input/output data of a simulation) or they are of more principle, general, eventually even fuzzy nature requiring a kind of translation for being understandable to the model. When it comes to the responding part of the dialogue the user needs to understand the simulation output for getting the answers s/he was looking for.

4.3 Needs for Supporting Model-User Interaction

Summarising needs and challenges in model-user interaction as discussed in this section, it can be concluded that model-user interaction must be a bidirectional dialogue covering all phases of the simulation project. Particular support is needed for:

- bringing in a user's simulation knowledge for simulation target definition, model building, experimentation and interpretation of simulation output;
- accessing simulation knowledge directly or indirectly stored in the products of a simulation project, i.e. model, experiments, results, reports and presentations.

Each of these aspects does not only refer to the user as expert directly involved in the particular simulation project. It also takes into consideration the need for sharing knowledge and experience gained throughout a project with others who might not have been involved in this particular simulation project.

To cope with those challenges new conceptual-procedural approaches are required as well as new types of model-user interfaces and support provided by the simulation tool or surrounding environments. The following section deals with the question for how those approaches, interfaces and support functionality might look like and presents proposals from literature and the authors' own work.

5 Approaches for Mediating between User and Model in a Simulation Project

To illustrate and demonstrate opportunities and chances for a more knowledgeable user-model dialogue typical knowledge management activities within a simulation project have been selected: accessing knowledge from a repository on one hand and using knowledge for planning, running and analysing experiments on the other.

For this it is necessary to take into consideration general simulation methodology, application-specific knowledge and project-specific intentions.

5.1 Methods and Tools for Accessing Simulation Knowledge

The necessity to comprehensively document a simulation project and clearly describe all details of a simulation model is commonly accepted and has led to a number of research and development activities in this area. Here, first focus was put on simulation model documentation standards.

An early feasibility study concluded that: (i) model documentation and specification should be accomplished through a top-down analysis; (ii) a language for model specification and description is required; and (iii) the description of model dynamics remains the major obstacle to simulation model description [12].

A taxonomy approach to this proposes a two-part documentation system [10]: A written record of all aspects of the simulation model development and operation is intended to serve as a communication vehicle among those working on and with the model and with this also as a quality control instrument. It should provide easily read information about the simulation model outsider interested in either learning more about the model or in possibly using the model. The second part, a classification system, should allow representing the simulation model in a numerical system similar to ISBN. This development culminated in the Modelling and Simulation Body of Knowledge (M&SBOK) as proposed by Ören [19].

All of these approaches do well represent general model building and implementation aspects (i.e. the methodology), but do not or hardly deal with details and specifics of the respective application area the model is settled down in. This criticism remains valid with later stages of research for consistent simulation documentation.

On the way towards a simulation model development environment Balci and Nance understand documentation as simulation model specification to achieve a valid dynamic representation [2]. With this they strengthen the principle of inseparability of documentation from specification, but they do not pay attention to the other role of simulation—being a collection of expert knowledge on the why and how of model development.

The challenge of gaining and proving a simulation model's validity and credibility initiated the most recent discussions on simulation model and/or project documentation [26, 3, 5, 20]. They are still quite often linked to software engineering views without any link to the simulation model's domain although a wider range audience of simulation documentation had already been identified [27, 6]: (i) the original model developer or simulation expert; (ii) a new engineer, simulation user, researcher interested in studying or re-using the model; and (iii) others who want to understand the main frame, but not necessarily all modelling details. This latter group might also include the customers of a simulation project, i.e. the domain experts. According to this, even different ways of model documentation ranging from UML (Unified Modelling Language) diagrams via IDEF0 (Integrated Computer

Aided Manufacturing DEFinition) conceptual documentation to animation means are suggested. But this way just the form of representation is changed whereas the background on assumptions, decisions etc. is still hidden. Since documentation deals with storing experience and communicating it or making it accessible to others, a language or transfer medium needs to be chosen that delivers the knowledge stored in the model in a way the target audience understands best.

Summarising the current state-of-research with regard to supporting knowledge explication in simulation projects, it can be stated that approaches to document domain specific and methodological simulation knowledge in parallel, a comprehensive approach for seamlessly integrating respective functionality into the simulation tool as well as a formalized documentation structure are still missing.

Furthermore, all participants in a simulation project, i.e. simulation experts and domain experts, need to be encouraged (and supported) permanently to provide background knowledge about his or her motivation for:

- going in one rather than the other direction;
- changing the model structure or parameters in a certain way;
- keeping a particular type of possible solution and abandoning others;
- looking for information, knowledge and support from one source instead of another.

This helps to identify (i) who knows what about the object of investigation, but also about the simulation project behind it, (ii) why something was decided in which way, (iii) which system configuration and which set of parameters worked how well together, (iv) what is represented in the simulation model, and (v) what are the limitations of its validity and usability. With this, the process of a simulation project could become a process of knowledge creation and acquisition at the same time.

This procedure will only work if being integrated into the "normal" simulation activities directly and as seamlessly as possible—with no or little extra effort. Techniques like structured documentation, continuous exchange or ongoing reflection and generalisation help to cope with this and to master complexity and dynamics to the benefit of both a certain simulation project in particular and simulation methodology in general. From the technical point of view comfortable interfaces and links to the simulation tools need to be developed. Additionally, an extended functionality for commenting and annotation is directly to be integrated into them. Ideally, the main part of the project knowledge to be documented is directly (and in this way hidden to the persons involved in the simulation project) taken from documents and models which are produced in the project anyway as well as from tools which are commonly used for experimentation purposes or statistical analysis of simulation outcomes.

5.2 Structured Documentation for Formalising Simulation Knowledge

As basis for continuous and consequent documentation of the simulation model and the entire simulation project a common and even formalisable documentation framework is proposed (Fig. 5). It is adjustable not just to the specific simulation problem, but especially to the current state-of-knowledge with each aspect and thus can simultaneously host descriptions of different levels-of-detail for different aspects. Here, a simulation project meeting is used as example for specifying knowledge and input to be documented: apart from categories with typical information on a meeting, such as location, time, participants etc., this framework allows collecting detailed knowledge about the particular simulation project. For the latter, subject-related knowledge is clearly separated from procedure-related knowledge with individual sub-categories for both knowledge aspects, i.e. a documentation and criticism part with each knowledge category. Whereas the documentation part of procedure-related knowledge represents the main aspects of a simulation-based problem-solving process in general, the documentation part of the subject-related

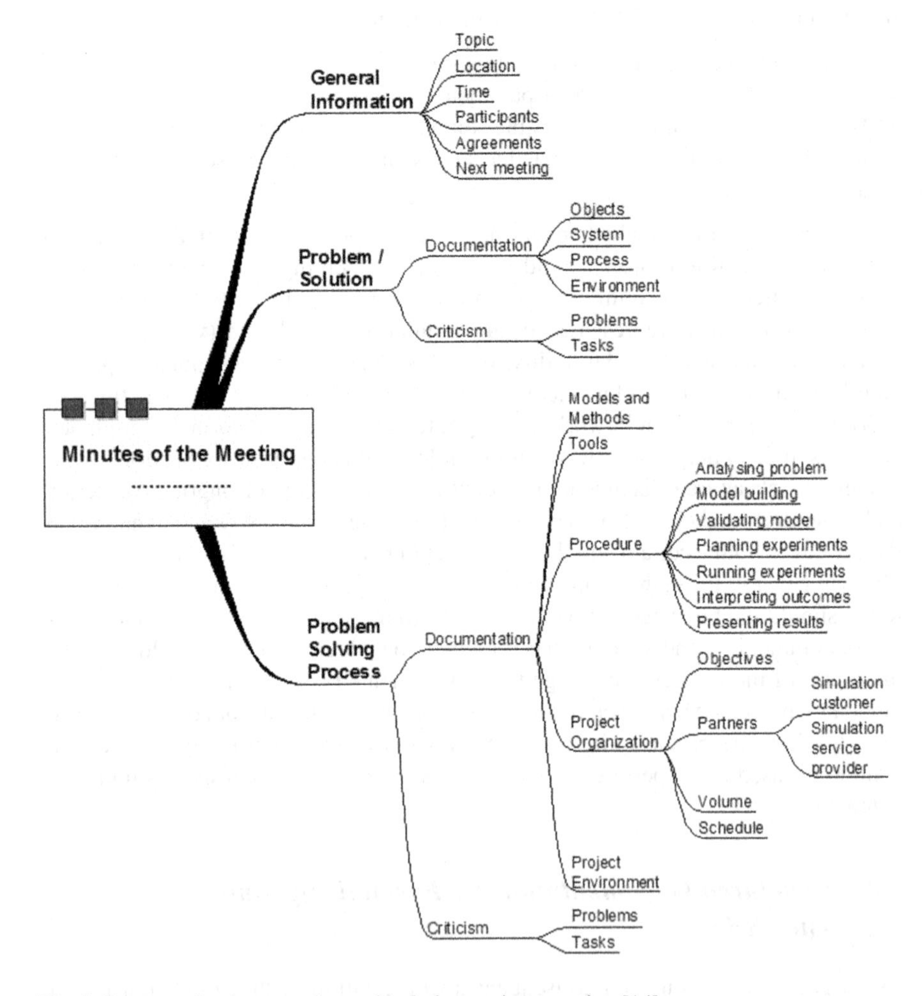

Fig. 5 Structure for documenting a simulation project meeting [14]

knowledge was specifically structured according to the application area which in our case is logistics. Consequently, on the first level it is further divided into object, system, process and environment categories which are suitable to completely describe a logistics problem and solution. If the proposed documentation framework is to be applied to any other application area of simulation, this specific part would have to be adapted to the relevant elements of problems and their solutions in this particular area.

Completing a copy of this template at each project meeting by filling in the particular contents and results, a growing set of documents with structured simulation knowledge is produced. In addition to this, further documents, written or oral communication outside the meetings and further external knowledge sources used within the project can be analysed in a similar way and thus increase the project-specific knowledge collection. From composing all of these individual documents step-by-step and according to project progress, a project-accompanying structured documentation of both the state-of-the-problem (and solution) and the state-of-the-problem-solving (including the state-of-development of the simulation model, methods and tools used or decisions taken) is derived.

At the same time the state-of-documentation resulting from each newly added source of knowledge also represents the particular state the project had reached at that point in time. With this, not only moment-specific representations of project knowledge, but also period-related representations of project progress become storable in a formalized way. Knowledge on the problem or solution and knowledge on the project procedure are always jointly documented. Because of this, not only a purposeful reflection of taken decisions is required, but also a clear explanation of all modifications to the model, procedure or solution initiated by those decisions. Together with the corresponding files containing the simulation model this would principally allow to return to any point in the problem-solving process and continue from there towards a different path whenever this seems to be necessary or appropriate.

Applying the approach for a structured documentation might offer a number of chances:

- To support an ongoing communication between the partners in a simulation project for achieving a common understanding of the logistics process and system to be investigated;
- To avoid any loss of information and knowledge in the course of the simulation project;
- To provide a kind of check list for data collection and a template for model design straight at the start of the project (What is to be taken into consideration for model building? Which information is required for the investigation?);
- To (automatically) create the draft of the simulation project report directly from the ongoing project documentation (including all assumptions, agreements, decisions).

Especially the vision of an automatically generated simulation project report is expected to address a particular need within the simulation community, because producing a report on the simulation project adequate to the processes and results is quite time-consuming and does not belong to the most-welcomed tasks within such a project. But the benefits that might be achievable by this are not only related to time-saving aspects and the increase of a simulation project's efficiency, but even more to provide a means for quality assurance.

5.3 Approaches for Benefiting from Users' Intuition and Experience in Simulation Experimentation

The impact of a person's knowledge and background on the design, level of detail and focus of the simulation model, i.e. on the way a simulation model appears and functions, was already demonstrated by the case study explained in an earlier section of this paper. Consequently, simulation needs to be seen in the context of human-centered processes and approaches for understanding simulation output need to match users' specific requirements. In recent literature this is being addressed by, for example, providing simulation modellers (or users) with methods and tools for automatic trace file analysis in order to better cope with large amounts of simulation output data. Those approaches mainly focus on formalising simulation outcome in the context of a certain application area.

In [11], for example, authors state that in tracing a simulation model a modeller finds himself in the situation where it is unclear what properties to ask to be checked by a model checker or what hypothesis to test. They assume that cyclic behaviour of model components is always good behaviour whereas all exceptions or disturbances in this behaviour indicate errors. Therefore, the aim is to provide support by automatically identifying and removing repetition from a simulation trace in order to pay particular attention on the non-returning, progressing part of a trace. This is to be achieved by automatic trace file reduction as it is assumed that modellers do not have enough background knowledge or experience to figure out interesting parts of the trace themselves.

In [29] authors assume that simulation usually aims to specify whether or not the concept of a material flow system meets formal requirements, but not how well it does it. This is said to be caused in limited methodological support and therefore strongly depend on the modelling/planning expert's experience and expertise. This is to be overcome by eliminating the user as weakest point through automatic analysis. For this an analysis tool is proposed that helps in identifying the concept's or the system's weak points, specifying their primary reasons and pointing out system immanent potential for performance increase.

Both approaches have in common the very much reduced role they give to the key actor(s) in any simulation project: the person who builds the simulation model and the person who uses the simulation model to run experiments. Instead they assume any result derived from simulation can directly and automatically be extracted from the trace file through statistical analysis, clustering or reasoning without any additional explanation by the simulating person. If this would be the case then any simulation model and any plan of experiments can be seen as objective representation of a particular part of reality and its problem situation. Any model building or experimentation activity no matter what background or intention one has would lead to the same model and to the same collection of simulation output. A particular simulation output always would lead to the same conclusions, i.e. simulation results, no matter what is being analysed by whom and how.

If this would be the case, why do simulation projects still require involvement of human resources of certain expertise? It is because simulation projects are not only sequences of formalising steps that can be fully represented by more or less complex logical algorithms, but also require intuitive problem solving, combining analysing steps and the need for creative thinking. Whereas the first can already be formalized or will be in future, the latter always remains linked to the person carrying out or contributing to or requesting simulation projects. Approaches to increase the degree of formalisation in simulation, no matter if they focus on automatic model generation or automatic trace file analysis and simulation result delivery, will always be limited by the impossibility of fully formalising the objectives and goals of a simulation. Therefore, any standardized or formalized approach for trace file analysis and interpretation of output without involving the model developer (and at the same time user in this particular case) would have failed or delivered very general (rough) results only. Instead, we propose an infrastructure and procedure combining formal analysis and human instinct of the simulation user (Fig. 6). The aim is to enable a true model-user dialogue referring to both aspects, the necessity for adjusting the simulation model to purposefully produce and collect useful simulation output data (according to simulation target definition) and the challenge for seeing behind the eventually huge amount of collected data to really get the message from the simulation output.

When the potential interests a simulation user might have in a simulation study are compared, one significant difference emerges: specific questions formulated by the user might directly be answered with concrete simulation output at data level; those usually fuzzy questions of principle from the more global user's point of view require interpretation and re-interpretation steps before being answered. Here, any question of principle has to be transferred to the data level by explaining it in detail and putting it in terms of concrete data (Fig. 6). As result of this process of interpretation a set of specific questions is defined with each of them providing a specific part of the overall answer in which the user is interested. Questions at data level correspond to results that can be delivered directly by the simulation even if minor modifications to the simulation model should be required [27]. This is the kind of study also current approaches for trace file analysis support [11, 29]. To derive an answer in principle to a question of principle the respective set of specific answers needs to be processed further. These steps of additional analysis and condensing can be understood as a process of re-interpretation to transfer results from data to user level.

All steps of interpretation and re-interpretation aim to link the user's point of view to that of the simulation expert. They not only require an appropriate procedure, but, even more importantly, an interpretative model representing the application area in which simulation takes place. This model needs to be based on knowledge and rules expressed in the user's individual expertise, but also in generalized knowledge

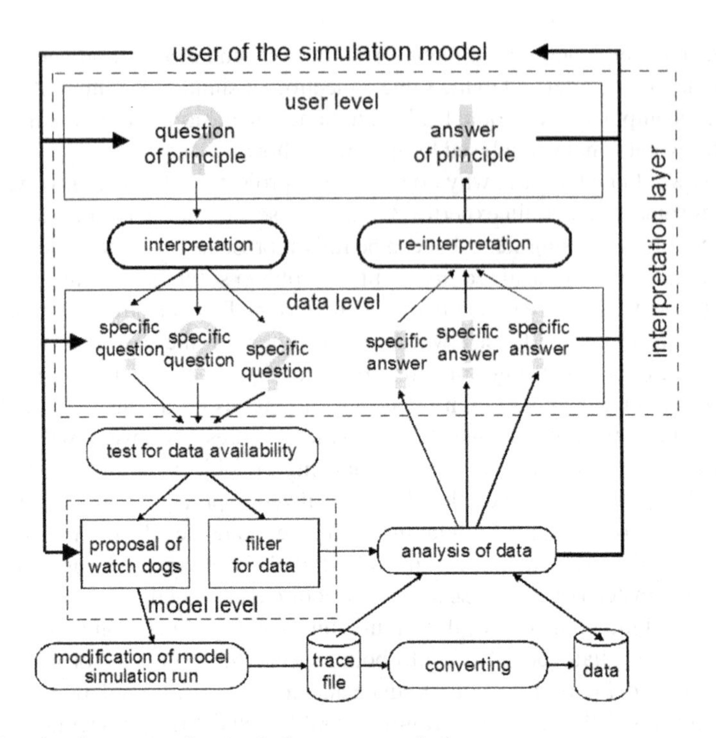

Fig. 6 User-data interaction for simulation output analysis

regarding design constraints or system behaviour and the experience of the simulation expert derived from prior simulations. As this knowledge might not only be of explicit nature, but also comprises implicit or tacit knowledge simulation users need to remain involved in the steps of interpretation and re-interpretation at least. Whereas explicit knowledge might be transferred into rules and algorithms, tacit knowledge cannot be separated from its owner and therefore requires direct involvement of the knowledge holder in the interpretation process. More specifically this means support is required for translating any question of principle into corresponding specific (data-related) questions as well as for deriving answers of principle from a number of specific (data-related) answers. Although a set of (standard) translation rules might be known, formalized and put into the rule base already, always further questions remain that are unknown to the rule base yet. Here, the domain expert needs support in (i) correctly formulating the right question and (ii) getting the full picture from the puzzle of available data and their analysis.

One approach for enabling this could be based on viewpoint descriptions. Viewpoint descriptions were introduced into model validation as a new kind of communication and interaction between the human observer of simulation results and the computer as the simulation model using authority that was called oracle-based model modification [26]. Here, the principle idea is that the user presents his or her observations (in the animation) as a viewpoint description to the computer that initiates a reasoning process. This results in definition and realisation of necessary changes to the simulation model in an ongoing user-computer dialogue. The main advantage of this concept lies in the reduced requirements for rule-base definition. Those aspects that easily can be formalized (e.g. typical quantitative observations or unambiguous logical dependencies) are translated into questions to the user (What is it s/he is interested in?) or various forms of result presentation (as figures or diagrams). Aspects that are non-imaginable yet or individual to the user or simply hard to formalize do not need to be included to provide meaningful support to the user. There is no need to completely specify all possible situations, views and problems in advance, because the person who deals with simulation output brings in additional knowledge, experience and creativity for coping with non-standard challenges. Even further, this way the rule-base continuously grows as it "learns" from all applications and especially from those that were not involved yet. On the other side the user benefits from prior experience and knowledge represented in the computer by receiving hints on what to look at based upon questions other users had asked or which were of interest in earlier investigations.

This approach helps in designing the interpretation layer for mediating between simulation user and simulation model or output no matter how many data have been gathered and how big the trace file grew. In the end, simulation results derived from running experiments by use of a particular simulation model are as good as they finally respond to the questions the simulation user is interested in. The challenge consists in knowing about questions a user in a specific project might have. Generally, a certain amount of (standard) questions can be pre-defined in correspondence with the application area and another set of questions might be defined by the user when starting into simulation modelling and experimentation. This might even lead to a specific focus in trace file generation and recording of simulation output data by purposefully introducing a cohort of observers to the model that directly correspond to the type and amount of data required for responding to questions already addressed by the user [27].

Effectiveness and efficiency of this interpretation process depends on the availability of the right data at the right level of detail. This quite often does not only depend on the simulation model and tool used for its implementation, but also on the opportunity to aggregate data in always new ways as it is not that exceptional that new questions arise in the cause of the simulation project when seeing results from previous experiments. In those situations it might either be necessary to re-run simulation with a modified observation concept or to aggregate or derive results from already existing simulation output in a different way [28].

Although being specific to a certain simulation project, respective analysis steps are possible to be pre-specified and also in the focus of approaches as presented by [24] and [29]. But beyond this, specific questions relevant in a certain simulation project might eventually even require, for example, to summarize (primary) objects as simulated into new (secondary) classes not simulated yet. In a transportation model with a number of trucks moving different types and different volumes of goods, for example, it suddenly might be of interest to know something about all those trucks arriving Tuesdays only. The simulation model itself knows trucks as one class of objects, but does not contain "Tuesday trucks" as a specific sub-class to this. This new class needs to be formed out of the situation and might then be added to the rule-base for trace file analysis, but cannot be pre-defined as simply not specified before. Consequently, any tool to support trace file analysis must allow and even support those interactions with the trace file which again goes far beyond formal statistical analysis.

In this context another question is related to the way the user's question is addressed to the data collection. In principle, there are three technologies for querying a database: (i) using a formal language like SQL, (ii) working with predefined search dialogues, or (iii) applying natural language. The first option requires specific knowledge and skills to formulate inquiries in the correct way; the second option might be limited in flexibility depending on the form's design and focus of the inquiry. Both options are therefore not really optimal solutions for supporting a real user-model dialogue. For this, the third option would be the ideal choice instead. This vision to allow the user asking questions to the simulation (or more precisely to the simulation output) in the own non-standardized way requires implementation of technology and logic for text or speech recognition and analysis. So-called natural language interfaces (NLI) to databases are subject to ongoing research [1, 22] with focus on "understanding" the human language in its complete version. Here, efficiency in translating and interpreting human inquiries to computer applications and databases will remain hence limited for quite some time. In contrast to this, natural language applications to a particular kind of databases like, for example, simulation output seem to be much more likely to come in shorter terms as those particular sets of data are characterized by a clear syntax and semantics.

From those discussions we can conclude that formal trace file analysis is just one important step for understanding the message of simulation results. The other one is the non-formal, more creative step of directly answering all questions that are of interest to the user (in our case the logistics expert). The precondition is to know (and understand) what the questions of the user are, but also the ability of the user to ask questions relevant to a particular problem. For the latter, the framework for trace file analysis and interpretation provides even further support: Typical questions no matter if they are of generic or specific nature help the user in identifying the problem or the questions to be asked or the aspects to be investigated. As discussed, this can be supported by the approaches for viewpoint description and defining observers or specifying analysis focus. Additionally, a pattern combining typical symptoms (i.e. visible situations or measurable characteristics) with the underlying problems causing those symptoms would be of huge benefit as this might also guide the user in truly understanding what happens in a specific system. Being able to ask those questions in natural language even simplifies the process and reduces the risk of misunderstanding, misinterpretation and misconception.

6 Conclusions

Human resources involved in a simulation project are the key factors for its success and efficiency. As discussed in the previous section it is always up to the simulation user to define objectives of any simulation and target functions of any experimentation. For this, detailed knowledge and understanding on the particular system/process to be investigated and problem to be solved is needed. Additionally, sound background knowledge on the domain and experiences in simulation-based problem solving is required. As this individual knowledge and experience belongs to the person carrying it and continuously develops and grows over time with each new simulation project, it can be separated from the person, i.e. externalized, to some extent only. Therefore, a mix of methods and tools for bringing in a user's knowledge and experience into the simulation project is needed:

- Formalize what can be formalized and incorporate this into simulation tools completed by a rule-based supporting system and an interface for its continuous improvement.
- Apply algorithms to routine problem-solving [11, 29].
- Enable a structured dialogue between the user and the tool by applying the concept of oracle-based simulation model validation [9].
- Provide support in structured documentation of problem, model, experiments, solution/findings and lessons learned [13].
- Use human intuition and tacit knowledge for all that cannot be formalized (yet).
- Allow the user to bring in his/her ability of flexible thinking for problems and questions that unexpectedly pop-up in the course of a simulation study [28].

A sophisticated framework especially helps to reduce routine work like statistics calculations through incorporated powerful analysis tools. In the same way creative thinking might be supported by proposing, asking, suggesting in a really interactive communication between the simulation user and the computer. Furthermore, it is crucial to initiate an ongoing learning and improvement process as basis of structured knowledge explication and gathering of experiences similar to what has been proposed by [6] for software engineering projects. Applying this approach to learning from simulation projects, a well-defined and well-structured documentation of simulation model, simulation runs and the entire simulation project with all its assumptions, agreements, and decisions has to be established (seamlessly and continuously). With this the process of a simulation project becomes a process of knowledge creation and acquisition at the same time without too much additional effort for all involved (Fig. 7).

To generalize research needs, the biggest challenge for properly handling modelling and simulation knowledge by applying knowledge management methods and tools consists in providing the right knowledge of the right quality and with the right costs at the right place and time. In other words, it is essential not to focus on the introduction of knowledge management technology and integration of software tools for storing and retrieving knowledge and information only. Instead the human resources running model building and simulation projects need to be put into the centre of gravity for giving them that kind and amount of support which is needed in a particular situation. Furthermore, the clue to the successful implementation of those knowledge management procedures is often an appropriate (supporting)

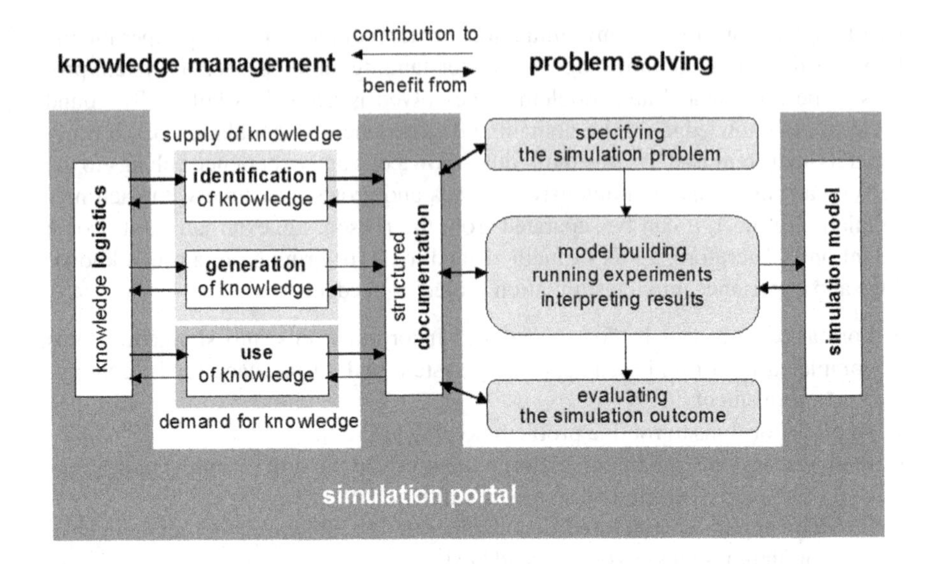

Fig. 7 Problem-solving and knowledge acquisition in the course of a simulation project

environment and climate in the organisation. Concerning this, there is a greater need for a cultural shift than for additional software tools and IT solutions.

Adopting a statement on human needs for computer technology by [25] the link between knowledge management and (simulation-based) problem solving can generally be described as follows: the old discussion about how to support problem solving is about what (software) tools can do; the new discussion about how to support problem solving is (and must be) about what kind of problem-solving support people really need. To understand those needs and translate them into a supportive infrastructure for a more knowledgeable user-model dialogue that is the challenge we are facing in research and development for applied simulation.

References

1. Androutsopoulos, I., Ritchie, G., Thanisch, P.: Natural language interfaces to databases—an introduction. Natural Language Engineering 1(1), 29–81 (1995)
2. Balci, O., Nance, R.: Simulation model development environments: A research prototype. Journal of the Operational Research Society 38(8), 753–763 (1987)
3. Balci, O., Ormsby, W.F., Carr, J.T.I., Saadi, S.D.: Planning for verification, validation, and accreditation of modeling and simulation applications. In: Proc. Winter Simulation Conference, pp. 829–839 (2000)
4. Beckman, T.J.: The current state of knowledge management. In: Liebowitz, J. (ed.) Knowledge Management Handbook. CRC Press, Boca Raton (1999)
5. Brade, D.: Enhacing modelling and simulation accreditation by structuring verification and validation results. In: Proc. Winter Simulation Conference, pp. 840–848 (2000)

6. Brandt, M., Ehrenberg, D., Althoff, K., Nick, M.: Ein fallbasierter ansatz für die computergestützte nutzung von erfahrungswissen bei der projektarbeit (a case-based approach for computer-based use of experience-based knowledge in projects). In: Proc. 5. Internationale Tagung Wirtschaftsinformatik (2001) (in German)
7. Bruzzone, A., Madeo, F., Tarone, F.: Modelling country reconstruction based on civil military cooperation. In: Proc. of European Modeling and Simulation Symposium, pp. 315–322 (2010)
8. Cuske, C., Dickopp, T., Seedorf, S.: JOmtoRisk. An ontology-based Platform for Knowledge-based Simulation Modeling in Financial Risk Management. In: Feliz-Teixeira, J.M., Brito, A.E.C. (eds.) Proc. European Simulation and Modelling Conference, pp. 79–86 (2005)
9. Helms, C., Strothotte, T.: Oracles and viewpoint descriptions for object flow investigation. In: Proc. of the 1st EUROSIM Congress on Modelling and Simulation, pp. 47–53 (1992)
10. Highland, H.J.: A taxonomy approach to simulation model documentation. ACM SIGSIM Simulation Digest 10(3), 19–23 (1979)
11. Kemper, P., Tepper, C.: Automated trace analysis of discrete-event system models. IEEE Transactions on Software Engineering 35(2), 195–208 (2009)
12. Nance, R.E., Roth, P.F.: Documentation of simulation models: prospects and problems. In: Proc. Winter Simulation Conference, pp. 722–723 (1997)
13. Neumann, G.: Projektwissen in der Logistiksimulation erschließen und bewahren: Auf dem Weg zu einer neuen Dokumentationskultur (gaining and storing project knowledge in logistics simulation: on the way towards a new documentation culture). In: Proc. of Simulation in Production and Logistics, pp. 341–350 (2006) (in German)
14. Neumann, G.: The role of knowledge throughout the simulation lifecycle: what does a simulation model know? In: Proc. of the 6th EUROSIM Congress on Modelling and Simulation (2007)
15. Neumann, G., Page, B.: Case study to compare modelling and simulation approaches of different domain experts. In: Proc. of I3M 2006—International Mediterranean Modelling Multiconference, pp. 517–522 (2006)
16. Neumann, G., Ziems, D.: Logistics simulation: Methodology for problem solving and knowledge acquisition. In: Proc. of the 6th Multiconference on Systemics, Cybernetics and Informatics, pp. 357–362 (2002)
17. Nonaka, I., Takeuchi, H.: The Knowledge-Creating Company: How Japanese Companies Create the Dynamics of Innovation. Oxford University Press (1995)
18. Ören, T.I.: A paradigm for artificial intelligence in software engineering. Advances in Artificial Intelligence in Software Engineering 1, 1–55 (1990)
19. Ören, T.I.: Body of Knowledge of Modeling and Simulation (M&SBOK): Pragmatic aspects. In: Proc. European Modelling and Simulation Symposium (2006)
20. Oscarsson, J., Urenda Moris, M.: Best modeling methods: documentation of discrete event simulation models for manufacturing system life cycle simulation. In: Proc. Winter Simulation Conference, pp. 1073–1078 (2002)
21. Page, B., Kreutzer, W.: The Java Simulation Handbook—Simulating Discrete Event Systems in UML and Java. Shaker, Aachen (2005)
22. Popescu, A.M., Armanasu, A., Etzioni, O.: Modern natural language interfaces to databases: Composing statistical parsing with semantic traceability. In: Proc. of COLING (2004)
23. Rego Monteil, N., del Rio Vilas, D., Crespo Pereira, D., Rios Prado, R.: A simulation-based ergonomic evaluation for the operational improvement of the slate splitters work. In: Proc. of European Modeling and Simulation Symposium, pp. 191–200 (2010)

24. Saanen, Y., van Valkengoed, M.: Comparison of three automated stacking alternatives by means of simulation. In: Proc. Winter Simulation Conference, pp. 1567–1576 (2005)
25. Shneiderman, B.: Leonardo's Laptop: Human Needs and the New Computing Technologies. The MIT Press, Cambridge (2002)
26. Spieckermann, S., Lehmann, A., Rabe, M.: Verifikation und Validierung: Überlegungen zu einer integrierten Vorgehensweise (validation and verification: towards an integrated procedure). In: Proc. 11th Conference Simulation in Production and Logistics, pp. 263–274 (2004) (in German)
27. Tolujew, J.: Werkzeuge des Simulationsexperten von morgen (tools of the simulation expert of tomorrow. In: Proc. of Simulation and Animation, pp. 201–210 (1997) (in German)
28. Tolujew, J., Reggelin, T., Sermpetzoglou, C.: Simulation und Interpretation von Datenströmen in logistischen Echtzeitsystemen (simulation and interpretation of data streams in logistics real-time systems). In: Engelhardt-Nowitzki, C., Nowitzki, O., Krenn, B. (eds.) Management komplexer Materialflüsse mittels Simulation. State-of-the-Art und innovative Konzepte. Leobener Logistik Cases, pp. 215–232. Deutscher Universitäts-Verlag, Wiesbaden (2007)
29. Wustmann, D., Vasyutynskyy, V., Schmidt, T.: Ansätze zur automatischen Analyse und Diagnose von komplexen Materialflusssystemen (approaches for automatic analysis and diagnosis of complex material flow systems). In: Proc. of the 5th Expert colloquium of WGTL—Wissenschaftliche Gesellschaft für Technische Logistik, pp. 1–19 (2009) (in German)
30. Zhou, M., Son, Y.J., Chen, Z.: Knowledge representation for conceptual simulation modeling. In: Proc. of Winter Simulation Conference, pp. 450–458 (2004)
31. Zülch, G.: Modelling and simulation of human decision-making in manufacturing systems. In: Proc. of Winter Simulation Conference, pp. 947–953 (2006)

Simulating Entrepreneurial Opportunity Recognition Processes: An Agent-Based and Knowledge-Driven Approach

Martin Ihrig

Abstract. This chapter describes how we abstract a complex phenomenon—opportunity recognition—and build a simulation model for studying strategic entrepreneurship in a knowledge-based economy. We develop an agent-based simulation tool that enables researchers to explore entrepreneurial strategies, their associated financial payoffs, and their knowledge creation potential under different environmental conditions. Opportunity recognition processes can be analyzed in detail both on a micro- and macro-level. To illustrate the modeling capabilities of the tool, we conduct some basic simulation runs and model three distinct entrepreneurial strategies—the innovator, the inventor, and the reproducer—and compare their knowledge progression and financial performance profiles. The software allows us to study the individual and the societal level effects that arise from competitive agent behavior in both national and international settings. It can account for international knowledge spillovers that occur in a globalized knowledge economy. The simulation can be used to conduct innovative research that will result in theory-driven hypotheses that can inform corporate and public-sector decision makers, which would be difficult to derive from empirical analyses.

1 Introduction

The academic field of entrepreneurship aims to develop theories to help us understand entrepreneurial opportunities and their formation [2, 3]. The concept of opportunity is central to the discussion of the entrepreneurial phenomenon [23]. Opportunities are emergent and tend to be one of a kind. Opportunity recognition—and the knowledge appropriation and development that define it—are usually complex and non-repeatable processes. This makes simulation modeling the appropriate

Martin Ihrig
Snider Entrepreneurial Research Center, The Wharton School, University of Pennsylvania, 418 Vance Hall, 3733 Spruce Street, Philadelphia, PA 19104, USA
e-mail: ihrig@wharton.upenn.edu

A. Byrski et al. (Eds.): Advances in Intelligent Modelling and Simulation, SCI 416, pp. 27–54.
springerlink.com

methodological approach for researching opportunity recognition [8, 9]. The unique simulation tool described in this chapter will help researchers to model different entrepreneurial actions and the contexts in which they take place. It thereby allows to conduct innovative research that will result in theory-driven frameworks and hypotheses that would be difficult to obtain from empirical analyses alone. Being able to explore distinct opportunity recognition strategies is the basis for advising entrepreneurs on their paths to success and governments on productive policy actions.

We first describe the three entrepreneurial actors that the literature puts forward and briefly explain our conceptual approach. We then present the simulation environment with which we build the application-specific model. The concrete parameterizations of that model are described in detail next. Following this, we show results of two virtual experiments that highlight the distinct modeling capabilities of the simulation tool. Finally, we build an extension of the model and simulate entrepreneurial activities in a global economy with international knowledge-spillovers.

2 Conceptual Background

The literature distinguishes three different actors with regard to the entrepreneurial process. First, we have the *inventor* whose sole task it is, according to Schumpeter [19], to "produce ideas", and who he contrasts with the *entrepreneur*, who is responsible for "getting new things done", i.e. "translat[ing] the inventions into practical, operational entities" [4]. Schumpeter [20] links the entrepreneur with innovation [16]—basically equating the entrepreneur with the *innovator* as the second actor. But, as Aldrich and Martinez [1] note, "innovation and entrepreneurship are not necessarily coupled". Analyzing economic development, Schumpeter [19] explains that the existence of successful innovative entrepreneurs encourage others to follow their example: "... the appearance of one or a few entrepreneurs facilitates the appearance of others, and these the appearance of more, in ever-increasing numbers". (p. 228) Aldrich and Martinez [1] talk about the "continuum from reproducer to innovator". Reproducers are "organizations whose routines and competencies vary imperceptibly from those of existing organizations in established populations" and "they bring little or no incremental knowledge to the population they enter, organizing their activities in much the same way as their predecessors" [1]. Hence, the third entrepreneurial player is the reproducer, so that we end up with three different entrepreneurial roles or strategies.

Our goal is to use simulation to explore the different ways of recognizing and realizing business opportunities and show that they can be meaningfully distinguished. Simulation modeling allows us to operationalize the theoretical concepts and processes discussed in the literature and analyze micro and macro effects dynamically. In particular, we are interested in creating a tool that allows researchers to compare payoffs of each entrepreneurial strategy in different environments, where one would expect to see distinctive knowledge progression and financial performance profiles. In addition, a simulation tool enables us to study the societal level effects that arise from competitive agent behavior. Simulation experiments will help researchers arrive at a better understanding of the impact of different entrepreneurial strategies in both national and international settings.

We take a knowledge-based approach when studying how entrepreneurs obtain their new venture ideas and develop them [14]. Based on Austrian economics, we consider knowledge, and in particular its appropriation, development and exploitation, as the basis for new venture creation. This is why we use the knowledge-based simulation environment *SimISpace2* [12] to model the different entrepreneurial strategies and the contexts in which they take place. *SimISpace2* is an agent-based graphical simulation environment designed to simulate strategic knowledge management processes, in particular knowledge flows and knowledge-based agent interactions. The simulation engine's conceptual foundation is provided by Boisot's work on the Information Space or *I-Space* [5, 6]. The *I-Space* is a conceptual framework that facilitates the study of knowledge flows in diverse populations of agents—individuals, groups, firms, industries, alliances, governments, and nations.

3 The Simulation Software *SimISpace2*

The purpose of *SimISpace2* is to improve our understanding of how knowledge is generated, diffused, internalized and managed by individuals and organizations, under both collaborative and competitive learning conditions. Through a user-friendly graphical interface, it can be adapted to a wide range of knowledge-related applications—in our case different entrepreneurial strategies of knowledge appropriation and development. Researchers can set up virtual experiments to explore how user-defined agents evolve as measured by *knowledge discovery* and growth of *financial funds*. During the simulation runs, agents compete and cooperate in performing different kinds of actions. The actions and interactions form the basis for the emergent properties of agents (e.g., stock of knowledge, financial funds, location, etc.) and of knowledge assets (e.g., diffusion, location, structure, etc.).

There are basically three sources in the simulation that a *SimISpace2* agent draws on to obtain or develop new knowledge. The first source is a global pool of knowledge that an agent accesses and extracts pieces of knowledge from. The second source of knowledge is from the knowledge portfolio that an agent already possesses, which can be internally developed into new knowledge through structuring and learning processes. Finally, an agent can obtain knowledge from other agents through certain actions such as scanning or trading. How different agents utilize these three sources ultimately depends on the complex attributes that the user assigns to each agent. Knowledge discovery affects the other principal indicator of an agent's development, its financial funds. Agents capitalize on knowledge discovery, and they grow their *financial funds* by executing several actions such as trading or exploiting. Financial funds are important to understanding an outcome of a simulation model, as they measure the success of agents and thereby present opportunity for analysis.

Before showing how we build the application-specific model—*SimOpp*—to operationalize opportunity recognition processes under different environmental conditions, we must review some of the basics of the *SimISpace2* simulation software. Ihrig and Abrahams [12] offer a comprehensive description of the entire *SimISpace2* environment and explain the technical details.

3.1 Basics

Two major forms of entities can be modeled with *SimISpace2*: agents and knowledge assets. When setting up the simulation, the user defines agent groups and knowledge groups with distinct properties. Individually definable distributions can be assigned to each property of each group (uniform, normal, triangular, exponential, or constant distribution). The simulation then assigns the individual group members (agents and knowledge items) characteristics in accordance with the distribution specified for the corresponding property for the group of which they are a member.

Knowledge in the simulation environment is defined as a "global proposition". The basic entities are *knowledge items*. Based on the knowledge group they belong to, those knowledge items have certain characteristics. All knowledge items together make up the *knowledge ocean*—a global pool of knowledge. Agents can access the knowledge ocean, pick up knowledge items, and deposit them in knowledge stores through the *scanning action*. A *knowledge store* is an agent's personal storage place for a knowledge item. Each knowledge store is local to an agent, i.e. possessed by a single agent. As containers, knowledge stores hold knowledge items as their contents. Stores and their items together constitute *knowledge assets*. Examples of knowledge stores include books, files, tools, diskettes, and sections of a person's brain. There is only one knowledge item per knowledge store, i.e. each knowledge item that an agent possesses has its own knowledge store. If an agent gets a new knowledge item (whether directly from the knowledge ocean or from other agents' knowledge stores), a new knowledge store for that item is generated to hold it.

The concept of a knowledge item has been separated from the concept of a knowledge store to render knowledge traceable. If knowledge items are drawn from a common pool and stored in the knowledge stores of different agents, it becomes possible to see when two (or more) agents possess the same knowledge, a useful property for tracking the diffusion of knowledge.

The separation between a global pool of knowledge items and local knowledge stores is particularly important when it comes to codification and abstraction (these only apply to knowledge stores, not to knowledge items). Knowledge items are held by multiple agents, and one agent's investment in codification or abstraction does not influence the codification and abstraction level of the same knowledge item held by another agent. Agents possess knowledge stores at a particular level of codification and abstraction. If the agent codifies its knowledge or makes it more abstract, the properties of the knowledge item itself—i.e., its *content*—are not changed, but it gets moved to a new knowledge store with higher degrees of codification and abstraction—i.e., its *form changes*.

SimISpace2 also features a special kind of knowledge: a DTI (knowledge *Discovered Through Investment*) is a composite knowledge item that cannot be discovered through scanning from the global pool of knowledge items, but only by integrating its constituent knowledge items into a coherent pattern. The software user determines which knowledge items will act as the constituent components of a DTI; and the only way for an agent to discover a DTI is to successfully scan and appropriate its constituent components and then to codify and/or abstract them beyond user-specified threshold values to achieve the required level of integration. Once these values are reached, the agent automatically obtains the DTI (via a *discover* occurrence triggered by the simulation software). Investing in its constituent components—i.e. scanning, codifying and abstracting them—is the main means of discovering a DTI. Specifying the values of different DTIs allows the user to indirectly determine the values of the networks of knowledge items that produce DTIs (networks that represent complex forms of knowledge). Once an agent has discovered a DTI item, it is treated like a regular knowledge item, i.e. other agents are then able to also scan it from the agent that possesses it (without the process of having to discover its child constituents).

To validate whether the basic theoretical relationships of the *I-Space* [5, 6] are properly implemented in the *SimISpace2* environment, several test cases were designed and run. The results all matched the expected *I-Space* behavior that the theory predicts. In particular, the special interactions between the structure, the diffusion, and the value of knowledge have all been taken into account. For this reason, any simulation model that will be implemented with *SimISpace2* will have the dynamics of a knowledge-based economy built in, and agent behavior can be analyzed accordingly.

3.2 Agents' Actions

To keep our model and the resulting analyses simple, we use only six of the twenty actions featured in the *SimISpace2* environment: relocate, scan, codify, discover, learn, exploit. Our virtual agents use those actions in each period of a run to accumulate new knowledge and develop it so as to discover DTIs. Agents can increase their financial funds by capitalizing on the knowledge they possess, in particular DTIs[1]. Based on different agent group behaviors, the increases in agents' individual financial funds and stocks of knowledge occur at different rates. Whereas all agents in our simulation will try to *learn* and *exploit* their knowledge (and thereby to grow their financial funds), agents will differ in their approaches to obtaining and developing knowledge in the first place. What follows is a concise review of the critical actions we assign for modeling knowledge appropriation and development.

[1] Agents' financial funds act as a measure of their success - the better the knowledge appropriation, development and exploitation strategy, the higher the funds will be. (Agents with zero financial funds 'die'.)

Scanning

An agent can scan for knowledge, randomly picking up knowledge items, either from the knowledge ocean or from other agents' knowledge stores. The probability of picking from the knowledge ocean (vs. from other agents) can be specified at the agent-group-level. While an agent can scan any knowledge item in the knowledge ocean, it can only scan knowledge items from the knowledge stores of other agents that fall within its *vision*. *SimISpace2* agents populate a physical, two-dimensional space (called *SimWorld*), and the vision property determines how far the agent can 'see', defined as being within a certain spatial radius from its current location.

A knowledge item that is successfully scanned is placed in a new knowledge store possessed by the agent, which picks up its codification and abstraction levels either from the knowledge group that the knowledge item belongs to in the knowledge ocean, or from those of the knowledge store where the agent found the item. Agents will only try to scan knowledge items they do not already possess, or not at that level of codification and abstraction.

The ease with which a knowledge item is scanned from another agent's knowledge store is some positive function of the store's degree of codification and abstraction [6]. Knowledge items in knowledge stores with higher codification and abstraction have a higher probability of being scanned.

Relocating

An agent can relocate within a certain distance of its position on the 100 by 100 *SimWorld* grid, with the *distance* moved being governed by the distance setting for the relocate action of its agent group. Relocating implies moving either closer to or further away from other agents or knowledge stores, and is thus relevant to scanning as it affects which other agents and knowledge stores the agent can see from its new position. As agents can only scan within the radius of their vision, they are only able to pick up knowledge from different areas by moving. Relocating agents leave their knowledge stores behind in the original location, but still retain access to them. (N.B.: When a new knowledge store is created, it is always assigned the same location as the agent that possesses it.)

Structuring knowledge (Codifying and Abstracting)

Codification and abstraction are separate actions that affect the knowledge stores (form) in which a given knowledge item (content) is held, although the agent must first possess a knowledge item in a store before it can perform these actions. Agents can create new knowledge stores at different codification and abstraction levels within the 0 to 1 range. The codification or abstraction levels of a newly created knowledge store are increased incrementally beyond those of existing stores. The knowledge item in the new knowledge store always remains the same—it is only the level of codification and abstraction of the knowledge store that changes. Stores with higher levels of codification and abstraction are both more likely to be scanned by other agents, and more valuable when exploited: however, the more diffused knowledge becomes, the less value agents can extract from it [6].

Learning

Before a knowledge item can be exploited, it has to register with an agent through a learning process. This can only apply to a knowledge item that an agent possesses. Its chances of successfully learning increase with the levels of codification of the knowledge store that holds it.

Exploiting

Agents can generate value for themselves by capitalizing on their knowledge, but only after it has been registered and internalized through learning. The exploiting agent's financial funds are increased by the value of the exploiting actions the agent undertakes, with the *exploit amount* calculation based on the user-set *base value* of the knowledge item involved. This takes account of the levels of codification and abstraction of the knowledge store holding the knowledge item, and of the level of diffusion of the knowledge item (the percentage of agents that possess the particular piece of knowledge in that period). The user can define an industry-specific table of revenue multipliers based on abstraction and codification levels. In the I-Space [6], the value of knowledge is some positive function of both its utility (the level of codification and abstraction) and of its scarcity (the level of diffusion). Therefore, typically, the higher the levels of codification and abstraction, the higher the revenue multiplier, i.e. more codified and abstract knowledge is worth more. More codified and abstract knowledge, however, is also more likely to be diffused, which erodes the value of knowledge. The calculations also allow for the effects of obsolescence, which also erodes value: obsolete knowledge is worthless. Whereas revenue multipliers depend on the codification and abstraction characteristics of a knowledge *store*, obsolescence depends solely on the properties of the knowledge *item* the store contains.

4 The Simulation Model *SimOpp*

We can now describe the *SimISpace2* model designed and built for the opportunity recognition context (hereinafter called *SimOpp*) and present the properties of the participating agent and knowledge groups [10].

4.1 Agents

Following Schumpeter [20] and Aldrich and Martinez [1] as described above, we create three *agent groups*: innovator, inventor, reproducer. The propensities to engage in (probabilities to choose and perform) particular *SimISpace2* actions vary from group to group based on the conceptual distinctions that have been made. In

Table 1 Distinct propensities to engage in actions (Activities 1 & 2)

	Scan		Relocate	Codify	Σ
	Ocean	*Others*			
Innovator	1		0	1	2
Inventor	1 +	*0.5*	*0.5*	0	2
Reproducer		*1 (0.5+0.5)*	*1 (0.5+0.5)*	0	2

total, agents engage in four activities[2]. One activity (1) is assigned for implementing the process of obtaining a new venture idea (first insight) and another activity (2) for implementing the further development of that idea. For the first activity, agents can arrive at an insight themselves through the 'Scan from the Knowledge Ocean' action, or obtain insights from a third party through the 'Relocate' and the 'Scan from Others' actions (since agents can move through the *SimWorld* and scan knowledge assets from other agents).

For the second activity, agents can further develop the idea themselves through the 'Codify/Abstract' actions, or they draw on others' knowledge resources, again via the 'Relocate' and 'Scan from Others' actions. There are two other relevant activities which all agents must perform if they are to be able to capitalize on their knowledge: those are implemented with the 'Learn' and 'Exploit' actions (activities 3 and 4 respectively)[3].

Table 1 summarizes the propensities of agents performing a particular action for activities 1 and 2—the activities that we use for modeling the distinct knowledge appropriation and development strategies. Whenever agents 'scan from others' (as opposed to 'scan from the knowledge ocean') the 'relocate' action is also enabled, and since the *relocate* and *scan* actions are treated as one activity, we assign a constant distribution of 0.5 to each of them (which adds up to 1 per activity)[4]. Looking at each agent group in turn and based on the description above, we can see what actions and properties agents have in common and what distinguishes them[5].

[2] One activity being one or more individual *SimISpace2* actions/occurrences.

[3] The numbering of the activities does not necessarily imply a particular order in which the actions are conducted in the simulation. Knowledge can only be learned once it has been obtained and can only be exploited once it has been learned. Which of the possible actions an agent chooses is determined randomly each period by the simulation software based on the distributions assigned for the propensities to engage in an action.

[4] We assume that agents divide their effort equally between moving through the *SimWorld* and scanning knowledge assets from other agents, so the weighting used for each action in this research is 0.5.

[5] Both propensity to learn and propensity to exploit (activities 3 and 4) are assigned constant distributions of 1 for all agent groups.

Innovator

Innovators perform four actions; they *scan* and they *codify*, and (as all the other groups) they learn and exploit. They can only scan from the knowledge ocean. This means that Innovator agents do not obtain any knowledge from other agents; they can conceive of a new venture idea themselves and further develop that idea independently.

Inventor

Inventors perform four actions—scan and relocate, and learn and exploit, but they do not codify. They can scan from both the knowledge ocean and from other agents. This means that Inventor agents can arrive at original insights themselves, but in contrast to Innovators, they lack the capability to further develop this knowledge. Therefore, they draw on outside knowledge from other agents.

Reproducer

Reproducers perform the same four actions - scan and relocate, and learn and exploit, but not codify. In contrast to the Inventors, they only scan from other agents, but not from the knowledge ocean. This means that Reproducer agents do not generate any knowledge themselves, they only source it externally.

There are ten agents in each agent group, and all agents start with financial funds of 100. The *relocate distance* and *vision* property are the same for all groups, but they change with each scenario under study (as will be explained below). Inventors and Reproducers are randomly spread out in the *SimWorld* (uniform distribution 0–100 for x and y location); Innovators are clustered together at the center (uniform distribution 45–55 for x and y location) as shown in Fig. 1—centrally located for when the other agents move around the *SimWorld* in the 1000 period runs.

4.2 Knowledge

The *SimOpp* model uses both *SimISpace2*'s basic and higher-level DTI knowledge types. We have three distinct basic knowledge groups: *Local*, *Entrepreneurial*, and *New Venture Idea Knowledge*.

Local Knowledge

Local Knowledge represents an agent's understanding of the local market and its culture. It starts at a high level of codification and abstraction (0.7) and has a base value of 5. A knowledge item's intrinsic *base value* is the starting point for calculating the *exploit amount*—the increase in an agent's financial funds after it has successfully performed an exploit action on a knowledge asset. Since the value of knowledge (in the *I-Space* [6]) is a positive function of its utility and its scarcity,

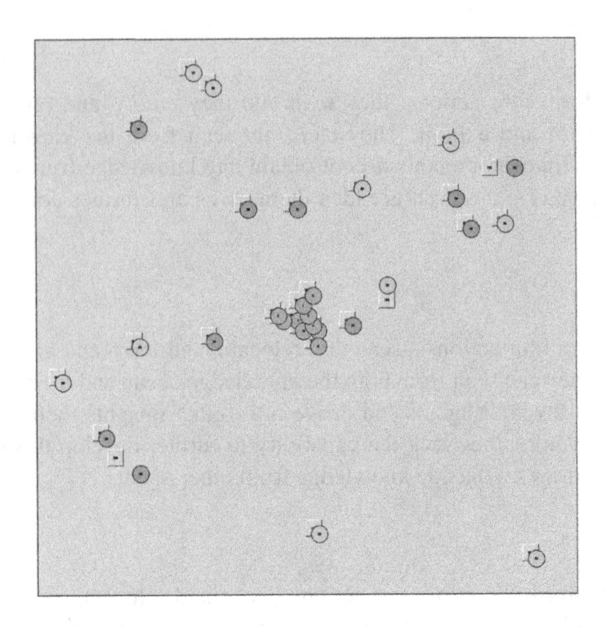

Fig. 1 *SimWorld* in period 1

both the levels of codification and abstraction of the knowledge item's store, and the item's level of diffusion, are included in the calculation.

Entrepreneurial Knowledge

Entrepreneurial Knowledge represents 'Know-*How*' [18]. Abilities like those to "sell, bargain, lead, plan, make decisions, solve problems, organize a firm and communicate" [22] are examples of knowledge items in this group. To this we add a 'creating the initial transactions' [24] set of skills like writing business plans, initiating sales, creating initial products and services, securing initial stakeholders and finances. Knowledge from this knowledge group starts at medium codification/abstraction levels (0.5) and has a base value of 10.

New Venture Idea Knowledge

New Venture Idea Knowledge represents the 'Know-*What*'. Knowledge items in this group are insights about a particular potential service or product offering. Knowledge from this knowledge group starts at low levels of codification and abstraction (0.3) but has a base value of 20.

There are ten knowledge items in each knowledge group, all of which have *obsolescence rates* of zero, *codification* and *abstraction increments* of 0.1, and no *per period carrying gain or cost*[6]. All agent groups are endowed with Local Knowledge and Entrepreneurial Knowledge (i.e. every agent in the simulation possesses all knowledge items from these groups), but they do not possess New Venture Idea Knowledge.

Opportunities

We use DTI knowledge to model opportunities. Once an agent possesses a knowledge item each from the Local Knowledge, Entrepreneurial Knowledge and New Venture Idea Knowledge groups, in knowledge stores with codification levels equal or greater than 0.6, it gains the corresponding DTI, i.e. the agent 'discovers' an opportunity (Fig. 2). There are ten DTIs, each based on a combination of the nth items of each basic knowledge group[7]. DTI knowledge items have high *starting levels* of *codification* and *abstraction* (0.8), high (compared to base knowledge) *base values* of 2500, *obsolescence rates* of zero, *codification* and *abstraction* increments of 0.1, and no per period *carrying gain or cost.*

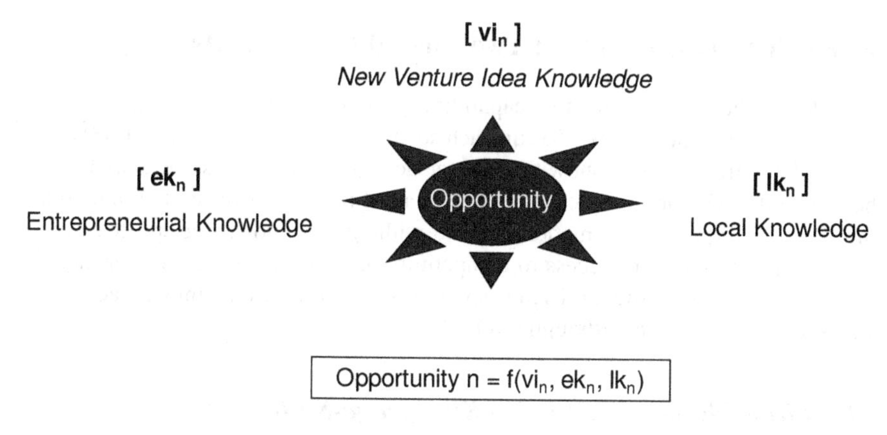

Fig. 2 Modeling opportunity in *SimOpp*

Agents obtain opportunities in different ways. The dynamics of this simulation mean that there are three.

[6] It is important to start with simple assumptions, switching certain settings off, and to only increase complexity slowly. This enables the researcher to gain a better understanding of the fundamental dynamics of the simulation, as it is easier to analyze results. While it is clearly beyond the scope of this chapter to test and explore the whole parameter space of *SimISpace2*, it is important to know about the different settings and *SimISpace2* options when it comes to designing future research efforts.

[7] For example, the underlying knowledge items for DTI 1, are knowledge item 1 of Local Knowledge, knowledge item 1 of Entrepreneurial Knowledge, and knowledge item 1 of New Venture Idea Knowledge.

Opportunity Construction. The classical way is to construct an opportunity[8]. An agent obtains all underlying knowledge items, structures them up to the specified codification threshold, and then gains the relevant DTI, i.e. the opportunity. As their prior stocks of knowledge, agents already possess Local Knowledge and Entrepreneurial Knowledge from the start of the run, and can obtain the missing New Venture Idea Knowledge item either directly from the knowledge ocean or from another agent's knowledge store. They then reach the required threshold either by codifying the knowledge themselves or by scanning it from another agent that already holds it at the required level.

Opportunity Acquisition. Agents can not only scan others' basic knowledge items, but also their DTIs. So, as well as being able to construct opportunities themselves, agents can acquire the knowledge about an opportunity directly by scanning it from another agent that carries the DTI in its knowledge store.

Opportunity Amplification. Agents can also develop and structure their opportunities further, either by increasing the codification levels of their DTIs themselves, or by scanning from other agents that possess higher codified stores of that DTI.

5 Putting *SimOpp* to Use: Two Virtual Experiments

To highlight the distinct modeling capabilities of our simulation tool, we now conduct two virtual experiments. We run each scenario for 60 runs, each of 1000 periods[9]. All the graphs below show the average across all runs, and some also display the standard deviation to indicate the significant difference between the individual lines. Virtual Experiment 1 models a world without any knowledge spillovers, i.e. an environment with zero access to competitors' knowledge; Virtual Experiment 2 a world with at least some knowledge spillovers, i.e. an environment with access to competitors' knowledge (although low).

5.1 Virtual Experiment 1: No Knowledge-Spillovers

For the first scenario we run, we take the setup described above and set both the vision property and the relocate property for all agent groups to zero. This means that agents can neither see other agents nor move away from their assigned positions.

[8] The occurrence type triggered in the simulation is called discovery, but we are not using this term yet because of the particular connotation it carries in the entrepreneurship literature.

[9] We also analyze 2000 periods sometimes to allow us to highlight certain trajectories. One period in the simulation is supposed to represent a particular length of 'real-world' time: the approximate conversion factor depends on the particular industry or environment being analyzed, although it could be estimated by looking, for example, at the particular duration of a specific process in the real world (e.g., it took x amount of years until Starbuck's business idea had been widely recognized in market y), and then identifying a graph that represents this process in the simulation (e.g., the knowledge diffusion curve) and checking on how many periods that run took.

With their vision set to zero, we expect Reproducers and Inventors not to acquire DTIs because of their inability to scan from other agents. They are unable (in contrast to Innovators) to construct opportunities in isolation: Inventors have unique insights but cannot develop them, while Reproducers lack access to new knowledge at all. So we expect the Innovator group to outperform the other groups in terms of financial funds, because it is the only group that can exploit DTIs. Fig. 3 displays the financial profile for each group over 1000 periods.

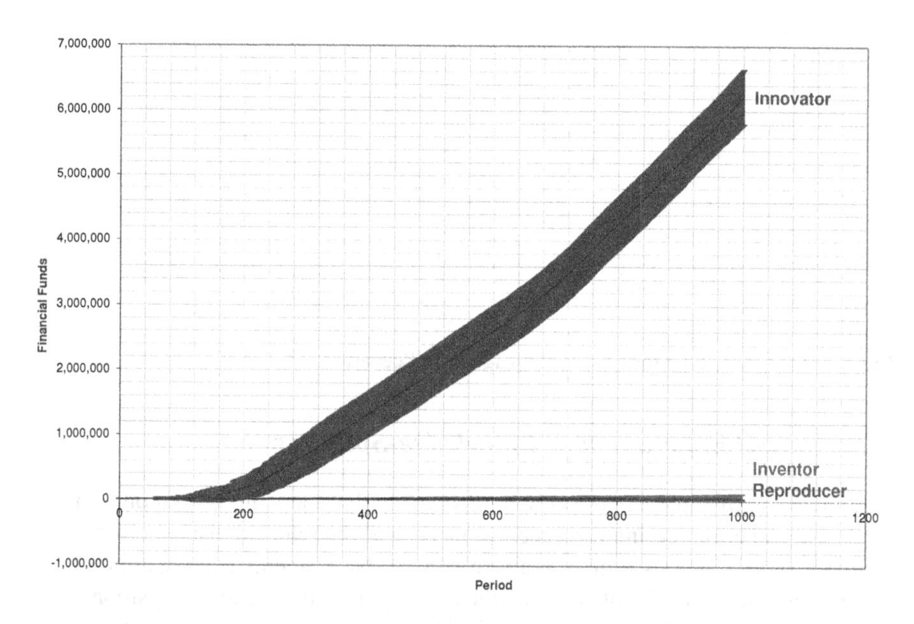

Fig. 3 Longitudinal Report Graph Financial Funds for Scenario 1

Indeed, the only group that really registers on the graph are the innovators—they are the only agents that obtain opportunities and capitalize on them accordingly, as Fig. 4 shows. The dominance of the innovator displayed in these graphs reflect much of the literature on entrepreneurship and opportunity recognition, which largely focuses on innovators, the classic 'heroes' of entrepreneurship who are supposed to be the entrepreneurs that make all the money, leaving no room for other players. But is this realistic?

In fact, it would be, if one assumed the inability of other market participants to obtain some of the knowledge an innovator holds, i.e. an environment with no knowledge spillovers—which is effectively the environment we model when we set the vision and relocate distances to zero. However, in the networked knowledge economy of the 21st century, appropriation is always possible, and appropriability is increasing. What might our model do if we adjust the settings to reflect this: to make players other than just the Innovators relevant?

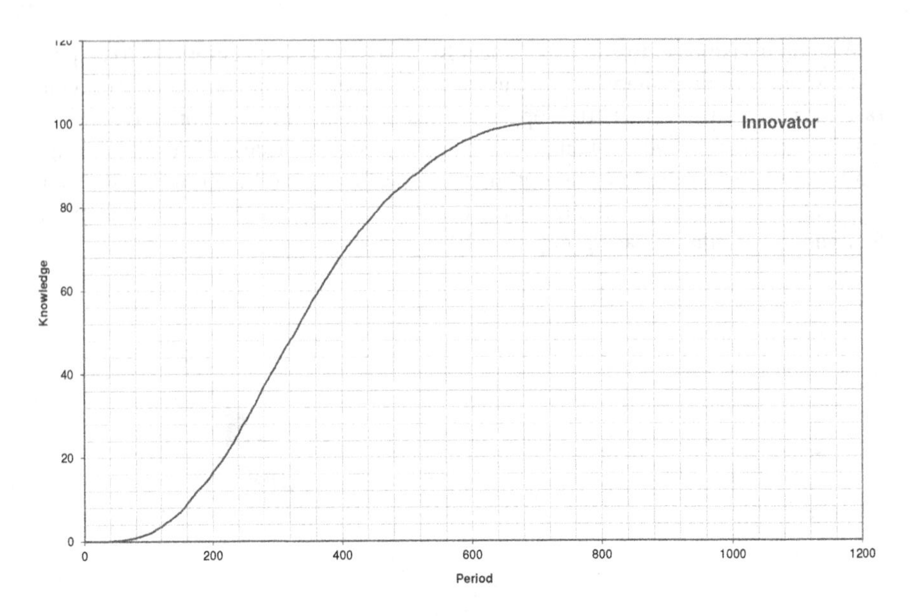

Fig. 4 DTIs obtained by Innovator group agents in Scenario 1

5.2 *Virtual Experiment 2: Knowledge-Spillovers*

Schumpeter [20] and Baumol [4] both recognized that there are indeed other players in the game. Increasing the vision property and the relocate distance from zero to just 5 should give us a more realistic picture. Now, agents can—within a limited radius—see other agents and can move away from their original positions—in short steps—around the 100 by 100 grid of the *SimWorld*. As an example, Fig. 5 depicts the area (black) that one (random) agent covered in one 1000 period run. We can now expect all agent groups to obtain at least some DTIs, because knowledge scanning is possible, even if limited. However, we cannot predict the specific effects this will have on the financial funds. Based on the distributions we have assigned to specify the properties of the three agent groups, we know their different general knowledge appropriation and development behaviors. But, again, how those different behaviors or strategies will play out in a population of agents in a knowledge environment, we do not know. We need the simulation to dynamically model the complex relationships among knowledge and agents across time to see how successful or not the different agent groups will be in terms of growing their financial funds and knowledge portfolios. Fig. 6 shows us the financial profiles for each group.

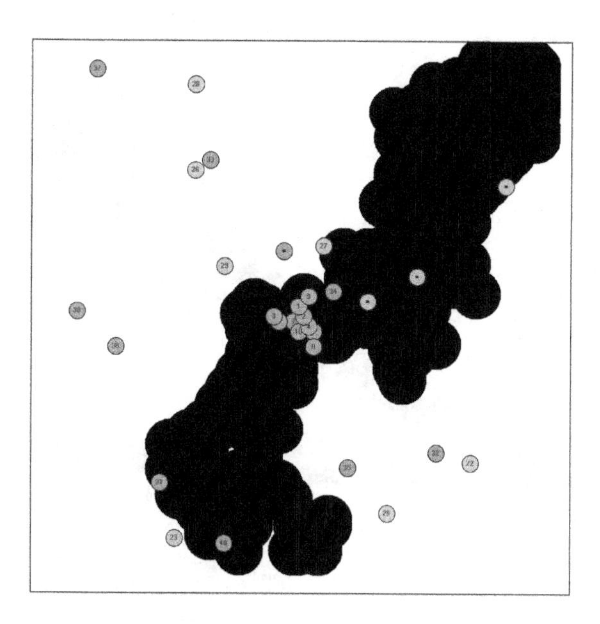

Fig. 5 *SimWorld* Report Period 1000

Whereas the sum of all financial funds (total generated by all agents)[10] does not change significantly between the first and the second scenario, we can now distinguish three different groups - based on the distinct opportunity recognition strategies of the three entrepreneurial types. With vision and relocate distance set to 5, Innovators perform better than Reproducers and Inventors, whose financial profiles overlap. Thus, when the assumption of inappropriability is relaxed, and agents can move in search of other agents' knowledge, the financial performance and performance volatility of the three entrepreneurial strategies—innovating, inventing, reproducing—will have distinct profiles resulting from the differences in their knowledge appropriation and knowledge development behaviors.

We can see that (compared to the Scenario 1 runs) the Innovator group 'loses' financial funds, which are picked up by the other two groups. This also illustrates the inner workings of the simulation. In both scenarios, Innovators follow the same pattern of actions (in particular, the number of their exploit actions does not change), but their financial funds in the first scenario are almost three times what they achieve in the second, demonstrating that these rents have been reduced by competition as a result of what we call the 'diffusion discount effect'. In the first scenario, fewer

[10] As a basic measure of the population's performance, we treat the total financial funds per period (the sum of all agent groups) as a proxy for GDP. Conceptually, it can be viewed as the market value of all final goods and services made within the *SimWorld* based on the exploitation of the agents' knowledge assets.

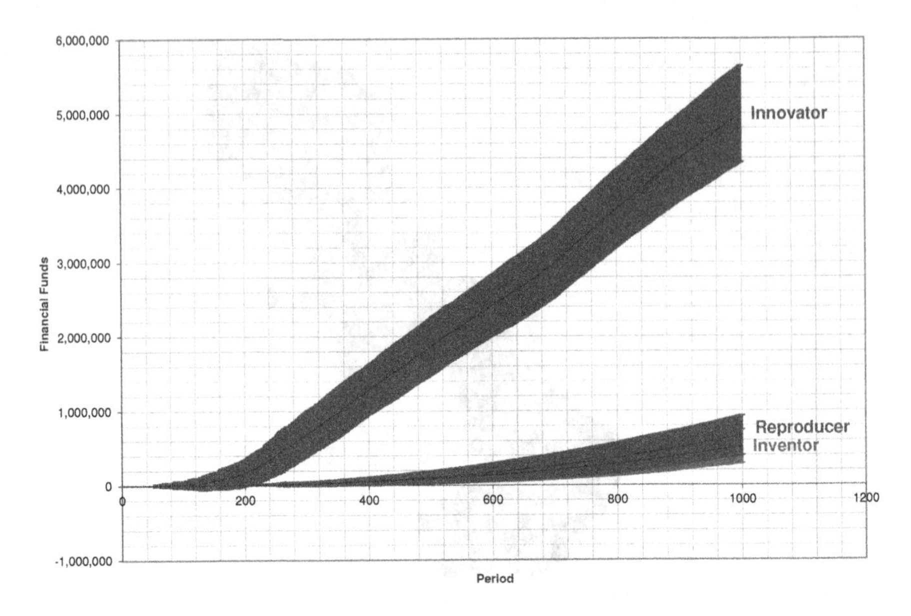

Fig. 6 Longitudinal Report Graph *Financial Funds* in Scenario 2

agents secure DTIs and those that do receive higher rents. More diffused knowledge is worth less, so innovators in the second scenario earn less (rent dissipation), because the total of rents is appropriated by and shared among other types of entrepreneurs. This showcases *SimOpp*'s distinct simulation and modeling capabilities, and together with the results described above, it can be seen as face validation of the model, showing that "the critical characteristics of the process being explored are adequately modeled and that the model is able to generate results that capture known canonical behavior" [7].

Simulation & Modeling Capability 1. *SimOpp* enables researchers to simulate the opportunity recognition process of different entrepreneurial strategies and their respective financial payoffs;

Simulation & Modeling Capability 2. *SimOpp* allows researchers to simulate the competitive agent behavior that results from different entrepreneurial strategies.

5.3 Opportunity Construction, Acquisition, and Amplification

The simulation allows us to look behind the groups' different financial performance profiles and to explore their comparative accumulation of DTIs (i.e., of opportunities). Fig. 7 shows the paths each group follows to obtain the ten DTIs (as there are ten agents in each group, the maximum is 100). The graph features the S-shaped

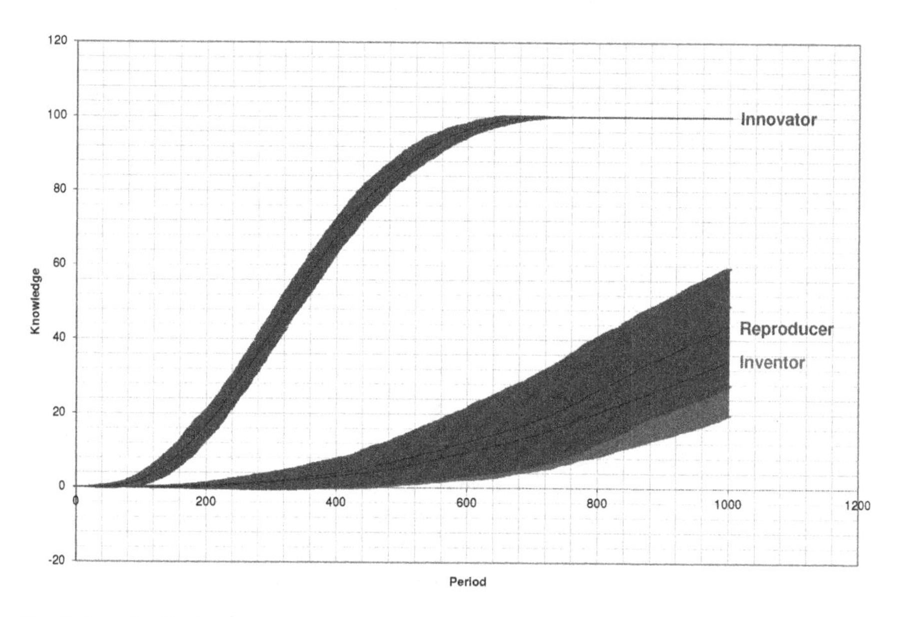

Fig. 7 Longitudinal Report Graph DTIs Obtained in Scenario 2

curves that are characteristic of diffusion processes [15][11]. The initial 600 periods are the most interesting to interpret.

The DTIs, with which we model opportunities, have a starting codification level of 0.8. Remember that the maximum level of codification is one and that the codification increment is set at 0.1. This means that up to three knowledge stores are obtainable per DTI per agent (with codification levels of 0.8, 0.9, 1.0), equivalent to 300 per agent group (see Fig. 8). As these DTI knowledge stores are the basis for exploit actions, the more stores an agent possesses the better they can perform.

Three different occurrences can let agents attain DTI knowledge stores, which means the results illustrated in Fig. 8 can be explained by three distinct 'actions' (ways with which entrepreneurs/agents can obtain and then develop opportunities). Fig. 9 shows the first one of these—the 'discovery' occurrence.

The graph depicts the outcome of the *Opportunity Construction* process described earlier. For a 'discovery' to happen, the underlying basic knowledge items have to be assembled and brought to the required threshold. We see that Innovators lead the process with Reproducers and Inventors in second place. The 'discovery' occurrence represents the classic entrepreneurial route of constructing one's opportunity step by step on the basis of one's idiosyncratic stock of knowledge [21]. The micro processes behind it are as follows: the missing basic knowledge group—new venture idea knowledge—can be obtained directly (in the cases of innovators and

[11] Research on diffusion of innovations [17] looks at the process of an innovation's adoption and implementation by a user. The main focus is on the market for and of an innovation. In contrast, we are looking at the knowledge dynamics that help *entrepreneurial agents* construct and attain opportunities

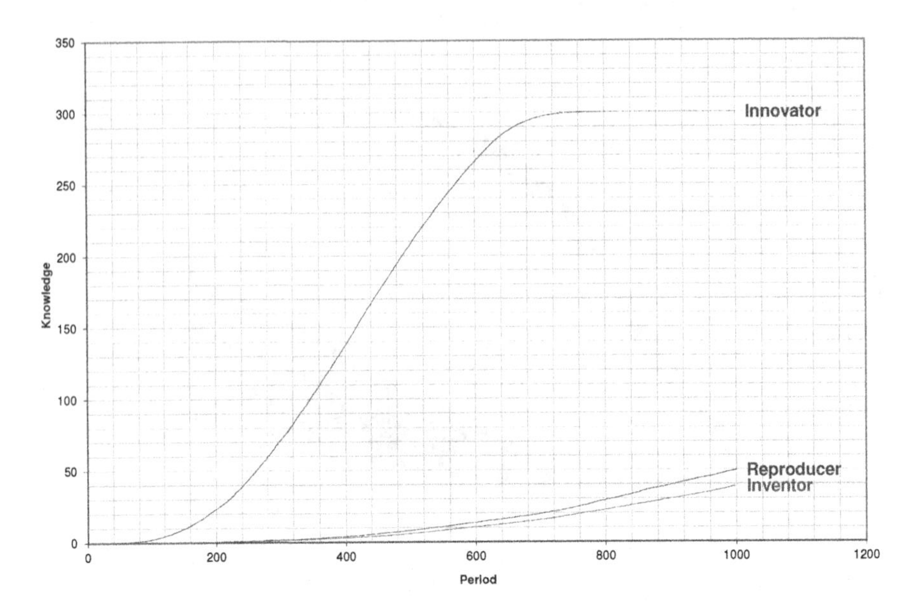

Fig. 8 Report Graph DTI Knowledge Stores in Scenario 2

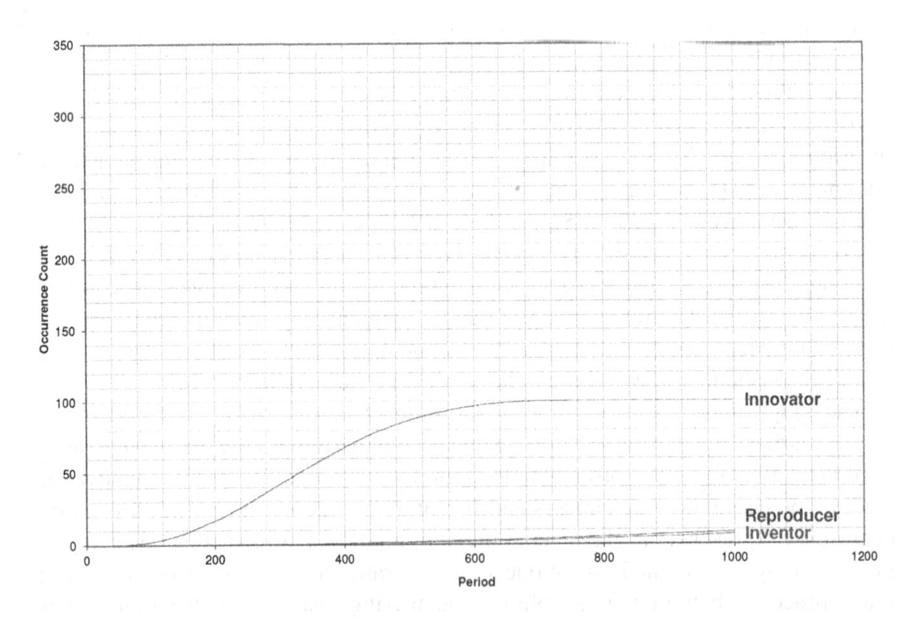

Fig. 9 Longitudinal Report Graph DTIs Discoverer in Scenario 2

inventors, by 'scanning from the knowledge ocean') or from another agent (as when Reproducers 'scan from others'). Bringing all the knowledge thus assembled up to the required threshold can also either be done directly (innovators 'codify') or by getting knowledge stores with higher codification levels from someone else (Reproducers and Inventors gain already-codified knowledge via 'scanning from others').

By doing everything directly, i.e. getting the idea and developing it themselves, the Innovator group comes top of the opportunity construction 'league' and the other two groups lag behind in terms of their ability to construct opportunities. However, these groups (Reproducers and Inventors) come first in the next graph that co-explains the total number of DTI knowledge stores: DTI scanning occurrences (Fig. 10).

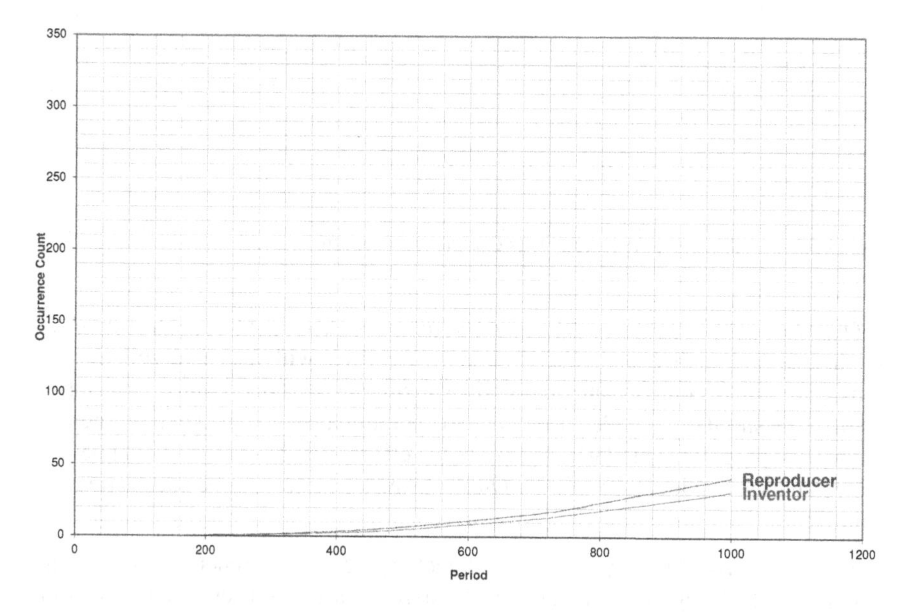

Fig. 10 Longitudinal Report Graph DTIs *Scanner* in Scenario 2

This graph shows what we described earlier as *Opportunity Acquisition*. The agents scan DTI knowledge stores from other agents, which allows them not only to obtain the basic knowledge about opportunities, but also the relevant further developed knowledge items—those DTI knowledge stores with higher levels of codification[12]. The 'production' of these can be observed in the next graph, which explains the number of DTI knowledge stores: DTI codification occurrences (Fig. 11).

Fig. 11 shows what we labeled earlier *Opportunity Amplification*. Only Innovators are able to further structure their knowledge about opportunities in our model.

[12] The Innovator group is not displayed in this particular graph as Innovators do not scan from other agents; they can only scan the basic knowledge from the ocean.

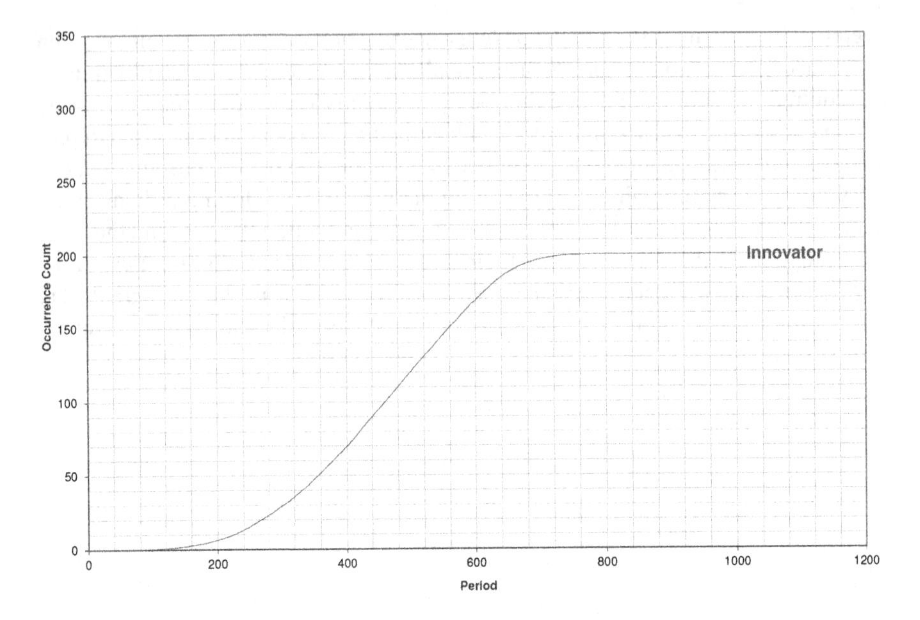

Fig. 11 Longitudinal Report Graph DTIs *Codifier* in Scenario 2

Opportunity Construction, *Opportunity Acquisition*, and *Opportunity Amplification* are distinct processes that, taken together (and also with their underpinning actions) give a more complete and fine-grained picture of what opportunity recognition entails. The simulation model enables us to operationalize different entrepreneurial actions and processes, and later analyze to the effects they have on financial performance and knowledge build up. *SimOpp*'s distinct features can be summarized as follows:

Simulation & Modeling Capability 3. *SimOpp* allows researchers to dissect the opportunity recognition process and to arrive at a more discriminating picture of entrepreneurial strategies. In particular, it facilitates the distinction between *Opportunity Construction*, *Opportunity Acquisition*, and *Opportunity Amplification*.

6 Extending *SimOpp*: Simulating *International* Opportunity Recognition and Knowledge Spillovers

Opportunity recognition processes are particularly interesting to study in an international context. To analyze the impact of the international transfer of knowledge in an increasingly global world, one has to construct a variant of *SimOpp* [11], which we do next.

6.1 Modeling the International Scenario

For this purpose, we must introduce an additional agent group—'National Innovators'—and an additional knowledge group—'Local Knowledge II'. National Innovators are our lead entrepreneurs from abroad. They were the first worldwide to have discovered a particular opportunity and implemented it in their country. Their label indicates that they are 'leading entrepreneurs' from the national setting where the knowledge originated [13]. With a couple of exceptions, the National Innovator group has the same simulation properties as the 'Innovator' group in the basic model (see Table 2).

Table 2 Settings for 'National Innovator' agent group (differences to 'Innovator' group in bold)

National Innovator: agent group that represents 'lead entrepreneurs from abroad'	
Knowledge agents possess:	• **Local II Knowledge** • Entrepreneurial Knowledge • New Venture Idea Knowledge ***[all knowledge groups are prior stock]***
Actions agents perform:	• **Abstract** • Learn • Exploit

As they are from a market, country or geographical region different to that of our three original groups, they have drawn on a different range of local knowledge in first constructing their opportunities—'Local Knowledge II' (which has the same simulation properties as 'Local Knowledge'). Since National Innovators are already 'ahead' in the entrepreneurial development process, and have already obtained all the required underlying knowledge, they are endowed with all knowledge groups from the start of the simulation: not only 'Entrepreneurial Knowledge' and 'Local Knowledge II', but also 'New Venture Idea Knowledge'. Since they already possess all relevant knowledge, the 'scanning' action is disabled for them. Coming from a different location and culture, National Innovators also have different cognitive predispositions, different ways of structuring knowledge. Therefore, we implement their knowledge structuring process with the 'abstraction' action rather than the 'codification' action[13].

When it comes to the opportunity construction process, National Innovators obtain their own set of ten DTIs 'abroad' (by combining knowledge items from the 'Entrepreneurial Knowledge', 'New Venture Idea Knowledge', and 'Local Knowledge II' groups, and bringing them to an abstraction level of 0.6). This set of DTIs,

[13] Note that in our model, we use codification and abstraction as two different ways of structuring knowledge, based on the region or culture the agent group is from. For this particular application, we do not make the conceptual distinctions between the two activities as described by Boisot [6].

called DTI II (as opposed to DTI I in the 'home' economy), cannot be constructed by our initial three groups, as their cognitive predispositions are different (they use the codification action, not the abstraction action to structure their knowledge). However, they can acquire DTI II items through scanning. Once scanned, they are in a position to capitalize on them, but the base value (of the DTI II items) is set to 20 and not to 2500 (the base value of DTI I items), since 'opportunities from abroad' do not (directly) constitute opportunities in the market of our three initial agent groups, so they are treated as additional new venture ideas instead (and are given the same base value of 20). To convert them into a 'real' opportunity, DTI IIs have to be adapted to the local region or market by combining them with the respective knowledge item of a 'Local Knowledge' group. This gives our three groups of entrepreneurs access to another set of ten DTIs - localized opportunities from abroad (DTI II in I) - which is given a base value of 2500 to reflect the special value of such opportunities. The opportunity construction processes are summarized in Fig. 12.

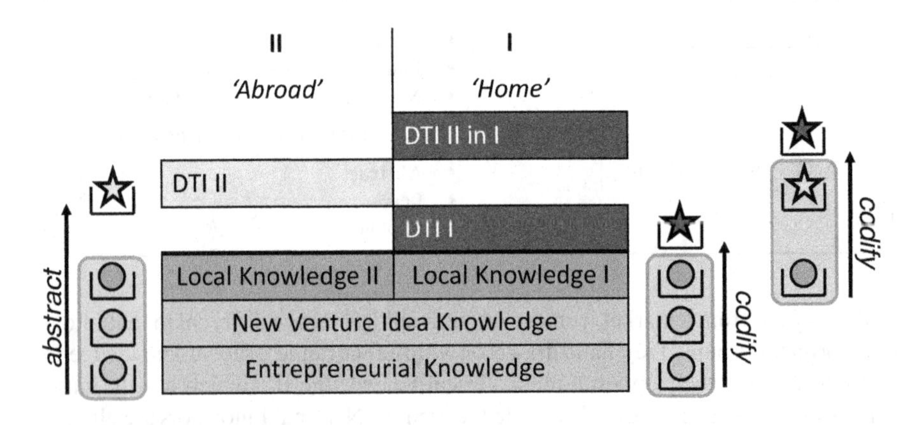

Fig. 12 Opportunity Construction: obtaining DTIs 'Home' vs. 'Abroad'

The *vision* and *relocate distance* settings for all agents remain at 5 (as in the base model). We do not use the 100 by 100 *SimWorld* grid to model the international scenario, but implement the international dimension differently, by having two different cognitive predispositions or ways of structuring knowledge, by introducing an additional local knowledge group, and by adding new sets of DTIs. We continue to use the vision and relocate distance properties in a more generic way, as in moving closer to or further away from knowledge sources, which may or may not be interpreted in terms of space/geography.

6.2 *The Base-Line Scenario*

Before running this new model, we have to establish a base-line set of outputs with which to compare the results of the simulation that incorporates the international dimension. We cannot just take the financial funds graph we produced in Scenario 2 above, because adding more agents and knowledge to the simulation will cause dilution effects, and an adjusted base scenario has to be created in order to later parse out the effect we are interested in, namely the effect of the availability of knowledge about opportunities from abroad. We make the following two adjustments. First, we run the base scenario with ten more (inactive) agents in order to account for the diffusion effect. The diffusion of a knowledge item, an important parameter that is part of the calculation of the exploit amount, is calculated by the number of agents that possess a knowledge item divided by the *total number* of agents. If we introduce our new 'National Innovator' group with ten agents, the increase in the total number of agents will have a direct, positive effect on the financial funds. Adding more agents in the simulation will increase the financial funds because knowledge is less diffused and thereby more valuable. Therefore, we take care of this effect in an adjusted base version. Second, we run the base scenario again with the 'Local Knowledge II' group and its ten additional knowledge items that we need for modeling the international context already added in. Increasing the numbers of knowledge items to be scanned, codified, learned and exploited also has a direct effect on the financial funds. The more knowledge items there are in the system, the longer it takes agents to reach the same amount of funds. It takes the agents more time to deal with the base knowledge before they can discover and exploit the more valuable DTIs, and therefore the financial funds graphs will shift to the right.

For our initial two scenarios above, we ran each scenario for 60 runs, each of 1000 periods. We now analyze 2000 periods, which will allow us to highlight certain trajectories. Because of the complexities of our simulation, the data storage and processing capacities required are extremely high. Adding more knowledge and agents and running the simulation for 2000 periods instead of 1000 limited us to only 20 runs (rather than 60) for the adjusted base scenario and the international scenario. However, running all the scenarios described above with only 20 instead of 60 runs showed no apparent difference between the results.

Fig. 13 displays the financial funds profiles of our three agent groups (in a closed economy) in the adjusted base scenario. The graph shows the average across all runs and also displays the standard deviation to indicate the significant difference between the individual lines. The outcome is consistent with Scenario 2 above. We can distinguish the different groups—based on the distinct opportunity recognition strategies of the three entrepreneurial types. Innovators outperform Reproducers and Inventors, whose financial profiles overlap.

6.3 Exploring Entrepreneurial Strategies in an International Environment

Having the adjusted base scenario in place, we can run our international case, which is now influenced by our lead innovator (the "National Innovator"), the additional knowledge group ("Local Knowledge II"), and the two more sets of DTIs. Based on the distributions we have assigned to specify the properties of the three agent groups, we know their different general knowledge appropriation and development

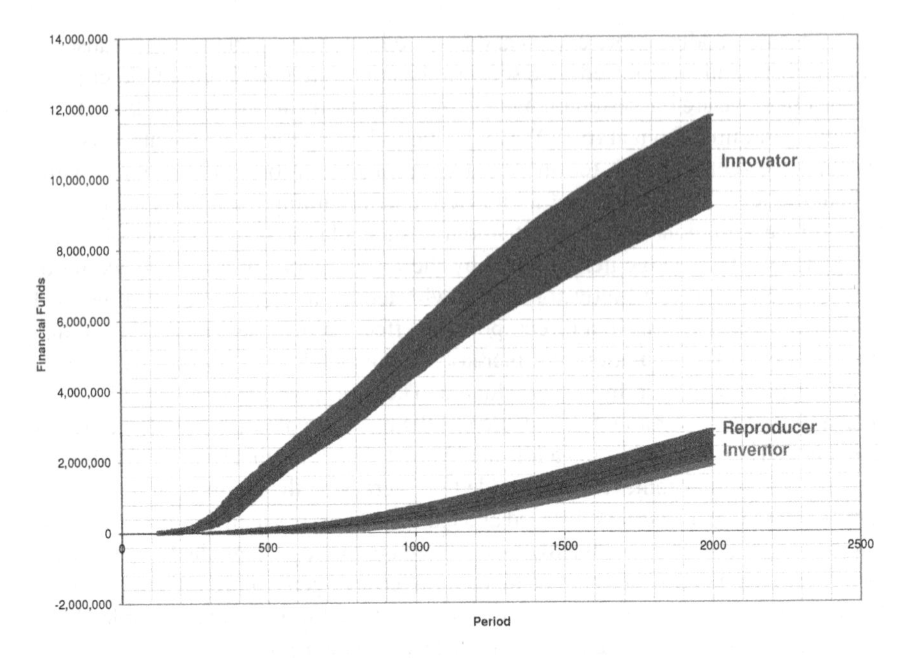

Fig. 13 Financial Funds Adjusted Base Scenario (closed economy)

behaviors, and we tested them in a closed economy (Fig. 13). How those different behaviors or strategies will play out in a population of agents in an international knowledge environment, however, we do not know. We cannot predict the specific effects this will have on the financial funds. We need the simulation to dynamically model the complex relationships among knowledge and agents across time to see how successful or not the different agent groups will be in terms of growing their financial funds and knowledge portfolios.

The financial performance of our three groups of entrepreneurial agents in an international world is displayed in Fig. 14. The financial performance of the National Innovator group is not displayed, because we are only interested in the effects on the original "Home" market and its participating entrepreneurs.

Whereas in the base scenario Innovators outperformed Reproducers and Inventors by far in terms of accumulating financial funds (Fig. 13), in the international case both groups get closer to the Innovator group. Scenario 2 above already highlighted the important effect of knowledge-spillovers; but when the international dimension is introduced, the outcome is even more revealing. The simulation experiment gives evidence that the payoffs resulting from different opportunity recognition strategies are changing in a globalised knowledge economy. For the development of entrepreneurship theory it is notable that in an international setting, when knowledge about opportunities is accessible from abroad, the financial performance profiles of the entrepreneurial actors can be significantly different - an insight that calls for further investigation in follow-on simulation studies using the SimOpp model and in future empirical research.

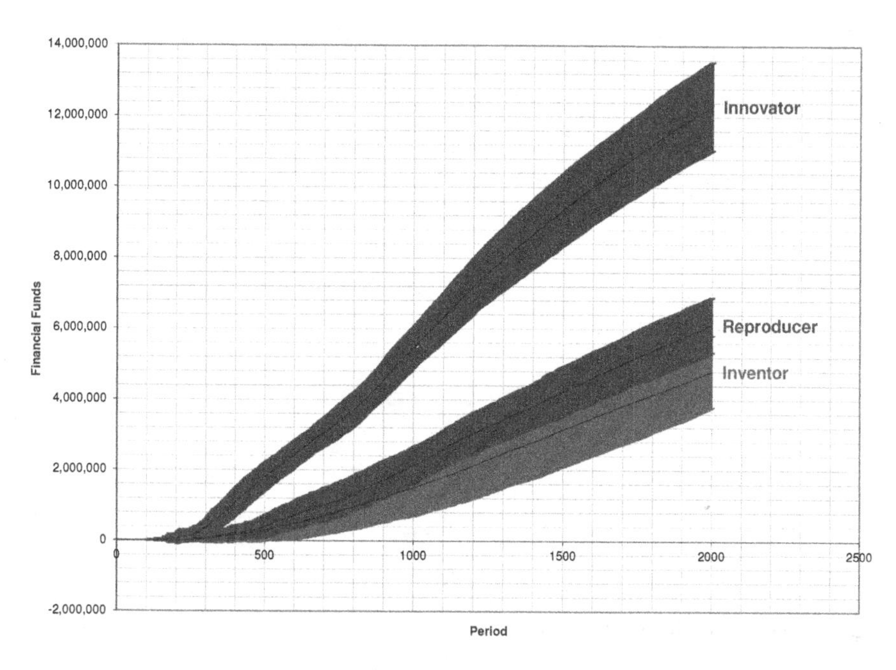

Fig. 14 Financial Funds International Scenario

An advantage of using simulation methods is that, in addition to performing analyses at the micro level—looking at individual agents and agent group behaviors—users also explore effects at the macro level. Not only can we assess the financial performance of different groups of entrepreneurial agents, but we can also see the societal (whole population) effects of their actions. We could analyze the total financial output (GDP) and also the "outgrowth of entrepreneurial opportunities"—the growth in the number of new products or services on the market—which society at large is interested in. An important result seems to be that the international environment leads to an influx of new opportunities to the market (as displayed in Fig. 15), which would constitute the basis for new innovation and employment. Reproducers and Inventors both have access to knowledge about opportunities from abroad and follow a similar knowledge trajectory in Fig. 15 (in contrast to Innovators, who just have access to 100 DTIs vs. 100 + 100 DTIs).

The international scenario lends itself to 'spatial' or geographic interpretations that have implications for regional policy initiatives and the individual strategies of transnational intra- and entrepreneurs. However, our new model also allows us to analyze industry dynamics, so that, instead of saying "Local Knowledge" and "Local Knowledge II" represent different cultures and regions, we characterize particular industries. Using ideas and opportunities from different industries can enable a company to get an advantage over its competition. Again, *SimOpp* can help analyze the particular entrepreneurial actions required to succeed in different competitive environments. So we can add an additional simulation and modeling capability.

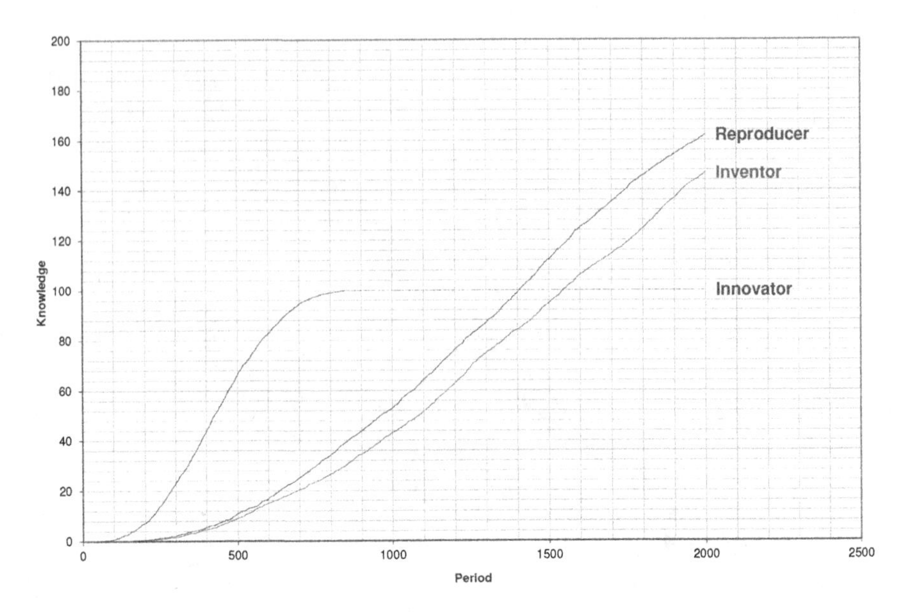

Fig. 15 DTIs Known International Scenario

Simulation & Modeling Capability 4. *SimOpp* allows one to model and simulate knowledge exchange between different cultural, spatial, and industry arenas and its resulting micro and macro effects.

7 Conclusion

We used the *SimISpace2* software to develop a model that can simulate the opportunity recognition process and competitive agent behavior and the respective financial payoffs associated with different entrepreneurial strategies. Dissecting the opportunity recognition process allows us to offer a more discriminating picture of entrepreneurial strategies, and we can simulate knowledge exchange between different cultural, spatial and industry arenas and analyze both the resulting micro and macro level effects. With its application specific model, *SimISpace2* provides researchers with a tool for analyzing strategic entrepreneurship in a knowledge economy and for deriving theory-driven propositions and hypotheses that can inform corporate and public-sector decision makers.

Entrepreneurship theory suggests three distinct groups of entrepreneurs - Innovators, Inventors, and Reproducers. Using our simulation model to test the validity of these distinctions, our results showed that the groups' different knowledge appropriation and knowledge development behaviors gave rise to divergent knowledge progression and financial performance profiles. Particularly in a global world, with rapidly expanding international travel and information and communication technologies, knowledge spillovers exist, both on a local and an international level.

We were able to extend our model to help researchers also analyze entrepreneurial strategies in an international environment. With access to knowledge about opportunities from abroad, competitive dynamics seem to change, which has both individual and societal level effects.

The focus of this paper has been the parameterization of the model, its face validation by conducting some basic simulation runs, and the illustration of distinct modeling capabilities. A research program that uses the tool to run and analyze a series of virtual experiments is now needed to fully leverage its capabilities to advance entrepreneurship theory. It is hoped that the simulation software we have built will be used for future simulation studies in the fields of entrepreneurship, strategy, international management and beyond, and that it will generate results with manifold theoretical and practical implications. It is not only academia that can utilize the tool, also corporate and public sector entities can exploit its benefits. There are numerous practical, real-world applications; for example, governments can study science, technology and innovation policies and knowledge-based firms can explore corporate entrepreneurship and innovation strategies.

References

1. Aldrich, H.E., Martinez, M.A.: Many are called, but few are chosen: an evolutionary perspective for the study of entrepreneurship. Entrepreneurship Theory and Practice 25(4), 41–56 (2001)
2. Alvarez, S.A., Barney, J.B.: Opportunities, organizations, and entrepreneurship. Strategic Entrepreneurship Journal 2(3), 171–173 (2008)
3. Alvarez, S.A., Barney, J.B.: Opportunities, organizations, and entrepreneurship. Strategic Entrepreneurship Journal 2(4), 265–267 (2008)
4. Baumol, W.J.: Entrepreneurship and a century of growth. Journal of Business Venturing 1(2), 141–145 (1986)
5. Boisot, M.H.: Information Space: A Framework for Learning in Organizations, Institutions and Culture. Routledge, London (1995)
6. Boisot, M.H.: Knowledge assets — securing competitive advantage in the information economy. Oxford University Press, New York (1998)
7. Carley, K.M.: Computational organizational science and organizational engineering. Simulation Modelling Practice and Theory 10(5-7), 253–269 (2002)
8. Davis, J.P., Eisenhardt, K.M., Bingham, C.B.: Developing theory through simulation methods. Academy of Management Review 32(2), 480–499 (2007)
9. Harrison, J.R., Lin, Z., Carroll, G.R., Carley, K.M.: Simulation modeling in organizational and management research. Academy of Management Review 32(4), 1229–1245 (2007)
10. Ihrig, M.: Investigating entrepreneurial strategies via simulation. In: Bargieła, A., Ali, S., Crowley, D., Kerckhoffs, E. (eds.) European Conference on Modelling and Simulation, Kuala Lumpur, Malaysia (2010)
11. Ihrig, M.: Simulating entrepreneurial strategies in a global knowledge economy. In: Burczyński, T., Kolodziej, J., Byrski, A., Carvalho, M. (eds.) 25th European Conference on Modelling and Simulation (ECMS), Kraków, Poland (2011)
12. Ihrig, M., Abrahams, A.S.: Breaking new ground in simulating knowledge management processes: Simispace2. In: Zelinka, I., Oplatková, Z., Orsoni, A. (eds.) 21st European Conference on Modelling and Simulation (ECMS 2007), Prague (2007)

13. Ihrig, M.: zu Knyphausen-Aufseß, D.: Discovering international imitative entrepreneurship: Towards a new model of international opportunity recognition and realization. Zeitschrift für Betriebswirtschaft Special Issue 1/2009 (2009)
14. Ihrig, M., Zu Knyphausen-Aufseß, D., O'Gorman, C.: The knowledge-based approach to entrepreneurship: linking the entrepreneurial process to the dynamic evolution of knowledge. Int. J. Knowledge Management Studies 1(1–2), 38–58 (2006)
15. Mahajan, V., Peterson, R.A.: Models for innovation diffusion. Sage Publications, Beverly Hills (1985)
16. Nafziger, E.W.: Entrepreneurship and economic development. Jai Press, Greenwich (1986)
17. Rogers, E.M.: Diffusion of innovations, 5th edn. Free Press, New York (2003)
18. Ryle, G.: The concept of mind. The University of Chicago Press, Chicago (1949)
19. Schumpeter, J.A.: The theory of economic development. Harvard University Press, Cambridge (1949)
20. Schumpeter, J.A.: The creative response in economic history. Addison-Wesley Press, Cambridge (1951)
21. Shane, S.: Prior knowledge and the discovery of entrepreneurial opportunities. Organization Science 11(4), 448–469 (2000)
22. Shane, S.: A general theory of entrepreneurship: the individual-opportunity nexus. Edward Elgar, Cheltenham (2003)
23. Shane, S., Venkataraman, S.: The promise of entrepreneurship as a field of research. Academy of Management Review 25(1), 217–226 (2000)
24. Venkataraman, S., van de Ven, A.H.: Hostile environmental jolts, transaction set, and new business. Journal of Business Venturing 13(3), 231–255 (1998)

Agent-Based Simulation in AgE Framework

Łukasz Faber, Kamil Piętak, Aleksander Byrski, and Marek Kisiel-Dorohinicki

Abstract. The chapter introduces AgE framework as a core for constructing agent-based simulation systems. Its features are described against other solutions that may be used in the area of agent-based simulation. The discussion focuses on technical issues—the support for agent-specific services as well as the mechanisms allowing for extensibility and flexibility of the configuration of simulation models and systems. The considerations are illustrated by a simple case study, which aims at showing the differences between AgE and several selected tools for agent-based simulation.

1 Introduction

For certain problems appropriate models may be built and desired research may be performed using available means—without specialized infrastructure (features of uniform motion may be explored using a battery-propelled toy car and a stopwatch, a toy car let down the slope may help in understanding motion with constant acceleration, etc.). However, the research on complex systems (e.g. factory assembly line) and phenomena (e.g. metal casting) often requires great effort that must be put on examining their determinants, defining conditions for experimental runs, and the observed features analysis. Therefore, appropriate strategies of performing experiments must be defined, requiring placing of sensors and reading them, statistical processing of the acquired data and computing their plausibility.

Unfortunately, paraphrasing Heisenberg's uncertainty principle, by introducing sensors into the system (e.g. reading the temperature during casting), the conditions of experiments are changed, and readings of these sensors are affected by their presence. Also 'live' system analysis suffers from constrained possibilities of repeating

Łukasz Faber · Kamil Piętak · Aleksander Byrski · Marek Kisiel-Dorohinicki
AGH University of Science and Technology, Al. Mickiewicza 30, 30-059 Kraków, Poland
e-mail: faber@student.agh.edu.pl,
{kpietak,olekb,doroh}@agh.edu.pl

A. Byrski et al. (Eds.): Advances in Intelligent Modelling and Simulation, SCI 416, pp. 55–83.
springerlink.com

the experiments with exactly the same conditions, which is required for statistical processing of acquired data.

Obviously, there exists a possibility of constructing mathematical models, which allow to formalize certain aspects of phenomena under interest, e.g. the process dynamics based on differential equations or Markov chains [24]. Yet such models are usually quite complex and have limited (though non-disputable) applicability in pure theoretical analyses (cf. [8, 23]). Computer-based models and simulations support the researcher with powerful means to create a virtual environment and conduct experiments with a possibility to repeat the simulations under different conditions, generating vast amounts of data ready to be processed statistically for presentation [28]. Thus, appropriately defined simulation may help in understanding systems or phenomena of interest, prior to performing real-life experiments.

As digitial computers are discrete-time machines *per se*, discrete-time simulation becomes a natural way of defining such models and implementing different supporting tools [3]. Since it was possible to generalize typical elements of such models (e.g. objects, events, actions), dedicated software tools became available, most of them proprietary, but several open-source too [22, 19]. At the same time the idea of agency (e.g. [29]) seems well suited for implementation of heterogeneous complex systems simulations, as the globally controlled algorithm is replaced with local perception and interactions among the agent, its neigbours in the environment and the environment itself. The resulting combination of the discrete and agent-based approach constitutes a base for a wide range of simulation systems (see, e.g., [20, 18, 16]).

In the course of this contribution, an agent-oriented framework *AgE* is presented as a tool supporting the construction of distributed simulation systems. The chapter begins with a short review of available discrete-event simulation frameworks. In the next section, the structure and principles of work of *AgE* are presented. The discussion of different features of *AgE* framework shows that it might be interesting and competitive when compared to other available open-source software. In the end, a case study depicting the application of selected frameworks to simulation of simple inter-agent interaction is presented.

2 Agent-Based Simulation Frameworks

Among a variety of problems (sociological, biological, etc.) requiring simulation approach there are complex processes observed in populations consisting of a huge number of different, possibly autonomous individuals. For such problems macro-scale models may be defined, which allow for understanding the dynamics of the emerged phenomena using appropriate mathematical apparatus, in order to perceive its certain features. Alternatively, mimicking the behaviour of single interacting entities, and observing the emerged phenomena in the whole population may be considered. Such entities, when situated in some environment, capable of perceiving the environment, interacting with themselves and the environment fall under the definition of autonomous agents [29].

Agency brings many improvements into the world of simulation, following the idea of decentralisation of control. Each agent may be autonomous, differently configured, utilising different means of discovering the features of the environment and its neighbours, utilising different algorithms and performing different actions in the system. Because of these, interesting added value may be taken for granted: building complex systems consisting of different, interacting beings will be natural when referring to agents as simulation objects.

As an example, one may consider economic modelling, which can, take into account new and, arguably, more precise characterisations of human beings. This way of modelling economic agents may become an alternative to more traditional mathematical models employed in economics. Those traditional descriptions of human beings normally exclude, for the sake of tractability, fundamental aspects such as qualitative descriptions of the agents' goals and intentions, beliefs and other attributes of human reasoning (e.g., bounded rationality) [26].

To recapitulate: the use of agents opens up new possibilities to introduce models very close to their real equivalents. At the same time the approach of agent-based simulation, because of its inherent logical decomposition and decentralisation of control, allows for building models featuring high flexibility and extensibility. Obviously all these features must be supported by dedicated software tools.

2.1 *Technical Issues*

In agent-based simulation the problem of synchronisation of autonomous entities, usually acting in parallel becomes of vast importance. Classical issues known from the parallel programming [1], such as deadlocks or starvation must be avoided, that becomes a non-trivial task when considering simulations consisting of hundreds or even milions of agents. Technical problems would arise if the agents were implemented as independent threads, especially in large systems, involving distributed processing. Far better seems to inverse the problem of synchronisation, by following so called *phase simulation* approach [21] in which there exists a synchronisation mechanism shared by a group of agents, which takes care of letting each regular agent (taking part in the simulation) do some activities (e.g., perform some query on the system state, change its state, register some action to be later performed by the synchronisation mechanism. In this way, complex parallel programming is changed into a kind of 'round robin' technique that does not pose problems in perceiving parts of the simulation as still autonomous (and acting in parallel, in, *nomen omen*, simulated way). Because of that, agent-based discrete-time simulation frameworks gained the attention of the authors.

Following the requirements of agency, other technical issues influencing the implementation should be considered, to make possible construction (or adaptation) of general-purpose agent-based simulation frameworks fulfilling the user's needs:

- life cycle control (means for definition of an agent being and managing its life cycle in the system),

- communication facilities (means for inter-agent and agent-platform communication making possible interaction between agents and the system, and among the agents, even if they are placed in remote locations, as different computation nodes),
- organisation (possibility of introducing some structures into agent organisation, e.g., groups or even trees of agents), making possible mimicking behaviour of real societes or implementing various divide and conquer-like algorithms),
- distributed computing (for simulations which require running on computer clusters so the total time of experiment is reduced),
- exetensibilty (low coupling on different levels of implementation, following e.g., reusable software components paradigm, that help in modularisation and generalization of the framework application, making possible easy exchange of algorithm, agent, environment parts so that the platform may be easily reconfigured for different experiments, repeating simulations in different conditions, focusing on configuration instead of programming),

One may also consider the platform code status—is the code open-source, up-to-date, cross-platform, making possible to quickly learn provided API, using as a support existing developer's blogs or fora.

2.2 Existing Solutions

There exists a plethora of multi-agent frameworks which may be used to support the construction of simulation systems. Some of them are oriented to specific kinds of simulation (see [19, 22]): e.g., simulating of movement of entities with 3D visualisation (see e.g., breve [17]), networking (see e.g, ns2/ns3 [9]), possibility of visual programming (see e.g. SeSam [27]). When looking for universal agent-based simulation frameworks (especially in open-source software domain), one should consider such products as Galatea [12], RePast [20], Mason [18], SystemC [6], SimPy[1]. Other ones are general-purpose agent-based programming frameworks (e.g. JADE [4]) that may be of course adapted to any kind of simulation. MadKit [16] should also be mentioned as a framework for simulating complex populations (following Agent/Group/Role paradigm [13]). An interesting, though not popular framework is Galatea [12] that makes possible utilising HLA [11] infrastructures for running the simulation as federation.

The below-mentioned frameworks were selected as the most promising examples of general-purpose agent-based simulation frameworks in the open-source market.

MASON

MASON is an agent-oriented simulation framework developed at George Mason University. The name refers to the parent institution, as well as derives from the acronym Multi-Agent Simulator Of Neighbourhoods (or Networks) [18]. Mason is advertised as fast, portable, 100% Java based. Multi-layer architecture brings

[1] `http://simpy.sourceforge.net/`

complete independence of the simulation logic from visualisation tools which may be altered anytime. The models are self-contained and may be included in other Java-based programs. Various means for 2D and 3D visualisation, and different means of output are available (PNG snapshots, Quicktime movies, charts and graphs, data streams).

Simulation in MASON consists of a model class `SimState` that composes random number generator and a `Schedule`. An object of `Schedule` class manages many agents, implementing `Steppable` interface, therefore an agent may interact with other ones and the environment by exposing predefined method that will be called by the `Schedule` [18]. The `SimState` may also manage the spatial structure of the simulation with a concept of `Fields` allocating different objects, thus constructing environment in which the agents may be situated.

Programming model of MASON follows basic principles of object-oriented design. An agent is instantiated as an object of a class, added to a scheduler and its `step` method is called during the simulation. There are no predefined communication nor organisation mechanism, these may be realized using simple method calls. There are neither ready-to-use distributed computing facilities nor component-oriented solutions.

First released in 2003, the environment is still maintained as an open-source project, distributed under Academic Free license (ver. 3.0). The current version (16.0) was released in the end of 2011.

RePast

RePast—Recursive Porous Agent Simulation Toolkit—is widely used agent-based modeling and simulation tool. Repast has multiple implementations in several languages and built-in adaptive features such as genetic algorithms and regression [20]. The framework utilizes fully concurrent discrete event scheduling, HPC version also exists [10]. In Repast 3, there are many programming languages interfaces (e.g., Java, Logo dialect, .NET languages, Lisp dialect, Prolog, Python). Logging and graphing tools are built-in. Dynamic access to the models in the runtime (introspection) is possible using graphical user interface. There are predefined libraries for different methods of modelling and analysis available, e.g., neural networks, genetic algorithms, social-network modelling, GIS support.

The implementation of a simulation system in RePast 3 is realized in a similar way as described for MASON. The class suitable for simulation should extend `SimpleModel` class, and contain appropriate implementation of `step()` function that will be called by the scheduler. It is to note that proper construction of the class' attributes allows to edit them using introspection mechanism supported by RePast GUI. The simulation may be even interrupted, available parameters may be changed and the simulation may be carried on.

Repast 3 consists of different implementation of the platform (Repast J—Java-based, Repast.NET—MS .NET and Repast Py—Python). It has been renowned for a long time, however, recently Repast 3 has been superseded by its next stage development called Repast Simphony (Repast S) bringing newly developed GUI, with some significant changes into the programming paradigm.

In Repast Simphony, graphical programming mode has been introduced, therefore a model may be constructed according to Rapid Application Development approach to software construction and the generated code may be easily integrated with Java and Groovy components. The execution environment supports exporting the simulation results to many popular external tools such as R, Weka or MATLAB. Sophisticated 2D and 3D visualisation features make possible integration with JUNG and GIS. The Repast Simphony inlcudes various libraries supporting genetic algorithms, neural networks, regression, random number generation and other mathematical tools.

In Repast Simphony, there has also been a new organisation concept called 'context' introduced. It generally consists of a group of unorganized agents (they may be organized using so-called projections) and may create a hierarchical structure (context can have many sub-contexts and so on). This idea affects the perception of agents in such way, that an agent in the sub-context also exists in the parent context, but the reverse is usually not true.

The latest (Simphony 2.0 beta) version of this open-source project, licensed according to 'new BSD' license, has been released in the late 2010.

MadKit

MadKit is a modular and scalable multi-agent platform written in Java aimed at modelling different agent organisations, groups and roles in artificial societies. Extensive GUI is available, as well as different programming languages for agents definitions (e.g., Python, Scheme, Jess, BeanShell) [16].

MadKit is built based on so called Agent/Group/Role organisational model [13] utilising plugin-based architecture. The architecture of MadKit is based on microkernels which provide only the basic facilities: messaging to local agents, management of groups and roles, launching and killing of agents. Other features (remote messages, visualisation, monitoring and control of agents) are performed by agents. Both thread based agents and scheduled agents for multi-agent based simulation may be developed. Different execution contexts are available: JSP, Applet, Console mode, GUI desktop, etc.

Threaded agents participating in simulation inherit from the `Agent` class and need to define at least the `live()` method that specifies their behaviour. This method represents the main loop of the thread. Non-threaded agents, inheriting from the `AbstractAgent` class, may also be used but they require a scheduling agent that can control their life cycle (e.g., which methods should be called at what time). Besides schedulers there is also one more kind of special agents, called *watchers*, inheriting from the `Watcher` class. They can use a set of *probes* to inspect parameters of other agents or the whole communities. All agents share also common life cycle calls: `activate()` run during agent initialisation and `end()` that is called at the end of the life-cycle.

Simulation does not require any particular structure or model to run. However, as mentioned earlier, it is possible to add an arbitrary scheduler or create complex agent communities and relations. Agents can locate other agents that handle a specific role (or many roles) or are members of a particular group. They can communicate with each other using these roles or group membership (i.e. using broadcast messages) and react to them.

MadKit provides an environment (in the desktop-like form) that allows for rapid creation and manipulation of agents with tools like a code editor, an agent designer or an agent observer that can, for example, log all exchanges of messages. There is also some support for binding GUI elements (graphical representations) to agents. User can replace parts of the environment and extend it with system agents that can use hooks in the micro-kernel to receive notifications about system events (publish-subscribe model).

The latest stable (4.2) version of this open-source project, licensed according to GPL license, has been released in 2008. Currently, alpha releases of version 5 are also available.

3 Agent-Based Simulation in AgE

AgE environment[2] is being developed as an open-source project at the Intelligent Information Systems Group of AGH-UST. *AgE* provides a platform for the development and execution of distributed agent-based applications—mainly simulation and computational systems.

Fig. 1 presents an overview of a system based on *AgE* platform. A user constructs the system by providing an input configuration in XML format. The configuration specifies the simulation structure and problem-dependent parameters. After the system start-up, the environment (agents and required resources) are instantiated, configured and distributed amongst available nodes where they start performing their tasks. Additional services such as name, monitoring, communication and topology service are also run on these nodes to manage and monitor the computation. The output of the simulation is problem dependent and may be visualized at run-time by dedicated graphical tools and interpreted by the user.

3.1 Structure and Execution of Agents

A simulation is decomposed into agents, which represent individuals or parts of or whole populations. Agents are structured into a tree with virtual root agent (as shown in Fig. 2) according to the simulation decomposition. The top level agents (called *workplaces*) along with all their children can be distributed amongst many nodes.

Agents can have named properties, which are features of an object, which can be referenced during run-time by its name in order to access, modify or even monitor

[2] `http://age.iisg.agh.edu.pl`

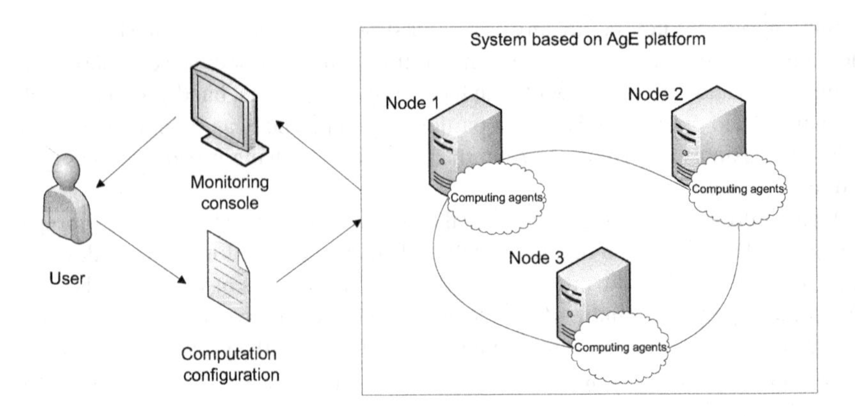

Fig. 1 AgE system overview

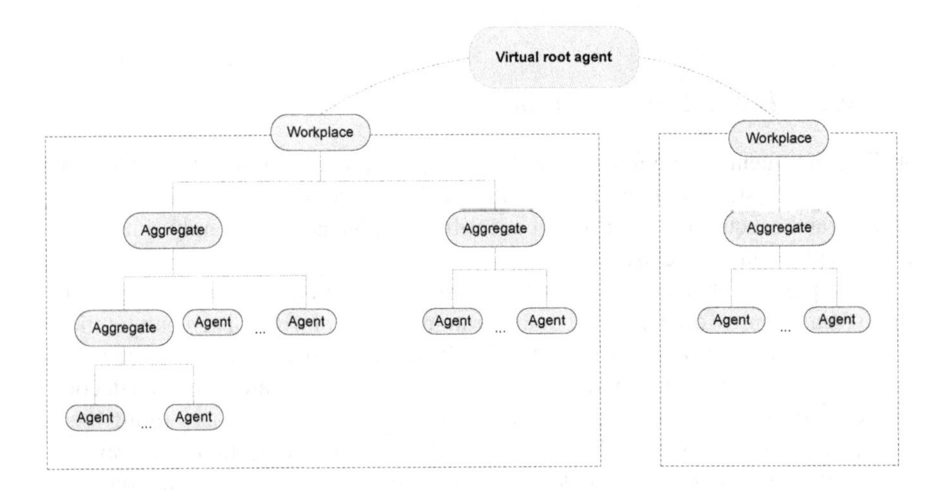

Fig. 2 Agent tree structure

its value. Properties are defined by annotating fields or methods of agents classes with dedicated Java annotations [7]. Each agent exists in an environment, defined by the parent agent, which provides a context of agent's processing. With the use of the environment, agents can communicate with their neighbour agents via messages, acquire specific information about them via queries, or even request them to perform specific actions.

It is assumed that all agents at the same level of the structure are being executed in parallel. The platform introduces two types of agents: thread-based and simple. The former are realized as separate threads so that the parallel processing is managed by Java Virtual Machine (similarly to JADE platform). Such agents can communicate and interact with neighbours via asynchronous messages. However, a large number of such agents would significantly decrease the performance of a simulation because of frequent context switching and raises synchronisation problems as discussed in Section 2. Therefore, following the concept of *phase simulation*, the notion of simple agents is introduced. The execution of simple agents is based on *steppable* processing which is to simulate pseudo-parallel execution of agents' tasks. Two phases are distinguished:

- Execution of tasks related to computation semantics in the `step()` method. In case of an aggregate agent all it's children perform their steps sequentially. While doing so, they can register various events, which may indicate actions to perform or communication messages, in the parent aggregate.
- Processing of events registered in an event queue. Since events may be registered only in agents that possess children, this phase concerns only aggregate agents.

The described idea of agents processing ensures that during execution of computational tasks of agents co-existing at the same level in the structure (agents with the same parent), the hierarchy remains unmodified, thus the tasks may be carried out in any order. From these agents perspective, they are processed in parallel. All changes to the agent structure are made by aggregates during processing of the events that indicate actions such as addition of a new agent, migration of an agent, killing an already existing agent, etc. They are visible for agents while performing the next step.

3.2 Actions

The environment of simple agents determines the types of actions which may be ordered by child agents. It also provides concrete implementations of these actions and thereby supplies and influences agents' execution. Thus actions realize the agent principle of goal level communication [5], because agent only lets the environment know what it expects to be done but it does not know how it will be done.

Simple agents request their parent aggregates to execute actions during an execution of a step of processing. Then, all of actions are executed sequentially (in order of their registration) by the aggregate after all children agents finished their operations.

Because some of the actions can significantly change the environment (for example removal or migration of an agent) so that the other actions would become invalid, the following phases have been introduced:

1. initialisation (`init`), when target addresses are verified,
2. execution (`main`), when the real action is executed,
3. finalisation (`finish`), for performing activities that could not be executed during the main phase (e.g. removal of an agent when other agents could refer to it).

All changes of agents structure that can influence execution of other registered actions are performed in the finalization phase. As a result, performing an action in the execution phase is safe. In the initialization phase actions can perform some preparetion activities that are required by other actions.

Two types of actions exist:

- Simple actions that can define only one task to be performed on only one agent.
- Complex actions — they are containers for other actions and can hold a tree-like structure. Actions wrapped by them are executed in a well-defined order and allows to create more complicated scenarios like an exchange of resources, when the separate component actions are required for getting a resource from one agent and for delivering it to another.

Most simple aggregate actions are defined as methods in a class of an aggregate agent and the default aggregate implementation provides some actions out-of-the-box:

- adding of a new agent,
- moving an agent to another aggregate,
- death of an agent,
- cloning of an agent.

Moreover, users can extend the platform with any actions they need. These actions can be created as strategies bound to the aggregate using the configuration of the platform. They allow to extend functionality of the platform in an easy way but have a downside of not having the possibility to refer to private members of the aggregate. Decision of how to execute such actions is made by the parent agent who resolves proper action implementation according to *Service Locator* design pattern [2].

3.3 Life-Cycle Management

The lifecycle of an agent consists of the following phases:

1. Construction — when a constructor of agent class is called.
2. Initialisation of the object dependencies and properties — when the `init()` method is called; at this point the agent has all its dependencies injected by the component framework based on dependency injection pattern mechanism. Also its properties are initialized using the component framework or by agent itself. For example at this stage, an agent generates an address
3. Initialisation of the environment — the moment when the parent of the agent calls the `setAgentEnvironment()` method. At this point the agent can use mechanisms that requires the existence of the local environment i.e. actions, queries, messaging.
4. Finalisation of the agent — the `finish()` method. The agent should finish its operation at this point.

Threaded agents additionally provide the `run()` method, called by the Java Virtual Machine after their dedicated thread was started. At this moment they can start the main loop of their execution.

The full lifecycle of the simple agents is shown in Fig. 3. Simple agents need to provide an implementation of the step. It is done in the `step()` method. This operation is called in an arbitrary order by the parent aggregate on every agent it contains. The actual execution from the point of view of the whole tree of agents is performed in the postorder way: firstly the aggregate lets children to carry out their tasks and only after they finished them it executes its own tasks.

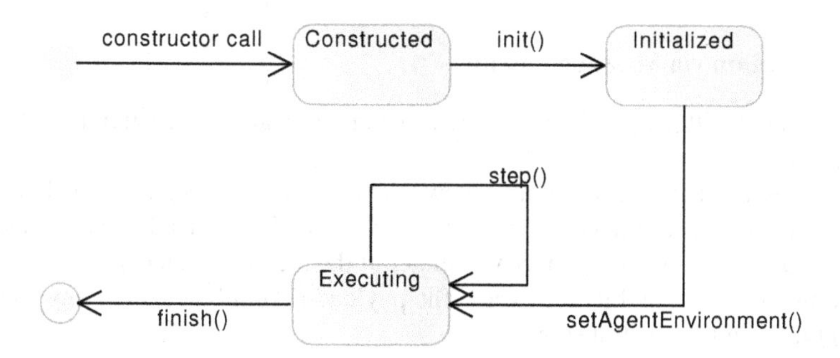

Fig. 3 A life-cycle of the simple agent shown as a state diagram

During the execution of the step, the simple agent usually needs to perform following actions:

- receive and send messages,
- execute queries,
- execute a part of the computation,
- order actions for the parent.

After iterating over all children, the aggregate needs to process the event queue. These events are usually actions requested by the children.

3.4 Communication Facilities

The platform allows for all agents to have a unique addresses, which allow for their identification and supports inter-agent communication. The particular property of being globally unique is guaranteed by a structure of the address. As shown in Fig. 4, the agent address comprises of three components: an UUID[3], a node address, and a name. Two former parts identify an agent in the platform instance and the last one is a replacement for an UUID provided for the user convenience (for usage in a user interface or logs).

An address is usually obtained by an agent during the initialisation of the component dependencies (see page 64 for the explanation of the agent life-cycle). It is done by requesting a new address object from the `AgentAddressProvider` instance that is a local component of a node.

[3] Universally Unique Identifier.

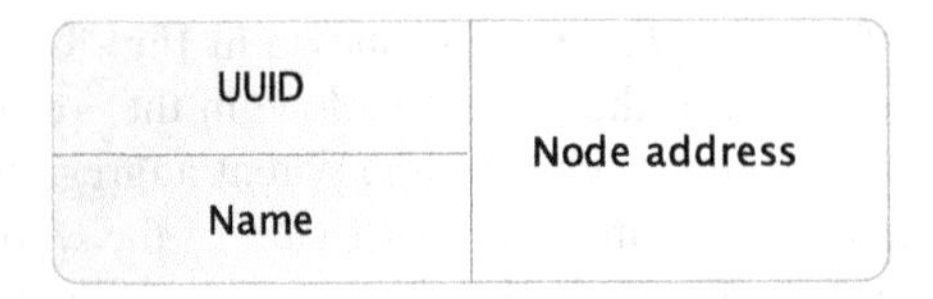

Fig. 4 Components of an agent address

Communication via Message Passing

Agents located within a single aggregate can communicate with each other via simple messages.

Interfaces used in messages are shown in Fig. 5. A message defined by the `IMessage` interface consists of a header and payload. The header, as defined by the `IHeader` interface must specify a sender of the message (usually the agent that created the message) and its receivers. The payload is simply a data of any (serialisable) type that is to be transported.

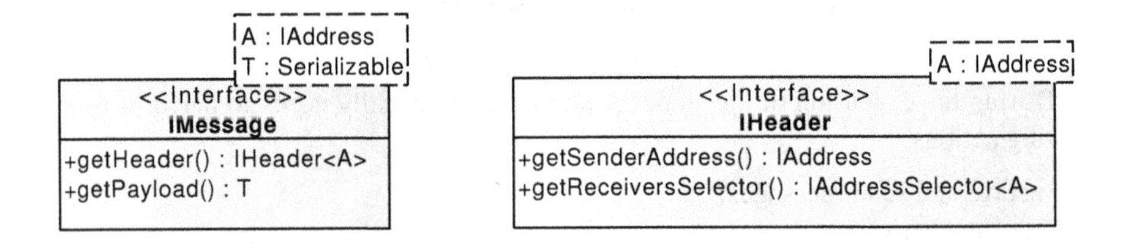

Fig. 5 Overview of interfaces used in messaging

Receivers are defined using selectors. They offer a possibility to define receivers with the *unicast*, *broadcast*, *multicast* or *anycast* semantics.

In the case of simple agents, sending and delivery of messages is performed by an aggregate agent. The sender adds a message event to its parent queue. The parent handles it by locating all receivers and calling a message handler on each of them. These messages are placed on a queue and can be received by the agent during its next step.

Thread-based agents use a similar queue of messages but are not restricted by the execution semantics and can inspect it at any point of time.

Query Mechanism

Queries offer a possibility to gather, analyze and process data located both in local and remote (from the point of view of the query executor) components.

The diagram in the Fig. 6 shows base classes and interfaces of the query mechanism along with their interconnections. The central point of this mechanism is the `IQuery` interface. It provides only one method: `execute()`. A query, as defined

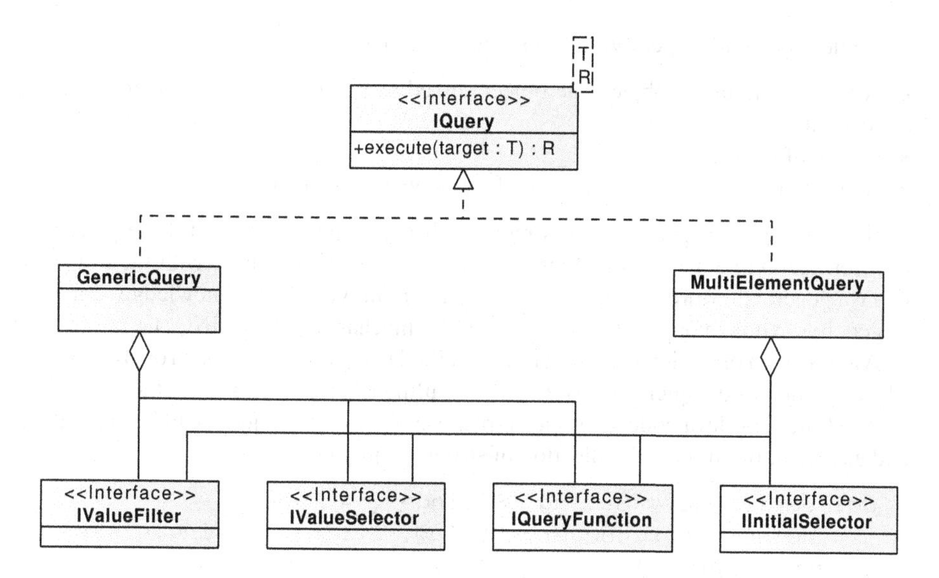

Fig. 6 Overview of queries base classes and interfaces

by this interface, is an action performed on a target object that leads to creation of query results. Specific implementations define a relationship between the target and results.

On the top of this interface and definition, a simple, declarative, yet extensible query language is built. Queries are implemented as (`GenericQuery` and `MultiElementQuery` classes in the diagram 6. It allows the user to perform tasks like: computation of the average value of some chosen properties from the agents in the environment, select and inspect arbitrary objects in collections and much more.

The following operations are defined:

- Preselection of values from the collection. It is only available if the query is performed over an iterable type instance. Its task is to select some (e.g. first ten or random) of objects without usage of the object-related information.
- Filtering by a value. This is an operation similar to WHERE clause in SQL.
- Selection of values. It can select specific fields from objects and it shows some similarities to the SELECT operation from SQL. If this operation is not defined then whole objects are selected.
- Functions working on an entire result set. They can remove, add or modify elements.

Operators are defined as implementation of specific interfaces (one for every operation, as shown in Fig. 6). They are presented to the user as static methods (e.g. `lessThan(),pattern()` etc.).

A query is built by specifying following properties:

- A type of the target object (the object passed as an argument to the `execute` method).
- A type of results.
- In the case of collections — a type of elements in a collection.

Such an exhaustive specification is required because queries rely on these pieces of information to control correctness of operators used by the user (with the usage of Java generics). Moreover, queries in AgE are built without the knowledge of the target object (it is in opposition to many similar mechanisms like LINQ[4]).

After that, an operation of the query is specified using aforementioned operations. The execution of the query is carried out by calling the `execute()` method.

The following Java code shows a simple example of how a query can be created and executed. In this case a collection of strings is queried.

```
CollectionQuery<String, String> q =
    new CollectionQuery<String, String>(String.class);
q.from(first(10))
    .matching(anyOf(
        pattern("li.*"),
        pattern("lorem[Ii]psum")));
Collection<String> results = q.execute(someList);
```

It can be noticed that queries definition uses the *fluent interface* pattern with specific operations being composed from static methods.

This approach of declaring a query without the knowledge of the target is additionally useful because it allows to execute a single query many times (possibly with caching the results or operations) or to delegate queries to be executed in another location. The query delegation is actually often used within the platform during the operation of querying an environment of a parent of an agent. This mechanism is essential for performing the migration of agents.

The other side of the queries mechanism is the extensibility offered to the user on many levels. It is possible to create completely specialized queries (by implementing the `IQuery` interface), extending the described declarative mechanism or even define in-line operators when creating a query. This elasticity of queries was very important because of performance requirements resulting from some applications of the platform. An approach was adopted, that the user is able to provide much faster solutions for his specific problems.

In some cases it is also useful to let know a queried object about a query being executed on it. For this reason the interface named `IQueryAware` was created. By implementing it any object can communicate to the query that it wants to be notified about some events related to the execution. Currently, two events are supported: initialisation and finalisation of the query.

The last part of the queries mechanism is caching. The platform offers a possibility to build a cache of query results. Its expiration time is based on the workplace

[4] `http://msdn.microsoft.com/en-us/netframework/aa904594.aspx`

step counter. This cache works as a wrapper to the query (and as such is an implementation of the `IQuery` interface). During the execution it checks whether stored results expired and possibly executes a real query replacing old results.

3.5 *AgE Component Framework*

The platform provides dedicated component framework, which is built on the top of an IoC container. It utilizes PicoContainer framework[5]—a popular open-source Java implementation of IoC container that can be easily extended and customized.

Both agents and strategies are provided to the platform as components. Their implementation classes can define named dependencies to other components (i.e. other agents, services or any other dependent classes) and simple properties that hold for example problem-dependent values. The dependencies definition for a component type, together with class's public methods (treated as component's operations) may be perceived as requirements closely related to component contracts (as proposed by Szyperski [25]).

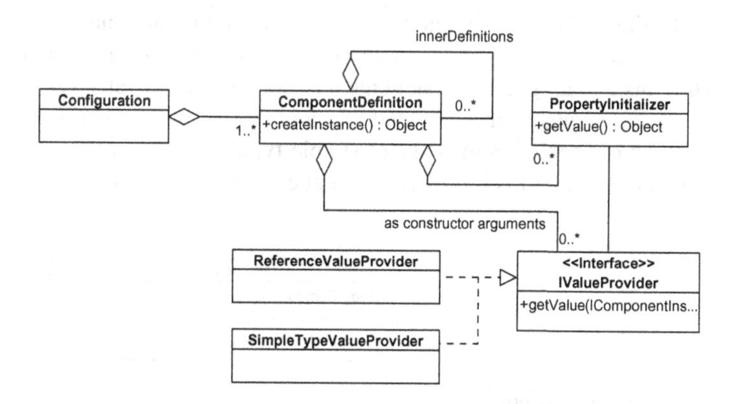

Fig. 7 Object configuration model

The process of assembling a system is divided into two main phases. In the first one, the input configuration is read from XML file with well-defined structure[6] and further transformed into object configuration model, structure of which is shown in Fig. 7. A `ComponentDefinition` instance describes a single component and contains data such as it's name, type (which is the name of a class) and scope, which is to determine if a new component will be created for each request (*protoype* scope) or only once during the first request (*singleton* scope). The definition also specifies the constructor arguments, which are implemented as `IValueProvider`

[5] `http://www.picocontainer.org`
[6] `http://agh.iisg.agh.edu.pl/age-2.3.xsd`

objects and used during constructor-based injection [14], as well as property initializers, responsible for initialising component properties with reference or simple values. Moreover, the definition contains `createInstance` method which creates a new instance of a described component with initialized dependencies (this process is described below). Component definitions may form hierarchical structures (via `innerDefinitions`). If a definition is a child of another one, it is said to "exist in the context of the outer definition" and is visible only for it's parent and other parent's children (siblings). Validation of the model, performed during processing of the input configuration, allows for detecting errors such as unresolved dependencies, non-existent components or incorrect property definitions.

In the next phase of system assembly process, a hierarchy of IoC containers is built according to a structure of component definitions. For each definition a dedicated adapter (`CoreComponentAdapter`) is created and registered in a container as shown in Fig. 8. Moreover, the adapter implements the interface, which defines methods for retrieving instances of components by name or type — `IComponentInstanceProvider`.

When a request for a component instance is directed to the container, it locates appropriate component adapter (using given name or type) and delegates the request further, to it. The adapter calls the associated component definition's `createInstance` method, which is responsible for creating a component instance. While instantiating a component the component adapter retrieves instances of dependent components from associated IoC container (or its parent), and the loop whole process starts again. In the case of simple types, a value is kept directly in a value provider object and is returned on a request. The whole process is repeated

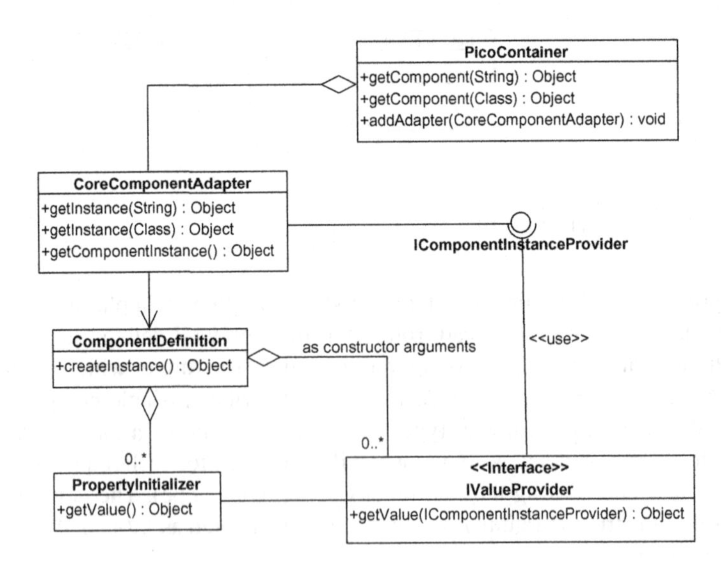

Fig. 8 Dependency Injection in AgE platform

until all dependencies are resolved and then the fully-initialized component instance is returned to the client.

The presented mechanism gives a possibility to build various structures of agents with their dependencies and initial properties values based on the input configuration.

3.6 Node Architecture

The simulation is executed in a distributed environment comprised of nodes connected via communication services. Each node is a separated process being executed on a physical machine. Nodes are responsible for setting-up and managing an execution environment for agents performing a simulation, as well as assuring communication and collaboration in distributed environment.

The main part of the node is a service bus that realize *Service Locator* design pattern. The bus is realized by *AgE component framework*, which utilizes IoC container to create and initialize an object that is a run-time instance of a service. Services are being registered in the container by the node boot-strapper or other services, based on component definitions, created using API or read from XML configuration file. A reference to a service instance can be acquired by service name or type via `IComponentInstanceProvider` interface.

The node distinguishes stateless and stateful services. The former offer functionality dependent only on parameters given in method call, therefore they does not hold any state and are always thread-safe. They are created on demand (at the first reference) and than their instances are cached in the container.

On the other hand, an instance of stateful service can hold data that influences its behavior. Such services implement `IStatefulComponent` interface, which introduces `init` and `finish` methods, called by the service bus while creating and destroying a service instance. Instantiation and initialisation is performed during node start-up. Stateful services can be also realized as threads, that are started in `init` method and finished asynchronously while destroying the service.

Fig. 9 shows an example node with registered services. The figure distinguish the main service (called *core service*), which constitutes an execution environment for agents, that provides functionalities such as global addressing schema, communication via message passing, query mechanism, life-cycle management.

This service also plays role of a proxy between agents and other services. Various services provide functionalities related to infrastructure (e.g. communication and configuration provider services), simulation (e.g. stop condition service) or external tools (e.g. monitoring service, which collects and stores simulation data for a visualisation application).

In one distributed environment particular nodes can have different responsibilities such as an end-user console, monitoring, management, and at last execution nodes. The role of a node is specified by the configuration of services plugged into its bus.

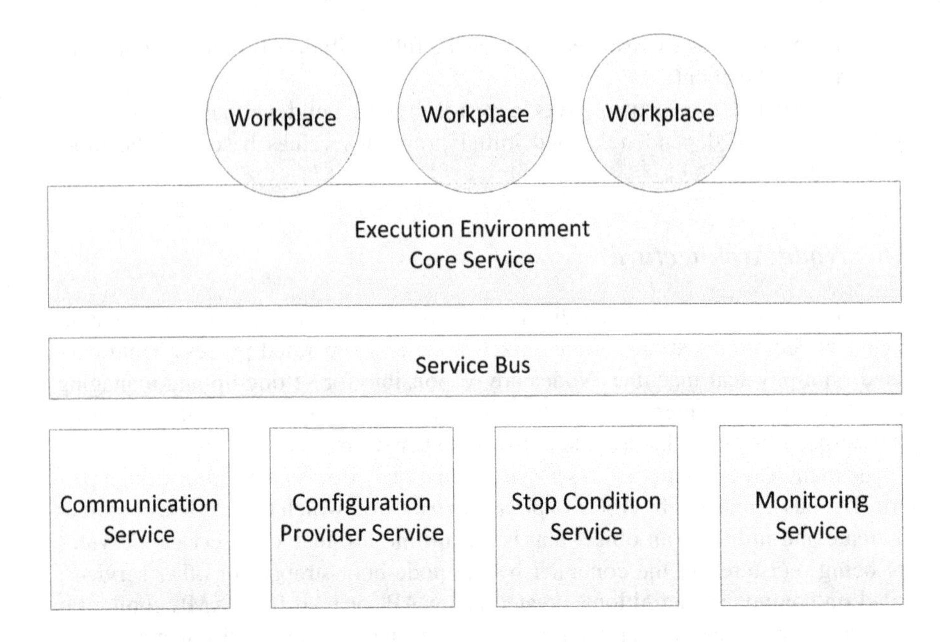

Fig. 9 Example node with services

Also, one can imagine a platform comprised only from a single node that works without any communication services[7].

3.7 Virtual Execution Environment

The platform introduces *a virtual execution environment* in distributed systems that allows for performing operations involving top level agents (workplaces) located on different nodes without their awareness of physical distribution. Such operations are executed by the core service according to *Proxy* design pattern [15]. The service uses the communication service to communicate with core services located on other nodes. This constitutes *a global name space* of top level agents in the distributed environment. In other words, the virtual execution environment can be perceived as a realisation of *virtual root agent*.

The name space of agents can be narrowed by introducing *agent neighbourhood* that defines visibility of top level agents. An agent can perform operations only on agents from its neighbourhood. The neighbourhood is realized and managed by a topology service (Fig. 10). This allows for creating virtual topologies among agents on the top of the distributed environment. Various topology strategies such as ring, grid, multi-dimensional grid can be applied in simulations.

[7] Such configuration is often used for test purposes.

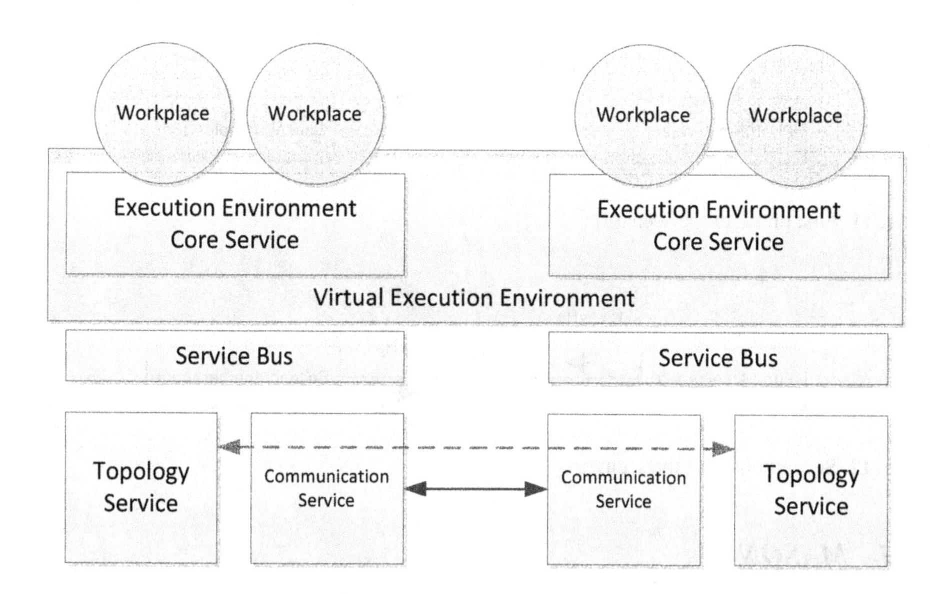

Fig. 10 Virtual execution environment

4 Implementation of Sample Case Study in Selected Platforms

To compare most basic mechanisms provided by chosen platforms a scenario needs to be defined, to be reviewed in every system. As 'most basic' we understand following elements (they were described in the section 2.1):

- life cycle of agents,
- organisation of agents,
- communication between agents (with possibly distributed data or interactions),
- level of coupling in the case of replacing an agent behaviour.

The (very simple) scenario that uses all of this elements can be described as follows:

1. Create and initialize a simulation with two agents.
2. Start the simulation. The agents should locate each other and start exchanging messages saying 'hello'.
3. Stop the simulation.
4. Change the implementation (behaviour) of the second agent (however, it still should send messages to the other agent).
5. Start the simulation again.

The diagrams 11 and 12 show interactions between these two agents before and after replacing the second one implementation.

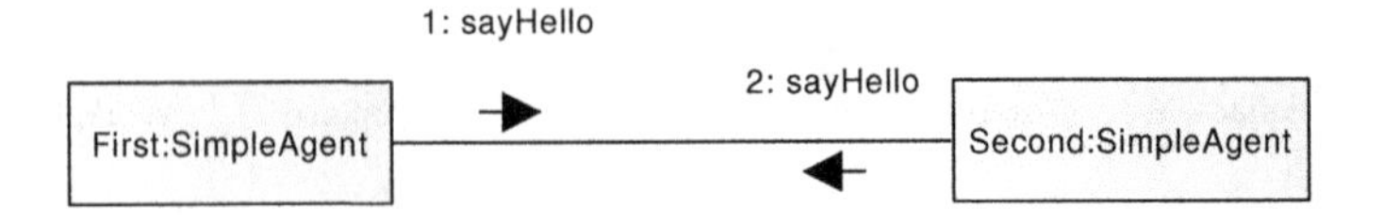

Fig. 11 First phase of the scenario.

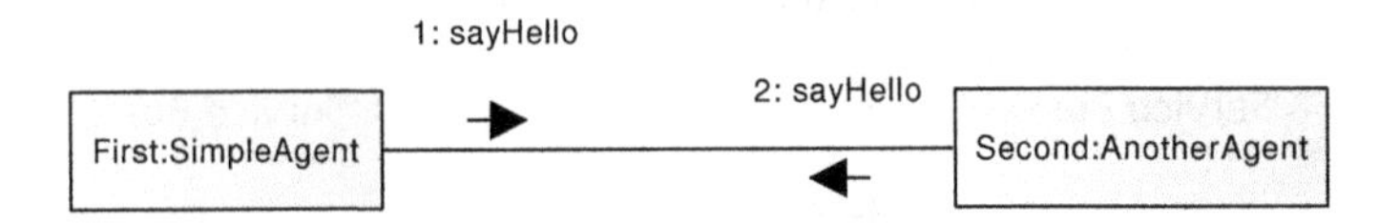

Fig. 12 Second phase of the scenario.

4.1 MASON

We have tested the scenario against the version 16 of the MASON.

Life Cycle of Agents

Simulation in MASON is created by inheriting from the `SimState` class. The execution of the simulation is started by calling the `doLoop()` method.

Agents are created either by implementing one of child interfaces of the `Steppable` interface, that requires a definition of the `step()` method. An instantiated agent needs to be scheduled to execute. It is done by calling one of methods of the `Schedule` instance located inside the simulation. For example, the following code schedules an agent to run repeatedly every unit of time:

```
Steppable agent = ...;
schedule.scheduleRepeating(agent);
```

Organisation

Agents may create a social network within a simulation. A social network, in MASON terms, is an undirected graph (or multi-graph) whose edges may be labeled and weighted. It opens possibilities to create arbitrarily complex relations between agents.

Moreover, MASON provides models for spatial organisation of agents. Using the `Continuous2D` class a plane for agents to be placed in could be created. Then, we could obtain agents to interact with on the base of their distance from the requesting agent:

```
Bag neighbours =
    space2d.getObjectsWithinDistance(myPosition,
    distance);
```

Where `myPosition` is a position of the agent and a distance is a number of the `double` type.

Communication

MASON does not provide any specific communication capabilities to agents. There is no distribution support either. Thus the developer needs to implement its own solution.

In the simplest case, communication can be realized with the usage of methods calls. It is a reasonable solution due to the fact, that the agent usually obtains a reference to its potential communication partners in some way (e.g. the `getObjectsWithinDistance` returns a bag of real agent objects). For example, saying 'hello' to all neighbours could be implemented as:

```
Bag neighbours =
    space2d.getObjectsWithinDistance(myPosition,
    distance);
for(int i = 0; i < neighbours.numObjs; i++)
{ if(neighbours.objs[i] != this) {
     ((Agent)neighbours.objs[i]).sayHello(this);}
}
```

A reference in the `sayHello` method can be used by the partner to respond to the call.

Distribution could be also added to the system but it would require the user to develop a complete solution (probably specific to the simulation) that would allow more-or-less transparent communication between separate MASON instances.

Coupling

A change of a behaviour of an agent is costly. MASON does not provide any built-in features to easily replace implementations of simulation objects (e.g. there is no configuration mechanism that would allow to define a specific implementation for a particular simulation run). Code (of an agent or the simulation initialisation) need to be changed and recompiled in order to use a new behaviour.

On the other hand, MASON provides utilities to build a GUI. One can make use of them to provide a way for the user to load new classes or change simulation parameters. However, it still needs to be programmed by the developer of the simulation.

4.2 RePast

We base this description on RePast Simphony 2.0 beta as this is a version currently recommended to use by developers.

Life Cycle of Agents

An agent in RePast is created as POJO — there is no requirements directed towards its implementation. To be able to run in the simulation it needs to have one of its method annotated with a scheduling plan. It is possible to use, for example, an active scheduling in the form of the `ScheduledMethod` annotation or the reactive one — the `Watch` annotation.

We would like to have an agent to execute in every step of the simulation since its beginning:

```
class Agent {
    ...
    @ScheduledMethod(start=1, interval=1)
    public void performStep () {
        ...
    }
}
```

The simulation is initialized and built by a `ContextBuilder` implementation that should be provided by the developer. It's implementation is dependent on the specific use-case but RePast provides a default version that behaves like a collection (i.e. `java.util.Collection`) of agents.

Organisation

Similarly to MASON, RePast allows for spatial organisation of agents on multidimensional continuous spaces or grids (however, there is no simple way to obtain neighbours of the object). Moreover, there is a graph organisation model that allows to construct arbitrarily complex networks of agents. It can be created in the context using:

```
NetworkBuilder<Object> builder = new
   NetworkBuilder<Object>("agents", context, true);
builder.buildNetwork();
```

And then, the agent, can obtain associated agents using (the graph is directed):

```
Network<Object> net =
   (Network<Object>)context.getProjection("agents");
for(Agent agent : net.getSuccessors(this)) {
    ...
}
```

Several other projections, like GIS[8], are also available.

[8] Geographic information system.

Communication

RePast does not offer any built-in messaging or interaction mechanisms beside plain method calls. As in MASON, it is possible to build a specific communication protocol for the particular simulation.

On the other hand, RePast has a support for the computation distribution using GridGain[9] or Terracota[10] plugins.

Coupling

RePast is similar to MASON also in this aspect. Agents depend on each other's interfaces and a simulation is constructed in a compiled builder. However, there is a support for specifying model parameters in configuration files (`parameters.xml`). Some components of a scenario may also be specified (and thus easily replaced) in the configuration.

4.3 MadKit

We used the version 4.2.0 of the project as it seems to be recommended release (although there are alpha releases of the version 5).

Life Cycle of Agents

Usually agents in MadKit are implemented by inheriting from the `Agent` class, representing a threaded agent. The main behaviour of such an agent can be provided by its creator in the `live()` method. Additionally, actions to be executed on agent initialisation and finalization can be specified by implementing `activate()` and `end()` methods.

In the discussed scenario a messaging in a do-while loop guarded by some condition could be simply implemented:

```
public class SimpleAgent extends Agent {
  ...

  public void live() {
    do
     {
       exitImmediatlyOnKill();
       ...
       pause(100);
     } while(someCondition);
  }
}
```

[9] `http://www.gridgain.com/`
[10] `http://www.terracotta.org/`

The `exitImmediatlyOnKill()` call is a way for an agent to check that it should stop its operation.

Furthermore, MadKit also allows to create agents with more general scheduling semantics thanks to `AbstractAgent` and `Scheduler` classes. The former is used as a base class for non-threaded agents and the latter is a scheduling agent that can execute methods of other agents according to the user specification, with the usage of so-called activators.

Organisation

MadKit offers a possibility for agents to organize into groups. An agent may want to become a member of some group and have a specific role within it. It can be done by firstly locating a specific group and then requesting a role with the `requestRole()` method or alternatively, by creating a completely new group. It can be done from the mentioned earlier `activate` method, called when the agent is started:

```
public void activate() {
  ...
  if (!isGroup("simple-exchange")) {
    createGroup(true, "simple-exchange", null, null);
  }
  requestRole("simple-exchange", "first-agent", null);
}
```

Firstly, the condition is checked, whether the group interesting for us exists and, if not, the agent creates it. Then, it tries to join this group with the role *first-agent*.

Communication

Mechanisms for the local messaging is provided directly in the base class of an agent as few versions of the `sendMessage()` method. In the discussed case we want to exchange messages between two agents. One can choose between two approaches: explicitly locating a communication partner and obtaining its address or by using its role name. The latter has an advantage of decoupling agents interactions from concrete agents instances — i.e. the partner can be easily replaced without introducing any change in the other agent.

```
Message message = new SpecialMessage();
sendMessage("simple-exchange", "second-agent",
   message);
```

Received messages are stored in a queue and can be obtained when needed by a call to the `nextMessage()` method.

It is also possible to use a distributed communication. However, this feature is not directly built into the system. MadKit follows an approach of the 'agentification' of services. The user has to introduce an agent that will be able to handle routing and forwarding of non-local messages. Such an agent needs to handle the *communicator* role in the *system* group. When a message that is non-deliverable within a local environment is sent, this agent will receive it and should forward it to another remote communicator.

Coupling

Coupling between agents is low, as most of the interactions is performed using a built-in messaging system that reduces a need to know each other agent interfaces. Most of the services are also built as agents (due to aforementioned 'agentification') and they are often identified by a specific role in the agent community and not by an interface. Thus, it is possible to change parts of the system as easily as casual agents.

MadKit GUI allows for simple changes to behaviour using a built-in code editor, so it is possible to stop the simulation, modify the agent code and rerun it. Similarly, agents can be added to or deleted from the simulation during runtime.

4.4 AgE

We have based this analysis on the version 2.5.0 (current stable) of the platform.

Life Cycle of Agents

As described on the page 64 the life cycle of an agent is well-defined. However, in most cases only the `step` method must be provided by an agent creator. This method is called once in every step of the computation.

AgE requires an agent to implement the `IAgent` interface. It also provides a default implementation of it in the `SimpleAgent` class that leaves only the aforementioned `step()` method unimplemented.

Organisation

Agents are organized in a tree structure. In this scenario three agents should be used: a workplace (that is a root of a tree) and two leaf agents on the second level of the tree. This structure is defined in the configuration file:

```
<agent name="workplace"
    class="org.jage.workplace.IsolatedSimpleWorkplace">
     ...
     <list name="agents">
         <agent name="agent1"
             class="org.jage.example.SimpleAgent">
             ...
         </agent>
```

```
        <reference target="agent1" />

        <agent name="agent2"
            class="org.jage.example.SimpleAgent">
            ...
        </agent>
        <reference target="agent2" />
    </list>
    <property name="agents">
        <reference target="agents" />
    </property>
</agent>
```

We define the workplace (which is represented by the built-in `IsolatedSimpleWorkplace` class) and, within it, a list of two agents of the type `SimpleAgent`.

Communication

AgE provides a specialized mechanism for sending messages. However, if one wants to send a message to a specific agent, he needs to obtain its address first (alternatively, an agent can send a broadcast message). It is usually done by querying the environment of the agent.

A sample query may look like this:

```
AgentEnvironmentQuery<SimpleAgent, IAgentAddress>
   query = new AgentEnvironmentQuery<SimpleAgent,
   IAgentAddress>(IAgentAddress.class);
query.select(agentAddress());
Collection<IAgentAddress> answer =
   queryEnvironment(query);
```

The part responsible for obtaining addresses of agents is the `agentAddress()` selector. The call to the `queryEnvironment()` method executes a query over the calling agent's environment (in which its partner is located).

Then, the agent can send a message:

```
Header<IAgentAddress> header = new
   Header<IAgentAddress>(getAddress(), new
   UnicastSelector<IAgentAddress>(receiverAddress));
Message<IAgentAddress, String> textMessage = new
   Message<IAgentAddress, String>(header, "hello");
sendMessage(textMessage);
```

At the moment AgE does not provide any means to distribute computation. However, work on this topic has already been started.

Coupling

In AgE, by applying separation of interfaces and providing communication and queries mechanism, agents, their dependencies and platform elements are very low coupled. Moreover, utilizing components techniques gives the wide opportunities to easily create flexible and customizable simulation systems. Particulary, agents structure, their behaviour or even simulation parameters can be easily changed by providing modified runtime configuration. Such changes do not require any code updates so that the platform components and libraries do not have to be recompiled. One can also imagine that multiple similar simulations can be run in batch-mode (one after another) without any user inference.

4.5 Summary

We have reviewed four different multi-agent frameworks and analysed facilities offered by them that allows us to create a simple simulation. Each of mentioned systems has its strengths and weaknesses in particular aspects of its usage in solving problems.

All of the systems have defined some kind of a life-cycle for agents. Some of them enforce it in a rather strict way (AgE, MadKit) and some provide very flexible mechanisms for scheduling an agent (RePast, MASON). Only RePast offers an easy way to schedule many methods from a single POJO class.

The second considered aspect was a possibility to organise agents in different ways. Both MASON and RePast have a good support for many (possibly co-existing in one simulation) organisational schemes (graphs, spatial neighbourhood, etc.). On the other hand MadKit requires following a strict Agent/Group/Role model but due to its flexibility one could possibly implement other arbitrary models with little effort. AgE also provides only one possible organisational model — a tree structure of agents.

The next aspect was communication and distribution. There is no defined communication schemes in RePast and MASON. The user has to rely on plain method calls or build their own mechanism. MadKit and AgE have chosen more strict approach and provide messaging facilities. Additionally, AgE has other means of communication available: queries and object properties. From described platforms only RePast provides an out-of-the-box way to distribute computation. In the rest of platforms user is required to come up with their own approach to this problem.

The last aspect was coupling. RePast and MASON both introduce a lot of interdependencies between agents (due to method calls as a messaging scheme) and with a platform (e.g. no built-in dependency injection). MadKit follows the scheme of agentification of services and also has rather low coupling between agents. AgE provides most advanced solutions regarding the decoupling and offers such facilities as external configuration, DI or easy runtime modifications.

5 Conclusions

In the course of the contribution, a review of selected Java-based, agent-oriented simulation frameworks was presented. The most important, in opinion of the authors, features of these programming environments were evaluated, such as agents life cycle management, inter-agent and agent-platform communication support, agents society organisation, distributed environment facilities, as well as the overall status of their development. The evaluation was conducted for the most popular general-purpose frameworks, such as MASON, RePast and MadKit, and proved that they had many advantages, such as overall maturity, sophisticated GUI or support for different scripting languages. However it was found that they all lacked either component-oriented features or built-in support for the low coupling of simulation structures. This leads to possibly inflexible designs, which makes it difficult to change isolated parts of the simulation system without altering their contextual dependencies.

Based on this scarce evaluation, the technical issues of AgE software environment were discussed. The platform seems to provide a well-suited execution environment for running agent-based simulations. It seems especially usefull when taking into consideration one of the most important issues of software engineering: flexibility and extensibility, which is achieved by the component-based design. Therefore it is very easy to extend the simulation prepared in AgE by changing its parts, parameters, upgrading them to the next versions etc.

The work on AgE development continues, in the near future implementation of user interface, monitoring facilities, experiment scheduling and persistence of results (among others) are planned to be completed.

References

1. Almasi, G., Gottlieb, A.: Highly Parallel Computing. Benjamin-Cummings Publishers, Redwood City (1989)
2. Alur, D., Crupi, J., Malks, D.: Core J2EE Patterns: Best Practices and Design Strategies. Prentice-Hall (2003)
3. Banks, J., Carson, J., Nelson, B., Nicol, D.: Discrete-Event System Simulation. Prentice-Hall (2009)
4. Bellifemine, B., Poggi, A., Rimassa, G.: Jade – a fipa-compliant agent framework. In: Proc. of PAAM 1999, London, pp. 97–108 (1999)
5. Bergenti, F., Gleizes, M.P., Zambonelli, F.: Methodologies and Software Engineering for Agent Systems. Kluwer Academic Publishers (2004)
6. Bhasker, J.: A SystemC Primer, 2nd edn. Star Galaxy Publishing (2004)
7. Byrski, A., Kisiel-Dorohinicki, M.: Agent-Based Model and Computing Environment Facilitating the Development of Distributed Computational Intelligence Systems. In: Allen, G., Nabrzyski, J., Seidel, E., van Albada, G.D., Dongarra, J., Sloot, P.M.A. (eds.) ICCS 2009. LNCS, vol. 5545, pp. 865–874. Springer, Heidelberg (2009)
8. Byrski, A., Schaefer, R.: Stochastic model of evolutionary and immunological multi-agent systems: Mutually exclusive actions. Fundamenta Informaticae 95(2-3), 263–285 (2009)

9. Carneiro, G., Fontes, H., Ricardo, M.: Fast prototyping of network protocols through ns-3 simulation model reuse. Simulation Modelling Practice and Theory 19(9), 2063–2075 (2011)
10. Collier, N., North, M.: Repast SC++: A Platform for Large-scale Agent-based Modeling. Wiley (2011)
11. Dahmann, J.: High level architecture for simulation. In: Proc. of First International Workshop on Distributed Interactive Simulation and Real-Time Applications, pp. 9–14 (1997)
12. Davila, J., Uzcategui, M.: Galatea: A multi-agent simulation platform. modeling. In: Proc. of Simulation and Neural Networks (MSNN 2000), Merida, Venezuela, pp. 216–233 (2000)
13. Ferber, J., Gutknecht, O.: A meta-model for the analysis and design of organizations in multiagents systems. In: Demaseau, Y. (ed.) Proc. of ICMAS 1998 Conference, Paris, pp. 128–135 (1998)
14. Fowler, M.: Inversion of control containers and the dependency injection pattern (2004), `http://martinfowler.com/articles/injection.html`
15. Gamma, E., Helm, R., Johnson, R., Vlissides, J.: Design Patterns: Elements of Reusable Object-Oriented Software. Addison-Wesley (1995)
16. Gutknecht, O., Ferber, J.: The madkit agent platform architecture. In: Agents Workshop on Infrastructure for Multi-Agent Systems, pp. 48–55 (2000)
17. Klein, J.: Breve: a 3d environment for the simulation of decentralized systems and articial life. In: Proc. of Artificial Life VIII, the 8th International Conference on the Simulation and Synthesis of Living Systems (2002)
18. Luke, S., Cioffi-Revilla, C., Panait, L., Sullivan, K., Balan, G.: MASON: A multi-agent simulation environment. Simulation: Transactions of the society for Modeling and Simulation International 82(7), 517–527 (2005)
19. Nikolai, C., Madey, G.: Tools of the trade: A survey of various agent based modeling platforms. Journal of Artificial Societies and Social Simulation 12(2) (2008)
20. North, M., Howe, T., Collier, N., Vos, J.: A declarative model assembly infrastructure for verification and validation. In: Takahashi, S., Sallach, D., Rouchier, J. (eds.) Advancing Social Simulation: The First World Congress, FRG. Springer, Heidelberg (2007)
21. Pidd, M., Cassel, R.A.: Three phase simulation in java. In: Proceedings of the 1998 Winter Simulation Conference, pp. 367–371 (1998)
22. Railsback, S., Lytinen, L.: Agent-based simulation platforms: review and development recommendations. Simulations 82, 609–623 (2006)
23. Schaefer, R., Byrski, A., Smołka, M.: Stochastic model of evolutionary and immunological multi-agent systems: Parallel execution of local actions. Fundamenta Informaticae 95(2-3), 325–348 (2009)
24. Stroock, D.: An Introduction to Markov Processes. Springer (2005)
25. Szyperski, C.: Component Software: Beyond Object-Oriented Programming. Addison-Wesley Longman Publishing Co., Inc., Boston (2002)
26. Tesfatsion, L.: Agent-based computational economics: Modeling economies as complex adaptive systems. Information Sciences 149(4), 262–268 (2003)
27. Ventroux, N., Guerre, A., Sassolas, T., Moutaoukil, L., Blanc, G., Bechara, C., David, R.: Sesam: An mpsoc simulation environment for dynamic application processing. In: CIT, pp. 1880–1886. IEEE Computer Society (2010)
28. Wainer, G., Mosterman, P.: Discrete-Event Modeling and Simulation: Theory and Applications (Computational Analysis, Synthesis, and Design of Dynamic Systems). CRC Press (2010)
29. Wooldridge, M., Jennings, N.: Intelligent agents: Theory and practice. Knowledge Engineering Review 10(2) (1995)

Software Environment for Online Simulation of Switched Reluctance Machines

Virgil Chindriş, Rareş Terec, Mircea Ruba, and Loránd Szabó

Abstract. Simulations are widely used in the study of both electrical machines and drives. In the literature several simulation programs can be found, which can also be applied in the study of the switched reluctance motors (SRMs). The real-time simulation tool developed by the authors is a novel approach as it enables also online simulations. The control of the SRM can be performed from the computer's keyboard (by energizing/un-energizing each coil from its corresponding key), from a built-in controller or from an external one, which can be connected via TCP/IP protocol. The program enables parameter changing and instantaneous value displaying during the simulations. The studied machine's characteristics can be plotted real-time, or after finishing the simulations. All the obtained results can be saved in external files for future data processing. The software tool proved to be very useful both in checking the design of a SRM, but also in setting up and verifying its proper control strategy. The online simulation program can also be practical in teaching electrical machines and drives, and it can be extended to be used in remote laboratories, too.

1 Introduction

The online (real-time) simulations are the one of the most advanced techniques used in studying both the electrical machines and drives [4, 5].

The online simulation of electrical machines and drives is an efficient and cost-effective approach to evaluate the behavior of both newly designed machines and controllers before applying them in a real system. Using such software tools new electrical machines and their controllers can be rapidly and easily tested [25, 28].

Virgil Chindriş · Rareş Terec · Mircea Ruba · Loránd Szabó
Department of Electrical Machines and Drives, Technical University of Cluj
RO-400114 Cluj, 28, Memorandumului, Romania
e-mail: {virgil.chindris,rares.terec}@mae.utcluj.ro,
{mircea.ruba,lorand.szabo}@mae.utcluj.ro

A. Byrski et al. (Eds.): Advances in Intelligent Modelling and Simulation, SCI 416, pp. 85–109.
springerlink.com

Therefore, online simulation techniques can also be very helpful where interactive changing of parameters immediately affects a change of the corresponding results [29].

The switched reluctance motors (SRM) are some of the most significant new developments in the field of electrical machines for variable speed drives. They have several attractive features, such as high output power, high starting torque, wide speed range, rugged and robust construction and low manufacturing costs [24].

Hence the online simulation tool for studying the SRM's performances to be presented in this paper could be of real interest for all the specialists working in these fields.

In the paper the simulation program will be detailed, both its algorithm and easy-to-use graphical interface. The usefulness of the program will be highlighted by the results of the simulations carried out on a typical structure of SRM.

2 The Switched Reluctance Machine

The SRM is a double salient electrical machine with a passive rotor [14]. In the SRM the torque is produced by the tendency of its rotor to get to a position where the inductance and the flux produced by the energized stator winding are maximized (variable reluctance principle).

The SRM's rotor and stator both have salient poles, as shown in Fig. 1.

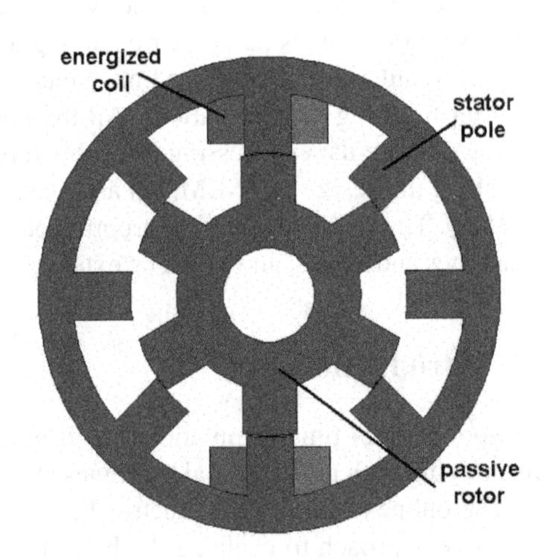

Fig. 1 The cross section of a SRM

The stator is formed from punched laminations that have been bonded into a stack. The rotor is made of conventional laminations without any kind of winding, excitation, squirrel-cage or permanent magnets [19].

The stator winding consists of coils placed on the stator poles. Typically a phase is created by two series or parallel connected coils placed on diametrically opposed poles of the machine. Each phase is independent and the machine's excitation is a sequence of current pulses applied to each phase in turn. The commutation of the SRM's phase currents must be synchronized precisely with the rotor position [24]. The SRM cannot be separated from the electronic supply device and its control [14].

The various advantages of the SRM make it an attractive alternative to the existing dc and ac motors in adjustable speed drives. The SRM drives can also deliver servo-drive performance equivalent to dc brushed motors. The rotor position sensing requirements, the need for an electronic converter and the higher noise and torque ripple, compared to other machines, are the main disadvantages of the SRM drives.

The most usual power converter for a four phase SRM is given in Fig. 2.

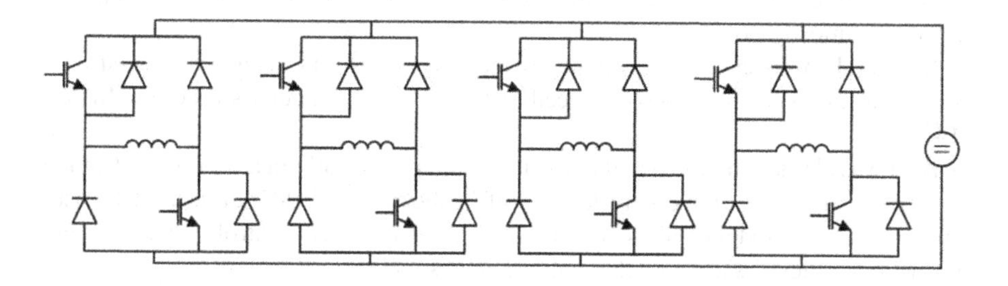

Fig. 2 The SRM's power converter

Each phase is fed thought a half H-bridge, hence the SRM is supplied only by unipolar current pulses. By opening and closing the transistors the correct current sequence required by the SRM is assured. The diodes have the role of guiding the reverse currents and by this advantaging the falling to nil of the phase current at each end of the conducting period. The bridges of the four phases are connected in parallel with the main power supply which ensures a constant dc voltage.

The most common control system of the SRMs is given in Fig. 3.

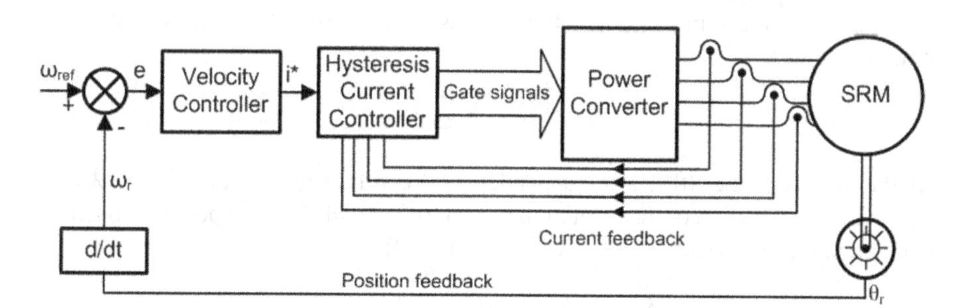

Fig. 3 The SRM's control system

The machine is equipped with a position sensor. The position is read by this sensor and the signal is sent to the main processor. Deriving the position in time domain will return the angular speed of the rotor. This is compared to a reference value. The error between the actual speed and the imposed one will be the input of the controller which will compute the current reference for the hysteresis comparator. Each phase is supplied within a certain rotor position range in order to maximize the developed torque. The hysteresis controllers will send the gate signals to the power switches from the converter.

3 Modeling and Simulation of the SRM

Unlike classical machines, the SRM is partially operated in magnetic saturation during the excitation cycle [23, 38]. Due to the saturation effect the mathematical models are mostly nonlinear.

In several approaches the flux linkage function of phase current and rotor position is used to calculate the torque produced and the overall characteristics of the SRM [14].

As it was already mentioned, the SRM cannot operate without a supply electronic converter and some sort of controller. Therefore the overall SRM's drive model must contain also the model of the switching devices and of their control system. Often the switching devices are modeled only as ideal switches.

Each phase of the SRM practically is a coil with a constant resistance R and a flux linkage λ, which is dependent of rotor position and phase current. The terminal voltage of the phase j can be expressed upon the basic voltage equation of the SRM's phases as [14]:

$$v_j = R \cdot i_j + \frac{d\lambda_j}{dt} \tag{1}$$

where the flux linkage is:

$$\lambda_j = L_\sigma \cdot i_j + \lambda_{m_j} \tag{2}$$

As the leakage inductance (L_σ) is usually not affected by saturation, it can be considered as constant and equal to all machine phases. In general terms, the magnetizing flux linkage through the phase j (λ_{m_j}) is composed of the phase's own magnetizing flux ($\lambda_{m_{jj}}$) and the sum of all the other phases' magnetizing flux linkage that goes through phase j:

$$\lambda_{m_j} = \lambda_{m_{jj}} + \sum_{k}^{k \neq j} \lambda_{m_{kj}} \tag{3}$$

As the phases of the SRM are independent, (3) can be simplified as: $\lambda_{m_j} = \lambda_{m_{jj}}$. Since λ_{m_j} is a function of the excitation current (i_j) and of the rotor position (θ), the magnetizing flux derivative can be expanded to [14]:

$$\frac{d\lambda_{m_{jj}}}{dt} = \frac{\delta\lambda_{m_{jj}}}{\delta i_j} \cdot \frac{di_j}{dt} + \frac{\delta\lambda_{m_{jj}}}{\delta\theta} \cdot \frac{d\theta}{dt} \tag{4}$$

where $d\theta/dt = \omega$ is the angular speed of the SRM.

Inserting (3), (4) and the above equation into (1) and taking into account that $\lambda_{m_{kj}} = 0$, there results:

$$v_j = R \cdot i_j + \left(L_\sigma + \frac{\delta \lambda_{m_{jj}}}{\delta i_j} \right) \frac{di_j}{dt} + \omega \frac{\delta \lambda_{m_{jj}}}{\delta \theta} \quad (5)$$

The magnetizing flux linkage $\lambda_{m_{jj}}$ is in fact the air–gap flux under the pole with coil j.

The torque produced by the phase can be determined from the flux linkage characteristics of the machine. In nonlinear magnetic circuits it is convenient to use virtual work or coenergy in order to determine the torque. The coenergy for the phase j of the machine is:

$$W_j' = \int_0^{i_j} \lambda_j di \quad (6)$$

and the torque is the partial derivative of the co-energy with respect to angle:

$$T_j = \frac{\delta W_j'}{\delta \theta} = \int_0^{i_j} \frac{\delta \lambda_j}{\delta \theta} di \quad (7)$$

The dynamic equation of the motion for the motor and its load is:

$$T = \frac{J}{p} \frac{d\omega}{dt} + B \frac{\omega}{p} + T_l \quad (8)$$

where J is the machine's moment of inertia, p is the number of pole pairs, B the viscous friction coefficient and T_l the load torque.

In order to predict the SRM's performance the above equations cannot be used without a relation that will express the flux linkage variation function of the rotor position and phase current. Such a relationship can be obtained analytically, by solving a magnetic equivalent circuit (MEC), or by using a numerical method to compute the magnetic field [14].

Most authors proposed analytic solutions to take account of the nonlinearity of the SRM magnetic characteristics [31, 34]. Some among them approximate from experimental data the inductance profile $L = f(\mu, I)$ of the SRM by trigonometric functions [13], or by using a cubic spline interpolation [27]. Unfortunately, these methods pose some problems since they involve extensive calculations and in several times they also require derivative computations. In spite of all of these disadvantages, the main defect of these methods resides in the accuracy of the obtained results.

An analytical representation of the flux linkage of an SRM as a function of current and rotor position, which takes into account also the machine's nonlinearities due to the double saliency of the construction and the magnetic saturation, has been presented in [17]. The position dependency on the flux linkage was proposed to be approximated by Fourier series representation.

In [12] a fast and effective analytical model is presented which uses the first five components of Fourier series in expression of current-dependent variable parameter's arctangent functions to express the whole flux linkage model of the SRM. The model can be built up with only 9 data points at 5 rotor positions. The electromagnetic torque could be further calculated from the proposed flux linkage model.

There have also been several attempts to produce accurate magnetic equivalent circuit models for SRMs [26, 30]. The models were used for the prediction of the motor characteristics, in cases mainly when other more precise methods could not be practical on their own.

There are different types of magnetic equivalent circuits developed for SRMs. Some of them only consider single-phase-at-a-time excitation [26], which cannot be used in simulating high-speed regimes of the SRMs. The modeling of multiple-phase-at-a-time excitation is more complicated, because it requires that the effects of saturation be spread throughout the machine [30], rather than concentrated at the pole tips, as in the case of MECs developed for the single-phase-at-a-time excitation case.

Two major problems arise when using the MEC based simulation approach: the computation of the air gap's reluctance, which is varying with the rotor's position, respectively the modeling of the iron core's saturation.

In [41] both problems are solved. The novel relationships given for the computation of the magnetic equivalent permeance of the air-gap solves the first problem. The saturation effect is taken into account by modifying the equivalent air-gap function with a saturation coefficient, computed based on a precise numeric field computation.

Several researchers are using SPICE, a wide-spread general-purpose circuit simulator program, to simulate SRMs. In [39] a variable reluctance model is expressed as a Fourier series (neglecting the saturation effect). By using this model the excitation voltage, the phase current and the torque can be accurately computed by using SPICE.

In [15] a circuit analysis model of the SRM using SPICE simulator is constructed on the basis of inductance-rotor position characteristics. The SPICE model consists of a main circuit and four subcircuits that express variable inductance, control signal, motor torque and equation of motion.

In the calculation model detailed in [16] the electric and magnetic circuit of SRM are separated. They are coupled by proper controlled sources. Using the SPICE model the dynamic characteristics of the SRM can be calculate very readily and accurately.

A much more precise modeling approach is by means of finite elements method (FEM) based two- or three-dimensional numeric field computation. Nowadays, this method is usually applied for designing purposes and for steady-state performance computation of diverse electrical machines. Therefore, the flux linkage and torque versus current and rotor position characteristics can be calculated easily during the design stage. Since the flux linkage characteristics are available, they can be used to apply any analytic model previously described [19, 14].

In [2] two models based on two-dimensional FEM calculated flux linkage characteristics is briefly discussed, and a model based on geometry data and an aligned flux linkage characteristic obtained via numeric field computation based analysis is introduced. The first proposed model is a direct one, since the flux linkage values are calculated function of phase current. In the other proposed model, the inverse one, the current is obtained as a function of flux linkage values. The first model is useful in the design of the SRMs, while the other one facilitates the study of the SRM's transient behavior, and also may be introduced in the SRM's control systems [14].

A major advantage of this approach beside its high accuracy is that the electromagnetic field analysis often can be coupled with the thermal FEM-based analysis, frequently inside the same program [43].

One of the most recent simulation techniques, the co-simulation, also can be used in the study of the SRM. In this approach each component of the drive system (the electrical machine, the power converter and the control unit) is simulated via different programs, each being the best fitted to the given purpose. An adequate software module links together the different simulation platforms [35].

In almost all of the cases the main program is built up in Simulink, the most frequently used program in simulating dynamic systems [33]. The electrical machines are modeled via advanced FEM-based numeric field analysis programs (as Flux 2D, ANSYS, MagNet, JMAG, etc.). The FEM model of the SRM is embedded into the main Simulink program. At every time step data is exchanged between the two coupled programs.

Such approach is presented in [11], where the SRM is simulated in Flux 2D and this is coupled to Simulink via the Flux-to-Simulink link. The machine's model is embedded in the Simulink program thru an S-type function. Thus the computing power of Flux 2D is joined by the advanced facilities of the MATLAB/Simulink environment. By using this coupled simulation tool the different working regimes of the SRM can be easily studied.

There are also links between the above mentioned field computation platforms and programs in the field of electrical drives and power electronics (CASPOC, SIMPLORER, PSIM, Portunus, etc.), which enable a in depth analysis of the entire SRM drive system [7].

Nowadays, applying artificial intelligent techniques to the electric machines increased along with boosting the speed of processors. Various studies have been published on modeling of SRMs by using intelligent techniques. Models based on fuzzy logic, adaptive neuro-fuzzy and artificial neural networks (ANN) approaches were used to develop optimized models that take into account also the nonlinearities of the SRM [20, 21, 40]. In the frame of these models the electromagnetic torque developed is practically estimated by means of learning the characteristics by advanced auto performing online measurements. Almost in all the cases these models were used in implementing the control strategies of the SRM.

Obligatory also the models used in simulating the drive of the SRM have to be mentioned. There are several kinds of drive models cited in the literature, both for current and voltage converters. These mainly use available simulation program packages, such as SPICE, CASPOC, SIMPLORER, etc. [42].

A generalized model of the SRM's drive was reported in [44]. The model uses the node voltages as variables, and the resulting matrix equations can easily be formed and modified by changing the respective rows and columns of the state matrix. By this approach it becomes possible to simulate any configuration of the drive system, and also the control dynamics effects can be studied [14].

As it can be seen, in the literature several simulation approaches for the SRM can be found. All of them have the drawback of obtaining the results only after completing running the simulation programs. Therefore the user can modify neither the parameters of the simulation and control, nor those of the machine.

4 The Simulation Program

The program is written by the authors in Borland Turbo Delphi 2006 [8, 1]. It is structured in 7 libraries: 3 for window forms, 3 for configurations and the last, for motor controllers. In total over 3000 lines of code (over 100 subprograms) were used.

To make the simulation to run in real-time, in addition to the main thread used for graphics, the program uses two additional threads [37]: for the motor simulation and for the built-in motor controller.

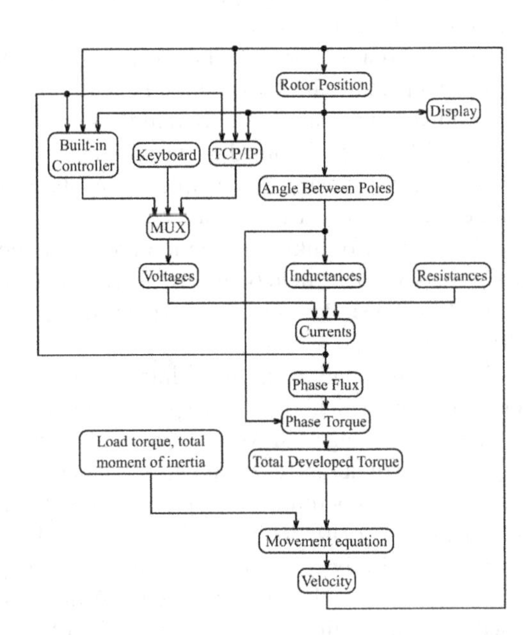

Fig. 4 The block scheme of the simulation program

The block scheme of the simulation program is given in Fig. 4. The tasks of the program at each stage are as follows:

1. Integrate velocity and get rotor's angle.
2. For every stator pole, compute the angle between the pole center and all rotor poles centers. Keep the smallest difference.

3. Assign an inductance to each stator pole based on the smallest difference between angles computed above.
4. The inductance values are forced to a fixed range through truncating.
5. Get voltages from either the built-in controller, or computer's keyboard (direct user control), or an external controller connected through TCP/IP.
6. Phase resistances are set from the GUI.
7. Compute the phase currents from phase voltages, phase inductances and phase resistances.
8. Compute the phase fluxes from phase currents and phase inductances.
9. Compute the phase torque from phase fluxes and the difference between stator pole angle and the closest rotor pole angle.
10. Sum up all the phase torques.
11. Compute the resultant torque from rotor inertia, load intertia, rotor torque, rotor friction and load friction.

If the user wants to connect the program to an external motor controller, through TCP/IP connection, the integrated server module generates one more thread [36, 9].

When the motor is simulated in real-time, the system requires a huge processing power. The program makes available the possibility to simulate the controller using an external simulator (see Fig. 5). This way the simulation speed increases.

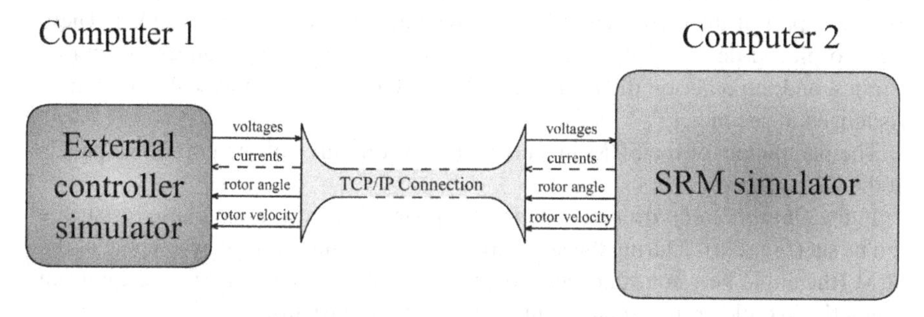

Fig. 5 Connection to the external simulator

The two simulators use a TCP/IP connection, where the SRM simulator is the server and the controller simulator is the client, running on different computers [18]. If the TCP/IP connection is slow, the current control can be simulated on the SRM simulator side, improving the simulation performance. This is the reason why the currents arrows are dashed in Fig. 5.

The block scheme of the simulation program is shown in Fig. 4, where the logical way as the program is built up can be followed. The easy use of the simulation program is assured by the user-friendly graphical user interface (GUI) given in Fig. 6 [22].

The GUI has three main parts. In the upper left corner the outline of the studied machine's cross section can be seen. In the right side, several pop-up panels can be

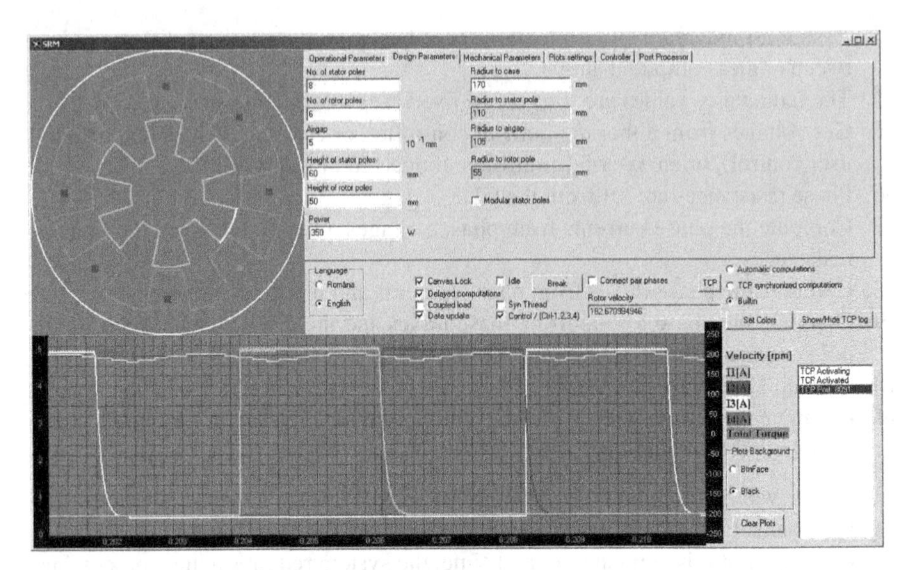

Fig. 6 The main GUI of the simulation program

opened for setting the parameters of both the machine and the simulation itself. In the lower part of the GUI, the results of the simulations can be plotted [10]. There are two plot scales, one at the left for currents, voltages, inductances and phase torques, and the other on the right, for velocity. On the horizontal scale the time is visualized in seconds.

The parameters of the SRM in study can be set in three of the pop-down panels of the GUI.

In the *Operational parameters* panel, the phase voltage and the phase resistance can be set (Fig. 7(a)). During the simulations, several data regarding each pole of the SRM (the angle between stator and rotor poles, phase voltage, inductance, current magnetic flux, electromagnetic torque generated) are visualized.

The main geometrical data of the machine can be set in the *Design Parameters* panel: the number of stator and rotor poles, the air-gap and the machine's main geometrical (Fig. 7(b)).

The mechanical parameters (the friction torque, the load torque, the moment of inertia of both the motor and of the load, etc.) can be introduced in the text fields of the *Mechanical parameters* panel (Fig. 7(c)).

As it can be seen, all the variables of the simulation program can be set up very easily in the GUI given in Fig. 6. By this, a high flexibility is assured for the program. By making several changes, practically any SRM structure can be easily simulated. The main parameters of the SRM's controller can be adjusted by very simple sliders in the Controller panel of the main GUI, as shown in Fig. 8.

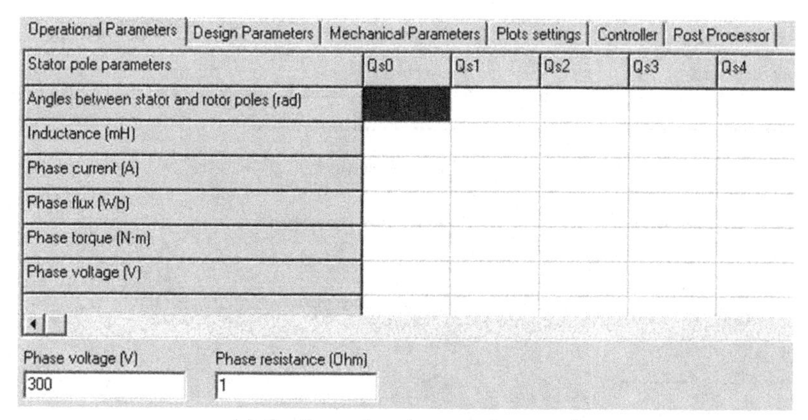

(a) the *Operational parameters* panel

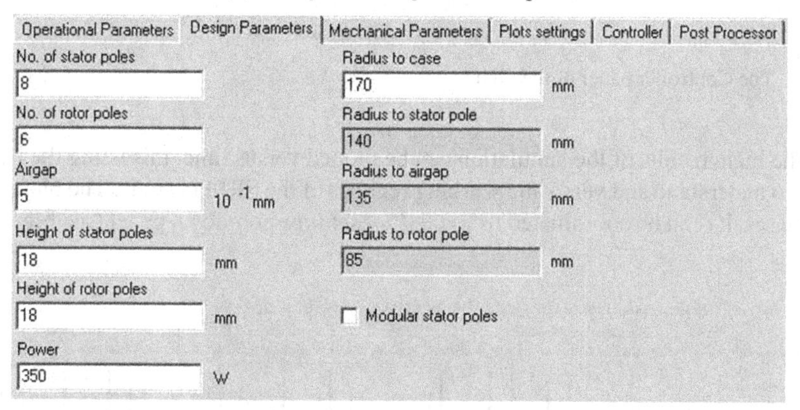

(b) the *Design parameters* panel

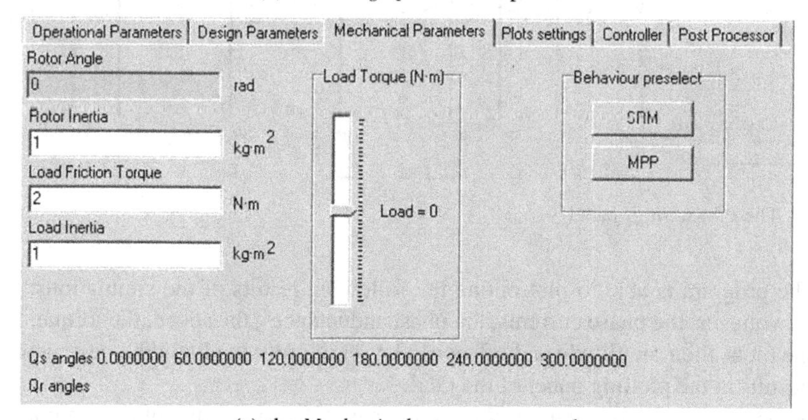

(c) the *Mechanical parameters* panel

Fig. 7 The pop-down panels of the GUI

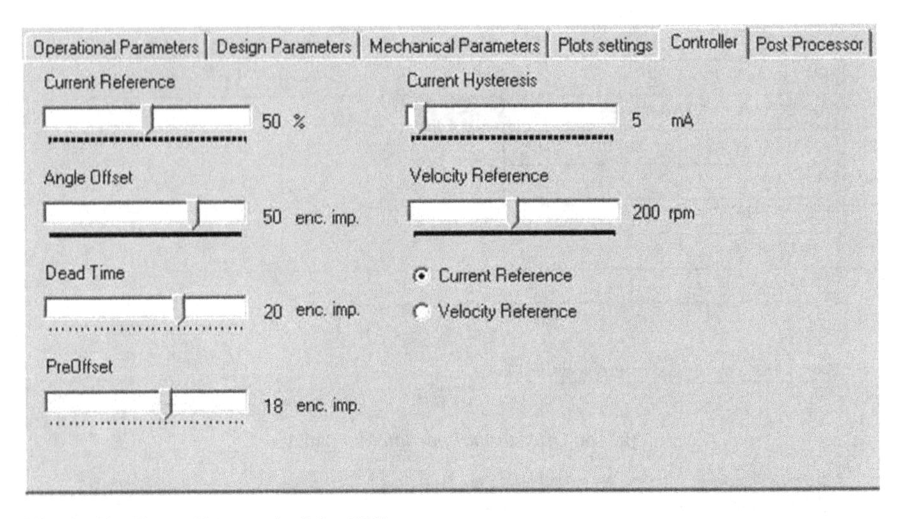

Fig. 8 The Controller panel of the GUI

The main results of the simulation can be plotted versus time, this being the best way to understand and verify the working regimes of the SRM in study. The plotting of the results can be coordinated from the *Plot settings* pop-down panel (see Fig. 9).

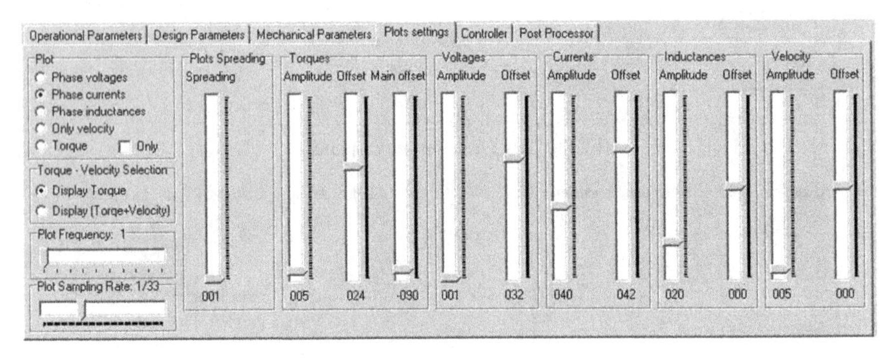

Fig. 9 The *Plot settings* panel

The program is able to plot online the following results of the simulations: the phase voltages, the phase currents, the phase inductances, the speed, the torque. For all the plots, their amplitude and offset can be set in order to obtain the best view of the results in the plotting panel of the GUI.

Although the program can display on its main GUI the results of the simulation in real time, it can also save these results in text files (see Fig. 10). By means of this feature the user can use other programs both for graphical and data processing of the obtained results.

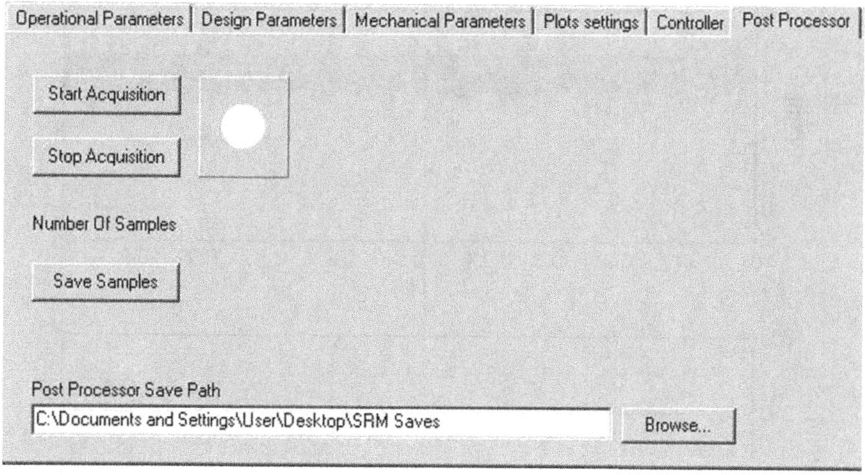

Fig. 10 The *Post processor* panel

5 Results of the Simulations

To emphasize the usefulness of the above presented program the results obtained by means of a simulation task performed for a sample motor will be given. The main data of the sample SRM are: stator poles number 8, rotor poles number 6, rated current 5 A, rated speed 200 rpm, rated torque 4 $N \cdot m$.

The simulations were performed for a 1 s time interval starting and steady state run when a ramp type speed profile was imposed. The results of the simulations were saved in text files and exported to MATLAB for advanced graphical processing.

The most important results obtained via the simulation (the torque and the speed of the SRM versus time) are given in Fig. 11(a). The phase currents are not plotted here. Due to the high frequency of the phase current's commutation, the current pulses could not be distinguished at this time scale.

As it can be seen in Fig. 11(a), the controlled SRM follows very strictly the imposed speed profile. In steady-state regime the speed of the machine is strictly constant. Also the inherent torque ripples of the SRM are clearly emphasized in Fig. 11(a). In order to study the waveforms of the phase currents, the results were also plotted for a shorter period (see Fig. 11(b)).

In the figure the typical waveforms of PWM controlled current can be easily observed. As the speed of the SRM is increasing, the frequency of the phase currents is also increasing due to the obligatory synchronization of the supply current pulses with the rotor's position.

For a more accurate view of the results zoomed plots are provided in Fig. 12.

Although the program basically is designed to deal with ideal cases, it can handle also faulty conditions of the machines. Diverse winding faults can be simulated by minor changes in the simulation's settings. This is very important since the SRM, as any other electrical machine, is subject to failures or faults [32].

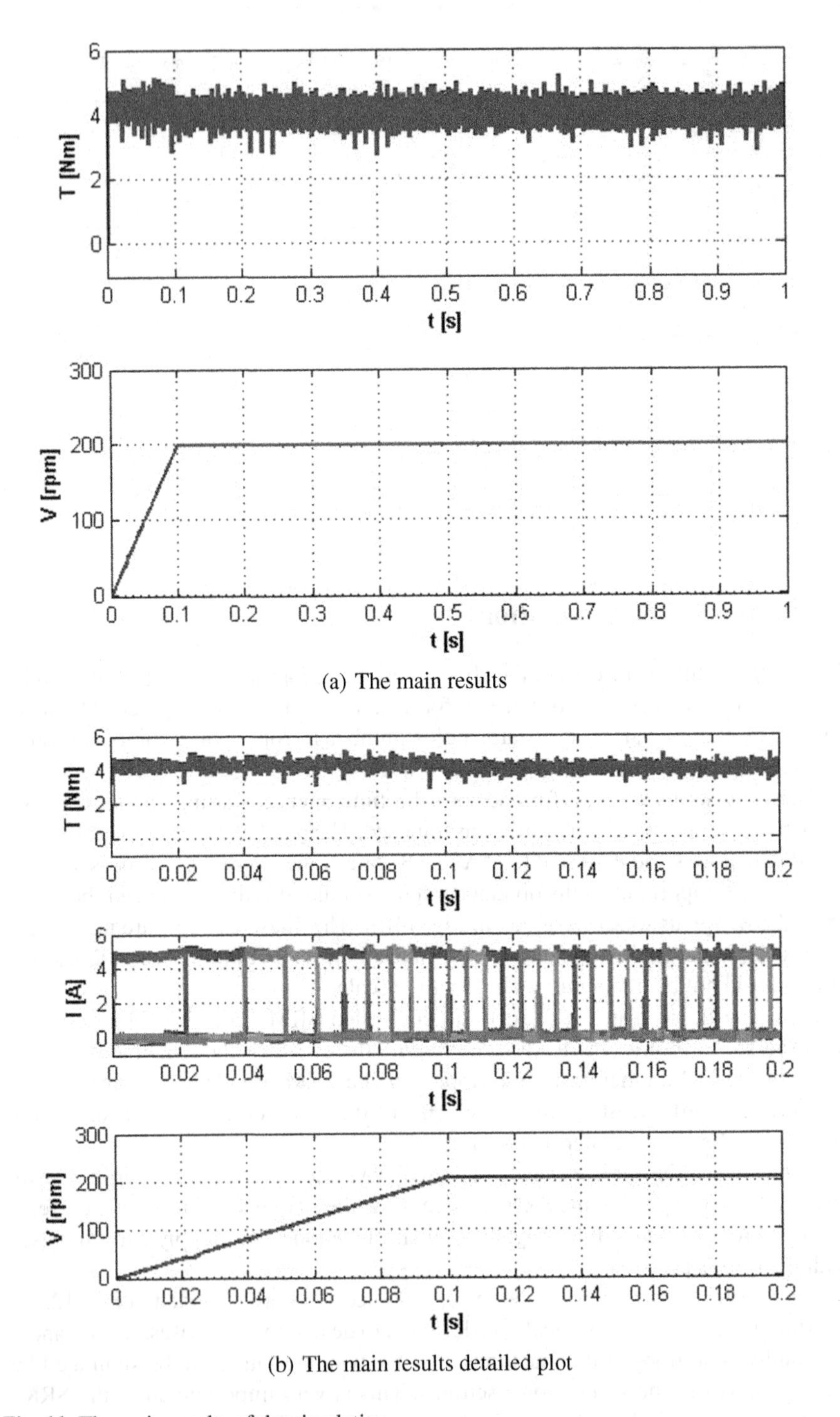

Fig. 11 The main results of the simulation

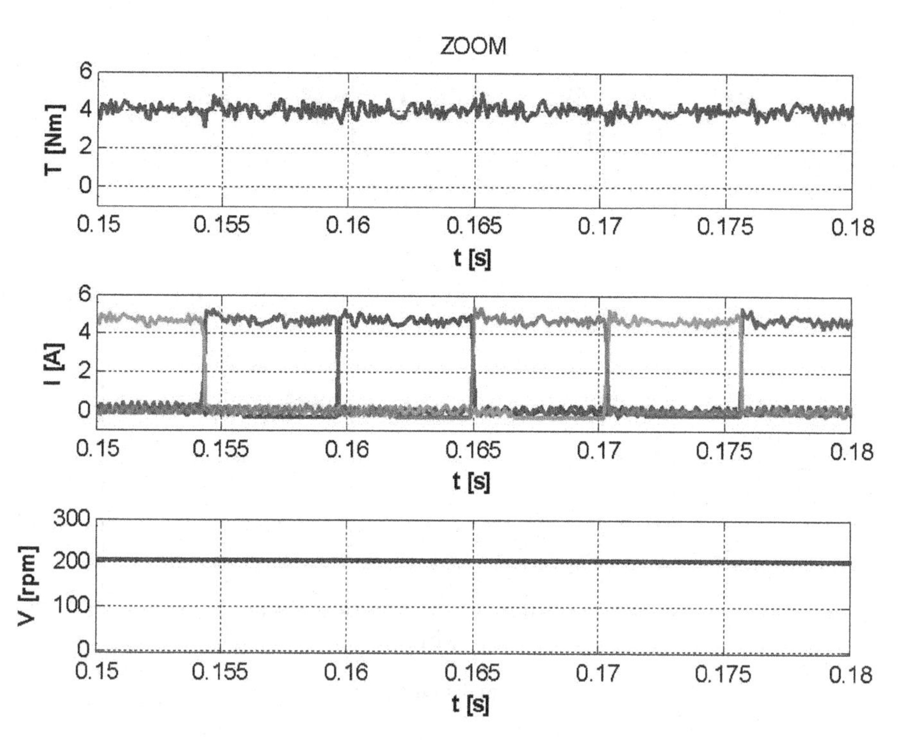

Fig. 12 A more detailed view of the simulation's main results

To point up also this feature of the program simulations for faulty conditions of the sample machine were also carried out. The simulation was performed for the same conditions as previously, but a coil of the SRM was considered opened. This is one of the most frequent machine faults. The main results are given in Fig. 13(a). For a better understanding of the results given in Fig. 13(a) a zoomed view can be seen in Fig. 13(b).

The lack of the currents through a coil of the machine can be clearly distinguished in both figures. As one of the two coils of a phase is opened the torque developing capability during that phase's work is diminished about with 50%. Therefore the SRM's torque will have much greater torque ripples as in the case of the healthy condition.

However also the fault tolerance capacity of the SRM in study is proved by means of simulations: the machine is able to continue its rotation also in spite of this winding error. Of course this has a price: less torque developed and torque ripples increased.

Upon all the results of the simulations presented above it can be concluded that all of them are perfectly in accordance with the theoretically expectations. This emphasizes both the precision of the simulation tool and its usefulness in studying the variable reluctance electrical machine's working principles and characteristics.

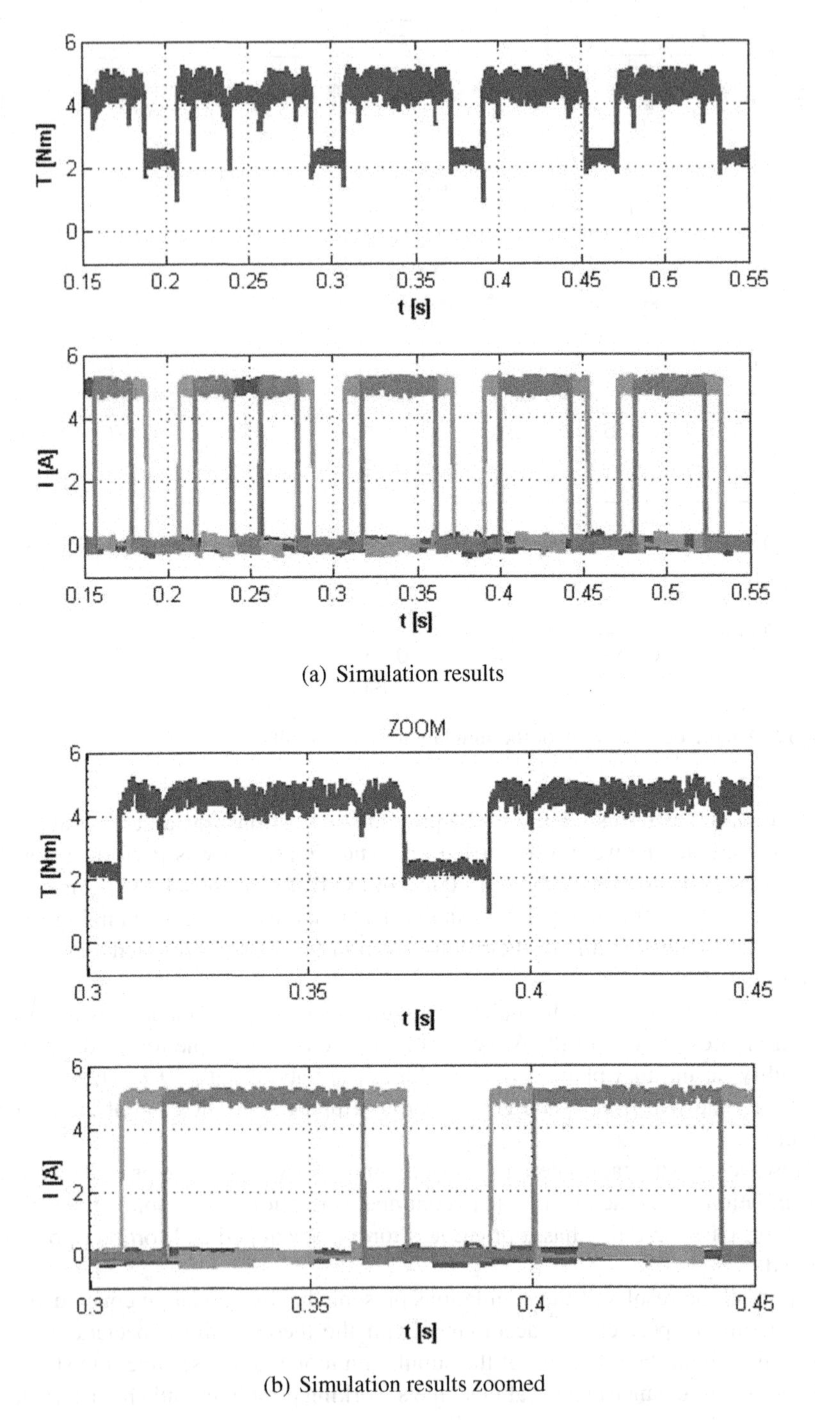

(a) Simulation results

(b) Simulation results zoomed

Fig. 13 Simulation's results for a faulty condition of the SRM

The online software tool can be also used in online testing of diverse control strategies developed for various variable reluctance machines. A demo of the simulation program can be tested at: `http://elbioarch.utcluj.ro/em/SRMdemo_sec.swf`.

6 Conclusions

In the paper an advanced simulation program, developed by the authors, of the SRM and its control system has been presented. It can be useful to show to the users how the SRM behaves in real time. Several simulations performed using this program proved its high credibility simulation qualities.

The main advantage of the proposed software tool is its capability to perform online simulations, offering the user the possibility to make real time changes even during the machine's simulation. By this, the influence of any parameter modification can be easily observed and studied.

The program can be used successfully both in research (for testing diverse SRM configurations and control strategies) and in higher education. By minor extensions of the SRM program or by structuring it as software components the simulation program can be also applied in remote (virtual) laboratories [6, 3].

Acknowledgements. This paper was supported by the project "Doctoral studies in engineering sciences for developing the knowledge based society—SIDOC" contract no. POSDRU/88/1.5/S/60078, project co-funded from European Social Fund through Sectorial Operational Program Human Resources 2007–2013.

A part of the paper was also supported by the Romanian National University Research Council (CNCSIS)—Executive Unit for Financing the Research, Development and Innovation in Higher Education (UEFISCDI), project type PN-II-RU, code TE_250, no 32/28.07.2010.

List of Symbols

λ, λ_m phase, respectively magnetizing flux
θ rotor angular position
ω angular speed
B viscous friction coefficient
i phase current
J moment of inertia
L_σ leakage inductance
p the number of pole pairs
R phase resistance
T torque
T_l load torque
v phase voltage

Appendix

Here two short parts of the simulation program's thousands line long code are detailed.

In the first code sequence the phase current of the SRM is computed in real time. The current is recomputed in a repeating loop at each time step by taking into account the phase voltage, respectively the resistance and inductance of the phase winding.

```
{...}
repeat
  {...}
  ic[i] := ic[i] + (E/R - ic[i]) / (L/R);
  {...}
until Terminated;
{...}

{ic - Instantaneous Current
E - Phase Voltage (DC Bus voltage of the power inverter, if
    the coil is energized. 0, otherwise.)
R - Phase Resistance
L - Phase Inductance
Terminated - simulation thread running status}
```

An important issue in the simulation program is the variation of the phase inductance versus the angular displacement of the SRM.

The inductance's idealized change during the rotor's movement is given in Fig. 14.

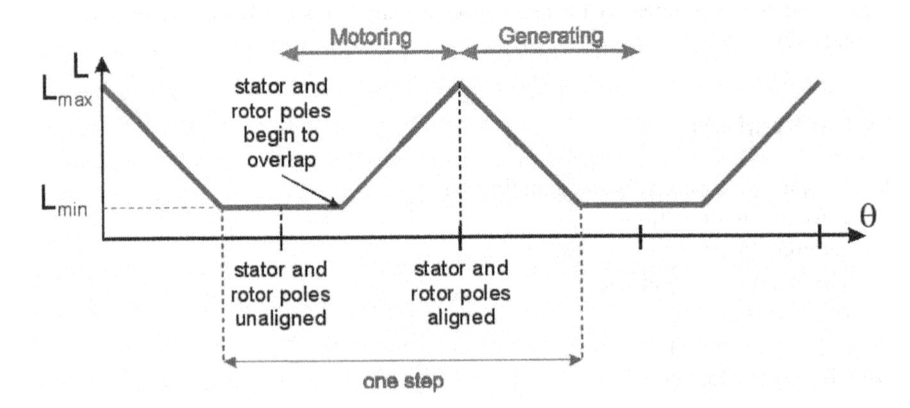

Fig. 14 The variation of phase inductance versus angular displacement of the rotor

As it can be seen in the absence of magnetic saturation the phase inductance varies linearly with the rotor position [23]. Hence it is very important to have the correct relative angular position of the rotor poles relatively to the poles of the stator.

In order to compute this position several angles are defined both on the SRM's stator and rotor, as shown in Fig. 15. The defined angles are stored in two arrays: *ang_ns*, respectively *ang_nr*.

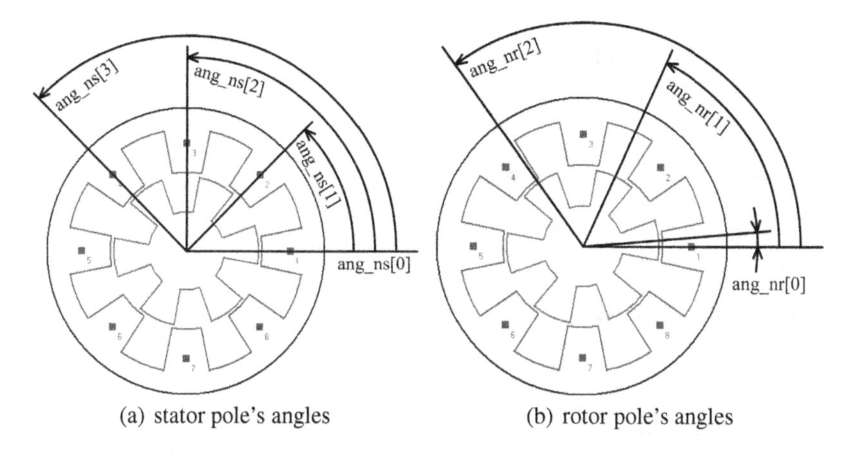

(a) stator pole's angles (b) rotor pole's angles

Fig. 15 The angles defined on the stator and rotor of the SRM

Next the implemented algorithm used to compute inductances based on the relative angles between a stator pole and each rotor pole is detailed. The closest rotor pole to the stator pole in discussion gives the minimum angle. This angle is directly used in the phase inductance computation, as it can be seen in the following program sequence.

```
type
  TFloat = Real;
var
  Qs, Qr: Integer; {number of stator and rotor poles}
  ang_ns: array of TFloat; {Statoric poles angles}
  ang_nr: array of TFloat; {Rotoric poles angles}
  LL: array of TFloat; {inductances for each stator pole}
  BetaS: TFloat; {pi / Qs}
  AngleSign: TValueSign; {torque direction (sign)}
  {...}

procedure ComputeAngleDifferences(i: Integer; var AngleSign:
    TValueSign; var MinAngle: TFloat);
const
  CComparisonError = 0.00001;
var
  j: Integer;
  dif: TFloat;
  angs, angr: TFloat;
  chd: Boolean; {changed}
```

```
begin
  AngleSign := 0;
  MinAngle := MaxInt; {(2^31 - 1)}
    for j := 0 to Qr - 1 do
  begin
    angs := ang_ns[i];
    angr := ang_nr[j];

    if angs >= pi then
      angs := angs - 2 * pi;

    if angs <= -pi then
      angs := angs + 2 * pi;

    if angr >= pi then
      angr := angr - 2 * pi;

    if angr <= -pi then
      angr := angr + 2 * pi;

    dif := Abs(angs - angr);
    chd := False;
    if dif >= pi then
    begin
      dif := Abs(dif - 2 * pi);
      chd := True;
    end;

    {Compare two real numbers with approximation.}
    if Math.SameValue(dif, 2 * pi, CComparisonError) then
      dif := 0;

    if MinAngle > dif then
    begin
      MinAngle := dif;
      AngleSign := Sign(angs - angr);
      if chd then {changed}
        AngleSign := - AngleSign;
    end;
  end;  {for}
end; {proc}

{compute inductances for each stator pole}
  thetaalign := (thetamin+thetamax)/2;
  for i := 0 to Qs - 1 do
  begin
    ComputeAngleDifferences(i, AngleSign, MinAngle);
```

```
    theta[i] := Abs(MinAngle);
    if theta[i] < thetamin then
      LL[i] := Lmin;
    if (theta[i] >= thetamin) and (theta[i] < thetaalign)
        then
      LL[i] := Lmin + 2*(Lmax-Lmin) / (thetamax - thetamin)
          * (theta[i] - thetamin);
    if (theta[i] >= thetaalign) and (theta[i] < thetamax)
        then
      LL[i] := Lmin + 2*(Lmax-Lmin) / (thetamax - thetamin)
          * (theta[i] - thetamax);
    if theta[i] > thetamax then
      LL[i] := Lmax;
end
```

The variation of the four phase inductances was visualized in the scope window of the GUI (see Fig. 16).

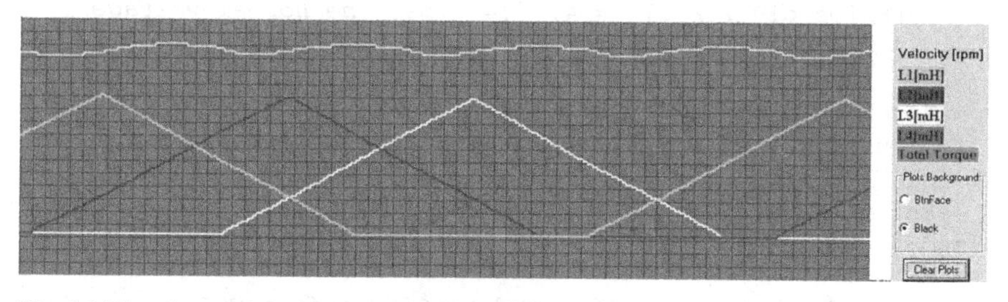

Fig. 16 The plots of the speed and the phase inductances vs. the angular displacement

As it can be seen, the obtained simulated waveforms variation of the phase inductances are in concordance with those plotted in Fig. 14 for the idealized case, when the saturation of the magnetic core is not taken into account. Finally the code sequence used for simulating the position controller of the SRM is given.

```
{This controller uses a 10-bit resolution virtual encoder
    and 4x multiplier.}
procedure PositionController(Qs, Qr: Integer; RotorAngle:
    TFloat;
  var AngleMaxEncCountInit: Integer; Hold: Boolean; var
      OutEE: array of Byte);
const
  pi_m2 = 2 * pi;
var
  angleMaxEncCount, AngleMaxEncCountTemp: Integer;
  ang_nr0_Positive: TFloat;
  DeadTime, PreOffset, AngleOffset: Integer;
  i, k: Integer;
  Qs_div_2: Integer;
  RotorInterval: Integer;
```

```
  StatorInterval: Integer;
begin
  Qs_div_2 := Qs div 2;
  RotorInterval := CEncoderMaxCount div Qr;
  k := Round(CEncoderMaxCount / (Qs_div_2 * Qr));
  ang_nr0_Positive := pi_m2 - RotorAngle;
  angleMaxEncCount := Abs(Round(CEncoderMaxCount *
      ang_nr0_Positive / pi_m2));
  if angleMaxEncCount > CEncoderMaxCount then
    angleMaxEncCount := CEncoderMaxCount - angleMaxEncCount;
  if angleMaxEncCount < 0 then
    angleMaxEncCount := CEncoderMaxCount + angleMaxEncCount;
  if Hold then
  begin {energize only one phase for incremental encoder
      reset}
    AngleMaxEncCountInit := AngleMaxEncCount;
    for i := 0 to Qs_div_2 - 2 do
      OutEE[i] := 0;
    OutEE[Qs_div_2 - 1] := 4;  {4 times the normal voltage
        in "Hold" state, without current control}

    for i := 0 to Qs_div_2 - 1 do
      OutEE[i + Qs_div_2] := OutEE[i]; {energize the pair
          coils}
    Exit;
  end;

  {get AngleOffset, DeadTime and PreOffset from GUI
      (controller tab)}
  AngleOffset := SRMMainForm.tbrAngleOffset.Position;
  DeadTime := SRMMainForm.tbrDeadTime.Position;
  PreOffset := SRMMainForm.tbrPreOffset.Position;
  if AngleMaxEncCountInit < 0 then
    AngleMaxEncCountInit := CEncoderMaxCount +
        AngleMaxEncCountInit;
  AngleMaxEncCountTemp := AngleMaxEncCountInit -
      AngleMaxEncCount + AngleOffset;
  if AngleMaxEncCountTemp < 0 then
    AngleMaxEncCountTemp := CEncoderMaxCount +
        AngleMaxEncCountTemp;
  Enc := AngleMaxEncCountTemp;  {global var used by post
      processor}
  StatorInterval := AngleMaxEncCountTemp mod RotorInterval;

  for i := 0 to Qs_div_2 do
    OutEE[i mod Qs_div_2] := 0; {reset all phase voltages}

  for i := 0 to Qs_div_2 do
  begin
```

```
    if (StatorInterval >= k * i - PreOffset) and
       (StatorInterval < k * (i + 1) - DeadTime) then
      OutEE[i mod Qs_div_2] := 1;  {energize only the
         appropriate phase}
  end;

  for i := 0 to Qs_div_2 - 1 do
    OutEE[i + Qs_div_2] := OutEE[i]; {energize the pair
       phases}
end;
```

References

1. Developer Studio 2006 Reference. Delphi Language Guide, C++ Language Guide, Together Reference. Borland Software Corporation, Scotts Valley, USA (2006)
2. Adurariu, E.P., San, L.S., Viorel, I.A., Tiş, C.M., Cornea, O.: Switched reluctance motor analytical models, comparative analysis. In: Proceedings of the 12th International Conference on Optimization of Electrical and Electronic Equipment (OPTIM 2010), Moieciu, Romania (2010)
3. Aldrich, C.: Learning online with games, simulations, and virtual worlds: Strategies for online instruction. Jossey-Bass Guides to Online Teaching and Learning. John Wiley and Sons, New York (2009)
4. Asghari, B., Dinavahi, V.: Permeance network based real-time induction machine model. In: Proceedings of the International Conference on Power Systems Transients (IPST 2009), Kyoto, Japan, pp. 1–6 (2009)
5. Bauer, P., van Duijsen, P.J.: Challenges and advances in simulation. In: Proceedings of the IEEE 36th Power Electronics Specialists Conference (PESC 2005), Recife, Brazil, pp. 1030–1036 (2005)
6. Bauer, P., Fedak, V., Hajek, V., Lampropoulos, I.: Survey of distance laboratories in power electronics. In: Proceedings of the 39th IEEE Power Electronic Specialists Conference (PESC 2008), Rhodes, Greece, pp. 430–436 (2008)
7. Bauer, P., Korondi, P., van Duijsen, P.J.: Integrated control — simulation design approach. In: Proceedings of the International Conference on Power Electronics, Drives and Motion (PCIM 2003), Nürnberg, Germany (2003)
8. Cantù, M.: Mastering Borland Delphi 2005. SYBEX, Indianapolis (2006)
9. Chindriş, V., Szász, C.: Real–time simulation of embryonic structures for high reliability mechatronic applications. In: Proceedings of the 16th International Conference on Building Services, Mechanical and Building Industry Days, Debrecen, Hungary, pp. 571–578 (2010)
10. Chindriş, V., Terec, R., Ruba, M., Szabó, L., Rafajdus, P.: Useful software tool for simulating switched reluctance motors. In: Proceedings of the 25th European Conference on Modelling and Simulation (ECMS 2011), Kraków, Poland, pp. 216–221 (2011)
11. D'hulster, F.: Switched reluctance motor (SRM) drive modelling using flux to simulink technology. Flux magazine 41, 10–11 (2003)
12. Ding, W., Liang, D.: A fast analytical model for an integrated switched reluctance starter/generator. IEEE Transactions on Energy Conversion 25(4), 948–956 (2010)
13. Henao, H., Capolino, G.A., Poloujadoff, M., Bassily, E.: A new control angle strategy for switched reluctance motor. In: Proceedings of the 7th European Conference on Power Electronics and Applications, Trondheim, Norway, pp. 613–618 (1997)

14. Henneberger, G., Viorel, I.A.: Variable Reluctance Electrical Machines. Shaker Verlag, Aachen (2001)
15. Ichinokura, O., Onda, T., Kimura, M., Watanabe, T., Yanada, T., Guo, H.: Analysis of dynamic characteristics of switched reluctance motor based on SPICE. IEEE Transactions on Magnetics 34(4), 2147–2149 (1998)
16. Ichinokura, O., Suyama, S., Watanabe, T., Guo, H.J.: A new calculation model of switched reluctance motor for use on SPICE. IEEE Transactions on Magnetics 37(4), 2834–2836 (2001)
17. Khalil, A., Husain, I.: A fourier series generalized geometry-based analytical model of switched reluctance machines. IEEE Transactions on Industry Applications 43(3), 673–684 (2007)
18. Kozierok, C.M.: The TCP/IP guide: a comprehensive, illustrated Internet protocols reference. No Starch Press, San Francisco (2005)
19. Krishnan, R.: Switched reluctance motor drives: modeling, simulation, analysis, design, and applications. CRC, Boca Raton (2001)
20. Lachman, T., Mohamad, T., Fong, C.: Nonlinear modelling of switched reluctance motors using artificial intelligence techniques. IEE Proceedings — Electric Power Applications 151(1), 53–60 (2004)
21. Lu, W., Keyhani, A., Fardoun, A.: Neural network-based modeling and parameter identification of switched reluctance motors. IEEE Transactions on Energy Conversion 18(2), 284–290 (2003)
22. Miller, F.P., Vandome, A.F., McBrewster, J.: Graphical User Interface. VDM Publishing House Ltd., Beau Bassin (2009)
23. Miller, T.J.E.: Switched Reluctance Motors and Their Control. Magna Physics (1993)
24. Miller, T.J.E.: Electronic Control of Switched Reluctance Machines. Newnes, Oxford (2001)
25. Mohan, M., Undeland, T.M., Robbins, W.P.: Power Electronics: Converters, Applications and Design. John Wiley and Sons, New York (2003)
26. Moreira, J.C., Lipo, T.: Simulation of a four phase switched reluctance motor including the effects of mutual coupling. Electric Machines and Power Systems 16(4), 281–299 (1989)
27. O'Dwyer, J., Vonhof, E.: Saturable variable reluctance motor simulation using spline functions. In: Proceedings of the International Conference on Electrical Machines (ICEM 1994), Paris, France (1994)
28. Ong, C.M.: Dynamic simulation of electric machinery: using Matlab/Simulink. Prentice Hall PTR, Upper Saddle (1998)
29. Perl, J.: Antagonistic adaptation systems: An example of how to improve understanding and simulation complex system behaviour by use of meta-models and on line-simulation. In: Proceedings of the 16th International Conference on Scientific Computing & Mathematical Modelling (IMACS 2000), Laussane, Switzerland (2000)
30. Preston, M.A., Lyons, J.P.: A switched reluctance motor model with mutual coupling and multi-phase excitation. IEEE Transactions on Magnetics 27(6), 5423–5425 (1991)
31. Radun, A.: Analytically computing the flux linked by a switched reluctance motor phase when the stator and rotor poles overlap. IEEE Transactions on Magnetics 36(4) (2000)
32. Ruba, M., Anders, M.: Fault tolerant switched reluctance machine study. In: Proceedings of the International Conference on Power Electronics, Intelligent Motion and Power Quality (PCIM 2008), Nürnberg, Germany (2008)
33. Soares, F., Branco, P.C.: Simulation of a 6/4 switched reluctance motor based on matlab/simulink environment. IEEE Transactions on Aerospace and Electronic Systems 37(3), 989–1009 (2001)

34. Strete, L., Husain, I., Cornea, O., Viorel, I.A.: Direct and inverse analytical models of a switched reluctance motor. In: Proceedings of the 14th Biennial IEEE Conference on Electromagnetic Field Computation (CEFC 2010), Chicago, USA, pp. 1–6 (2010)
35. Szabó, L., Ruba, M.: Using co-simulations in fault tolerant machine's study. In: 23rd European Conference on Modelling and Simulation (ECMS 2009), Madrid, Spain, pp. 756–762 (2009)
36. Szász, C., Chindriş, V.: Fault-tolerant embryonic network development for high reliability mechatronic applications. International Review of Applied Sciences and Engineering 1(1), 61–66 (2010)
37. Szász, C., Chindriş, V., Szabó, L.: Modeling and simulation of embryonic hardware structures designed on FPGA–based artificial cell network topologies. In: Proceedings of the 23rd European Conference on Modelling and Simulation (ECMS 2009), Madrid, Spain (2009)
38. Torrey, D.A., Lang, J.H.: Modelling a nonlinear variable-reluctance motor drive. IEE Proceedings B — Electric Power Applications 137(5), 314–326 (1990)
39. Tsukii, T., Nakamura, K., Ichinokura, O.: SPICE simulation of SRM considering nonlinear magnetization characteristics. Electrical Engineering in Japan, 50–56 (2003)
40. Ustun, O.: A nonlinear full model of switched reluctance motor with artificial neural network. Energy Conversion and Management 50(9), 2413–2421 (2009)
41. Viorel, I.A., Forrai, A., Ciorba, R.C.: On the switched reluctance motor circuit-field model. In: Proceedings of the International Conference on Electrical Machines (ICEM 1996), Vigo, Spain, pp. 94–98 (1996)
42. Wen, D., Deliang, L., Zhuping, C.: Dynamic model and simulation for a 6/4 switched reluctance machine system assisted by maxwell SPICE and simplorer. In: Proceedings of the International Conference on Mechatronics and Automation (ICMA 2007), Harbin, China, pp. 1699–1704 (2007)
43. Wu, W., Dunlop, J.B., Collocott, S.J., Kalan, B.A.: Design optimization of a switched reluctance motor by electromagnetic and thermal finite-element analysis. IEEE Transactions on Magnetics 39(5), 3334–3336 (2003)
44. Yao, R., Stiebler, M., Liu, D.: A generalized simulation for switched reluctance motor variable–speed system. In: Proceedings of the International Conference on Electrical Machines (ICEM 1998), Instanbul, Turkey, pp. 1686–1691 (1998)

A Hybrid Emulation Environment for Airborne Wireless Networks

Marco Carvalho, Adrián Granados, Carlos Pérez, Marco Arguedas,
Michael Muccio, Joseph Suprenant, Daniel Hague, and Brendon Poland

Abstract. Airborne networks are among some of the most challenging communication environments for protocol design and validation. Three dimensional and highly dynamic in nature, airborne network links are generally difficult to simulate accurately and often expensive to recreate in field experiments. The physical characteristics of the data links in these kinds of networks are not only influenced by the usual environmental factors and interferences, but also by the frequent changes in the relative angles and positions between antennas, shadowing effects of the airframe, non-uniform ground effects and other factors. In this work we summarize our research for the development of mLab, a hybrid emulation environment for airborne networks that combines theoretical path loss models with statistical data-driven models to create a low-cost, high fidelity, specialized emulation environment for the development, evaluation, and validation of protocols and algorithms for airborne wireless networks.

1 Introduction

The majority of the mobile ad hoc network (MANET) research is based on computer simulations, emulation network environments and field tests. While some important

Marco Carvalho
Florida Institute of Technology, 150 W University Blvd., Melbourne, FL 32901 USA, and Institute for Human and Machine Cognition, 15 SE Osceola Ave., Ocala, FL 34471, USA
e-mail: mcarvalho@fit.edu

Adrián Granados · Carlos Pérez · Marco Arguedas
Florida Institute for Human and Machine Cognition, 15 SE Osceola Ave., Ocala, FL 34471, USA
e-mail: {agranados,cperez,marguedas}@ihmc.us

Michael Muccio · Joseph Suprenant · Daniel Hague · Brendon Poland
U.S. Air Force Research Laboratory, 26 Electronic Parkway, Rome, NY, USA
e-mail: {michael.muccio,joseph.suprenant}@rl.af.mil,
{daniel.hague,brendon.poland}@rl.af.mil

A. Byrski et al. (Eds.): Advances in Intelligent Modelling and Simulation, SCI 416, pp. 111–129.
springerlink.com

theoretical work does exist in the literature, they generally constitute less than 25% of the body of work in field, as noted by Kurkowski et al. [12] in their review of five years of published research. Thus, wireless modeling and simulation play an important role in MANET, and airborne networks research.

In most wireless simulation environments, theoretical data link models such as Free Space or Two Ray Ground [15] are used to recreate the effects of the physical environment [6, 9]. The relative position of nodes, as well as their mobility patterns and physical interface settings, were normally treated as inputs to mathematical models used to calculate the quality of individual links. In most simulation tools and frameworks [1, 4, 7, 8, 2, 13] mathematical link models are then used to calculate the rates of drop or corrupt data-rates necessary to reflect specific scenarios and operational settings. However, the quality of the simulation is directly affected by the accuracy of such models [17], as well as the simulation and experimental approaches [11, 14, 12].

Not surprisingly, the alternative approach for validation and development is to rely on field experiments, which is usually costly, time consuming and very difficult to replicate. In the case of airborne networks the issue is even more complicated. Three dimensional and highly dynamic in nature, airborne network links are generally difficult to simulate accurately, and often expensive to recreate in field experiments. The physical characteristics of the data links in these kinds of networks are not only influenced by the usual environmental factors and interferences, but also by the frequent changes in the relative angles and positions between antennas, shadowing effects of the airframe, non-uniform ground effects and other factors.

From a protocol design perspective, an accurate emulation of these kinds of link-effects is important to enable the experimental validation of laboratory results, and to better understand the behavior of the protocol. However, popular non-commercial wireless network simulators [1, 3, 4, 10] rely solely on common theoretical radio propagation models that are often constrained to specific environments and communication parameters, lacking of models to emulate the environmental and external effects that influence the characteristics of wireless links.

In this context, empirical data-driven radio propagation models have been previously proposed [5, 16, 19] to create more reliable emulation environments for different target scenarios and varying wireless technologies. Hence, we have developed mLab, a hybrid emulation environment for airborne networks that combines theoretical path loss models with statistical data-driven models to create a low-cost, high fidelity, specialized emulation environment for the development, evaluation and validation of protocols and algorithms for airborne wireless networks.

In the following sections, we present the design and architecture of mLab, and also discuss the integration of mLab and EMANE (the Extendable Mobile Ad-hoc Network Emulator) to provide additional link enforcement mechanisms as well as to facilitate the reuse and sharing of mLab's statistical link models. Finally, we give a detailed description of our approach for generating data-driven corrective models using the datasets from the Protocol Emulation for Next Generation Wireless UAS Networks (PENGWUN) project, an internal effort of the Air Force Research

Laboratory (AFRL). We conclude this chapter, by presenting the experimental results of the tests and evaluation of our proposed statistical data-driven models.

2 The mLab Hybrid Emulation Testbed

The mLab hybrid emulation testbed is one of the components of the PENGWUN project. The architecture of the mLab (Fig. 1) testbed has four components: the hardware infrastructure, the modeling component, the coordination component, and the enforcement component. In particular, the last three components are responsible, respectively, for the modeling of wireless links and node behaviors, the estimation of the communications topology and the characteristics of each communication link, and the enforcement of policies (at the lowest possible levels) for wireless link emulation.

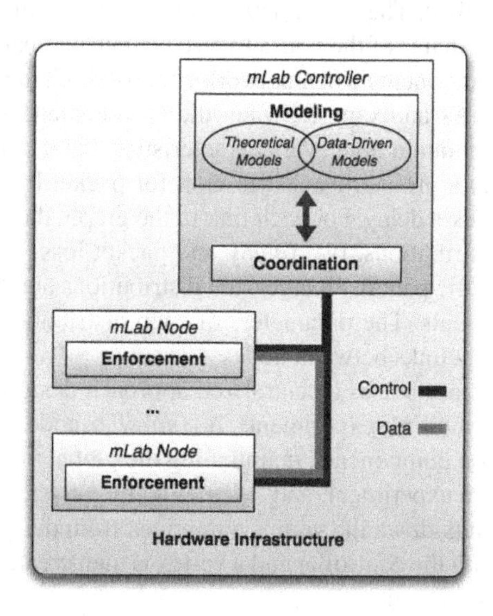

Fig. 1 Conceptual mLab architecture with enforcement components.

The emulation environment is recreated on top of a hardware infrastructure that must provide the necessary resources for the emulation, model enforcement and system coordination. There are several approaches for creating the hardware testbed and most often, deployments tend to include fixed infrastructures with sets of nodes dedicated to the emulation itself, and a few nodes dedicated to system configuration and control. Our initial design for mLab relied on a fixed set of nodes connected through two independent networks (one for data and one for control), with one platform node (the controller node) acting as the modeling and coordination element. In this configuration the user interfaces, and all other emulation nodes may be directly or remotely connected to the controller. The only requirement is that network latency at the physical level must be significantly smaller than the required emulated latencies, and the capacity at the physical level must be significantly larger than the capacity of the emulated system.

Testbed nodes have two interfaces, one connected to a data-only high bandwidth switched network and the other one connected to a control network, which is used for the control and monitoring of link behavior and network policies. The current version of mLab replaces the physical infrastructure by a virtualized environment where the controller can be dynamically connected with emulation nodes running as virtualized machines in a cloud environment. This new design for the physical infrastructure is functionally similar to the original concept.

The modeling component is responsible for modeling the physical links. As a modular component, it can be replaced to represent any arbitrary physical link using theoretical propagation models, data-driven statistical models, or a combination of them. The coordination component holds a model of the network topology represented as a graph structure. Each vertex in the graph maps to a specific host in the testbed. The coordination component continuously monitors the position and other attributes of the nodes to re-calculate and parameterize the models and enforcement components when network conditions change.

As an example, when the enforcement is performed by virtual drivers at each emulation node, the characteristics that define a communications link are only two, (1) a probability distribution for packet loss and (2) a probability distribution for packet delay. For each link in the graph, the coordination component maintains four distributions, two (delay and packet loss) for each direction. As nodes move, and traffic pattern changes, the distributions are then adjusted based on the physical link models. The parameters of each distribution are then used to configure the actual data links between nodes.

mLab uses a centralized approach designed to support the concurrent execution of multiple experiments. A *controller* node encapsulates the modeling and coordination components, maintaining the global view of the network topology for each active experiment, and calculating the physical link characteristics. Whenever a node shuts-down the vertex is removed from the graph. Upon startup, each node registers with the controller and a vertex is then created in the graph with the unique identifier of the node.

The enforcement component is a distributed software component that exists both at the controller node and at each host in the testbed. It provides coordination of policies related to current link conditions and communication constraints, and provides the enforcement capabilities (at the lowest possible levels) of packet loss and delays for the emulation. In our current implementation, the enforcement infrastructure is provided through customized virtual network drivers, directly interfacing with the controller node, or through EMANE, which provides pluggable Medium Access Control (MAC) and physical (PHY) layers for emulation of heterogeneous mobile ad-hoc networks.

3 mLab—EMANE Integration

EMANE is an infrastructure used for emulation of simple as well as complex heterogeneous mobile ad-hoc networks [13]. It supports pluggable MAC and PHY layers that allow for the emulation of commercial and tactical networks with multiple tiers and varying wireless technologies. It also supports multiple platforms (Windows, Linux and Mac OS X) and provides mechanisms to set up small as well as large-scale emulations using centralized, distributed or virtualized deployments. EMANE provides a modular architecture with well-defined APIs to allow independent development of emulation functionality for different radio models (network emulation modules), boundary interfaces between emulation and applications (transports), and distribution of emulation environmental data (events).

In EMANE, each node is represented by an instance of an emulation stack, which is comprised of three components: transport, Network Emulation Module (NEM), and Over-the-Air (OTA) Manager. This set of components encapsulate the functionality necessary to transmit, receive, and operate data routed through the emulation space. In particular, the transport component is responsible for handling data to and from the emulation space and interfaces with the underlying operating system using different platform-dependent mechanisms such as TUN/TAP (Linux, Mac OS X) and WinTap (Windows). On the other hand, a NEM provides emulation functionality for the MAC and PHY layers, including CSMA, TDMA, queue management, and adaptive transport protocols for the MAC layer, and waveform timing, half-duplex operations, interference modeling, probability of reception, out-of-band packets, and more for the PHY layer. Finally, the OTA Manager provides the necessary messaging infrastructure to deliver emulation radio model data to all nodes participating in the emulation. Control messages from the OTA Manager are distributed to each NEM through a multicast channel. Fig. 2 shows a diagram of the EMANE emulation stack.

The NEM and its corresponding MAC and PHY components are responsible for the enforcement of link characteristics. Hence, packets in the emulation space contain a special EMANE identification header and it is up to each layer to decide whether a packet is dropped or delayed before passing it to the upper layers. Each NEM requires a base configuration that specifies the network emulation model and the parameter values to configure the capabilities provided by the MAC and PHY components of the model. This configuration dictates the behavior of the emulation stack, which can be modified through the dissemination of emulation environmental data, such as location information or path loss, to emulate the dynamics of the network (e.g. node movement). This data can be distributed to NEMs in real time from the EMANE Event Service using a multicast channel, similarly to how the OTA Manager distributes control messages.

The type of environmental data that can be used to change communication constraints and how this data affects the behavior of the emulation depends on the configured network emulation model of the NEMs. For example, the Universal PHY Layer, a common PHY implementation for the various MAC layers that can be used with EMANE, may accept Location events specifying the latitude, longitude and

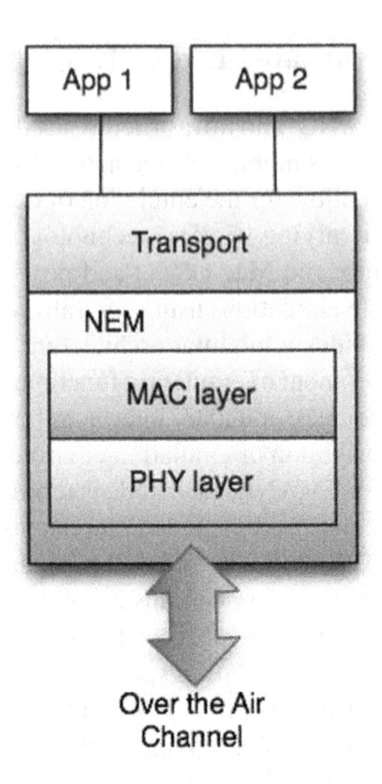

Fig. 2 EMANE emulation stack.

altitude of the NEM. When the Universal PHY Layer receives a Location event, the PHY implementation automatically computes the path loss using the most recent location information for all NEMs and the selected propagation model (Free Space or Two Ray Ground).

Alternatively, the Universal PHY Layer also supports path loss as a type of emulation environmental data that can be utilized to change link emulation conditions between NEMs. In this case, path loss computation is performed by a third-party outside of EMANE (e.g. mLab). When the Universal PHY Layer receives a *Pathloss* event containing a variable list of path loss and reverse path loss values between the receiving NEM and one or more NEMs in the emulation, the PHY implementation changes the characterization of the links to other NEMs based on the specified path loss values.

The integration between mLab and EMANE takes advantage of EMANE's network emulation models and event functionality to specify communication constraints and enforce link characteristics between mLab emulation nodes. In particular, EMANE's Universal PHY Layer's modes of operation (location or path loss) are used by mLab to perform link characterization using either EMANE's theoretical propagation models, mLab's theoretical propagation models, or the statistical data-driven models derived from the PENGWUN datasets that we have developed as part of this effort.

In mLab, emulation modes may be configured with multiple network interfaces over one or more different media (channels). In the case of EMANE, a medium corresponds to a network emulation model, such as IEEE80211abg, which emulates

IEEE 802.11 MAC layer's Distributed Coordination Function (DCF) channel access scheme and IEEE 802.11 Direct Spread Spectrum Sequence (DSS) and Orthogonal Frequency Division Multiplexing (OFDM) signals in space, or RFPipe, a model that provides low fidelity emulation of a variety of waveforms.

In order to integrate EMANE's capabilities into mLab, we had to provide mLab with the ability to automatically generate and deploy EMANE's XML configuration files, as well as the ability to generate EMANE *Pathloss* and *Location* events. Additionally, we also extended EMANE with a simple feedback mechanism that allows mLab to receive notification for changes in path loss, and implemented a TxPower event that allows mLab to dynamically control the transmitter power settings of a NEM.

```
<?xml version='1.0' encoding='utf-8'?>
<scenario name="MyScenarioDefinition"
          description="A small example scenario.">

  <!-- Defines a medium that uses EMANE's IEEE 802.11abg model -->

  <medium id="ename80211abg">
    <enforcer classname="us.ihmc.mlab.emane.EmaneEnforcer">
      <model classname="us.ihmc.mlab.emane.Emane80211EnforcerModel">
        <param name="mac.rtsthresh" value="100"/>
        <param name="phy.txpower" value="15"/>
        <param name="phy.frequency" value="2417000"/>
        <param name="phy.pathlossmode" value="freespace"/>
      </model>
    </enforcer>
  </medium>

  <!-- Defines nodes -->

  <node id="1" label="UAV">
     <interface medium_id="emane80211abg"
                address="192.168.0.1"
                netmask="255.255.255.0" />
     <position lat="29.187617746963323"
               lon="-82.13927366751467"
               alt="400.0"
               roll="0.0"
               pitch="0.0"
               yaw="0.0"/>
  </node>

  <node id ="2" label="GN">
     <interface medium_id="emane80211abg"
                address="192.168.0.2"
                netmask="255.255.255.0"/>
     <position lat="29.18655204125978"
               lon="-82.13563653875342"
               alt="80.0"
               roll="0.0"
               pitch="0.0"
               yaw="0.0"/>
  </node>

</scenario>
```

Fig. 3 Example of mLab XML Scenario File

3.1 Generation and Deployment of EMANE's XML Configuration Files

When the mLab controller loads a scenario file (Fig. 3), the EMANE enforcer component in the controller generates the necessary EMANE configuration files for the physical nodes that will take part in the emulation. Then, these XML configuration files are deployed to create and configure a NEM with the given network emulation model (e.g. IEEE80211abg MAC and Universal PHY Layer) to emulate the desired medium. An XML scenario may contain one or more medium definitions, each of them with specific model configuration parameters. Each medium is given an ID, which is then used to associate the node's network interface with a particular network emulation model.

A subset of the most relevant configuration parameters of each EMANE's network emulation model can be specified through the definition of the medium in the mLab scenario file. Parameter names that start with the prefix "phy." map directly to the configuration parameters of EMANE's Universal PHY Layer, which is used by both the IEEE80211 and RFPipe models as their PHY component. Parameter names that start with the prefix "mac." map directly to the selected EMANE MAC model. A list of the configuration parameters that can be specified through the XML scenario file is shown in Table 1.

3.2 Link Characterization

mLab performs link characterization by generating the appropriate EMANE *Pathloss* or *Location* events for each of the configured network emulation models (Fig. 4). For example, each time there is change in node position, mLab recalculates path loss values using mLab's theoretical or data-driven propagation model implementations and generates an EMANE *Pathloss* event to change the link conditions in the emulation environment. Instead, if we are leveraging completely on EMANE's capabilities, mLab generates Location events after changes in position to have EMANE dynamically compute path loss values using its own theoretical propagation models (this is the default behavior of the system).

3.3 Path Loss Feedback

As a consequence of doing link characterization using EMANE's own propagation models, mLab can no longer determine if any two given nodes can reach each other directly, and therefore it is unable to provide the required information to MView (mLab's visualizer) so that the link can be displayed depending on the connectivity between NEMs (Fig. 5). To solve this problem, we extended EMANE with a feedback mechanism that allows it to notify changes in link characteristics (i.e. path loss) back to mLab via UDP messages.

One of the challenges we encountered when doing the necessary changes to support this feature in EMANE was that, by design, path loss computation only occurs

Table 1 Subset of the most relevant configuration parameters that can be specified in a mLab scenario file

Name	Description	Default	80211abg	RFPipe
phy.bandwidth	Center frequency bandwidth in KHz.	20000	✓	✓
phy.antennagain	Antenna gain in dBi.	0	✓	✓
phy.systemnoisefigure	System noise figure in dB.	4.0	✓	✓
phy.pathlossmode	path loss mode of operation (path loss, freespace, 2ray).	freespace	✓	✓
phy.txpower	Transmit power in dBm.	15	✓	✓
phy.frequency	Transmit center frequency in KHz.	2417000	✓	✓
mac.enablepromiscuousmode	Determines if all packets received over the air will be sent up the stack to the transport layer.	off	✓	✓
mac.rtsthresh	Minimum packet size in bytes required to trigger RTS/CTS for unicast transmissions.	0	✓	
mac.datarate	Datarate/burstrate (Kbps) of the waveform being emulated.	1000		✓
mac.delay	Delay (usec) that is to be included in the transmission delay.	0		✓
mac.jitter	Jitter (usec) to be included to the transmission delay.	0		✓

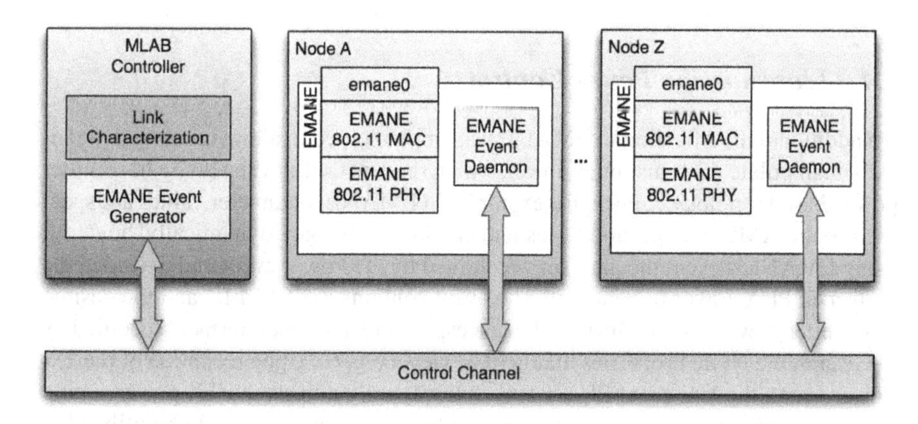

Fig. 4 mLab - EMANE Integration

Fig. 5 Lines represent links and indicate connectivity between nodes (NEMs)

upon reception of data packets, which means that EMANE cannot provide any path loss information when there are no transmissions. To solve this problem, we had to modify EMANE so that path loss is also computed when *Location* events are received and handled by the Universal PHY Layer. In this case, EMANE sends a message to mLab containing the new values for path loss, which then estimates the connectivity between two nodes and passes the information to MView so that links can be properly visualized.

3.4 Transmission Power Control

An additional modification to EMANE was implemented to allow mLab to dynamically manipulate the transmitter power, which renders useful when doing transmitter power-based topology control, for example. In EMANE, parameter values are specified in the XML configuration files and cannot be changed dynamically, however, using EMANE's event mechanism we defined the TxPower event and extended the Universal PHY Layer to handle it. The event contains the NEM ID and the desired transmitter power level (dBm) for the corresponding network interface, specified by the user in the Node Properties dialog of MView (Fig. 6). Upon reception of the TxPower event, the Universal PHY Layer automatically adjusts the TX power setting to the desired value. It is worth noting that the same approach could be utilized to change other configuration parameters if the necessary events are made available.

The main advantage of this approach is that it is clean and fits naturally into EMANE's architecture; however, one of the disadvantages is that user-specified settings (e.g. transmitter power) can be overridden by the model's MAC layer through

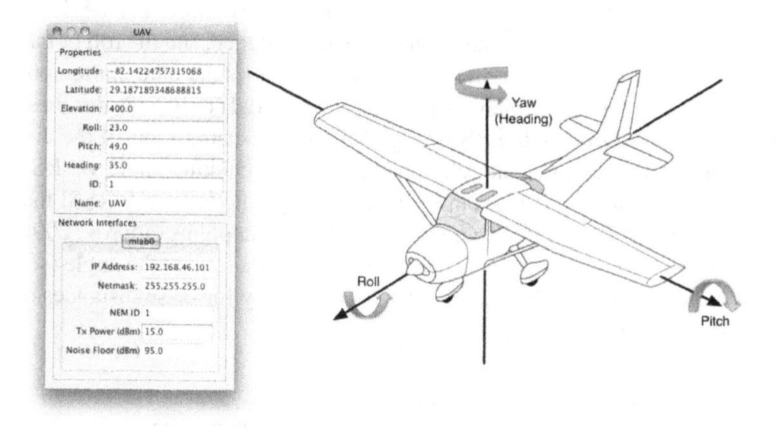

Fig. 6 The Node Properties dialog in MView

the MAC-PHY Control Messaging API, which provides the MAC component with the ability to override default PHY configuration parameters. If this is the case, then the TX power level specified by mLab using the TxPower event will have no effect.

One possible solution would be to extend the MAC layer to also handle the TxPower event, however, it will require modifying every possible model that provides its own MAC implementation. Therefore, we assume that when transmitter power control is required, the MAC of the selected model has not been configured to override PHY's configuration parameters, in particular the transmitter power.

4 Statistical Link Modeling

As part of mLab, we have developed and validated a statistical data driven model extracted from PENGWUN datasets to predict the link characteristics and network dynamics of an airborne network. In most cases, theoretical path loss models only consider distance between nodes for estimating the received signal strength indicator (RSSI) of the wireless link. However, in airborne communications signal is also attenuated when the ground and aircraft antennas lose line of sight when the aircraft is banking. Hence, we have used the PENGWUN datasets to create models for estimating the RSSI of wireless links in airborne networks based on distinct parameters.

The PENGWUN datasets consist of measurements of different platform, network and communication parameters such as GPS positions of ground nodes and aircrafts, heading, elevation and bank angle of the aircraft, RSSI and packet loss. These measurements are taken during actual field tests and saved to log files that are later imported into a database for further processing. This post-processing includes the automatic consolidation and synchronization of logged data from source and target nodes that is used in the generation of the statistical data driven models.

In the PENGWUN datasets, the variables that mostly affect the RSSI are the distance, because the signal gets attenuated with the distance, and the roll or banking of the airplane, because the ground antenna and the airplane antenna temporarily lose line of sight when the airplane is banking. Our models only consider these two dimensions when estimating the RSSI. Fig. 7 shows the scatter plots of the distance vs. RSSI and Figure 8 shows the relationship between banking angle vs. RSSI for one of the flights. The distance seems to be a better predictor for the RSSI than the angle, according to these graphs, but notice that at certain points (approx. 200m and 400m) the RSSI varies considerably. If at these points we look at the angle (see Fig. 9), this high variability can be explained. The variability for these distances occurs when the banking angle changes from 0° to 40°.

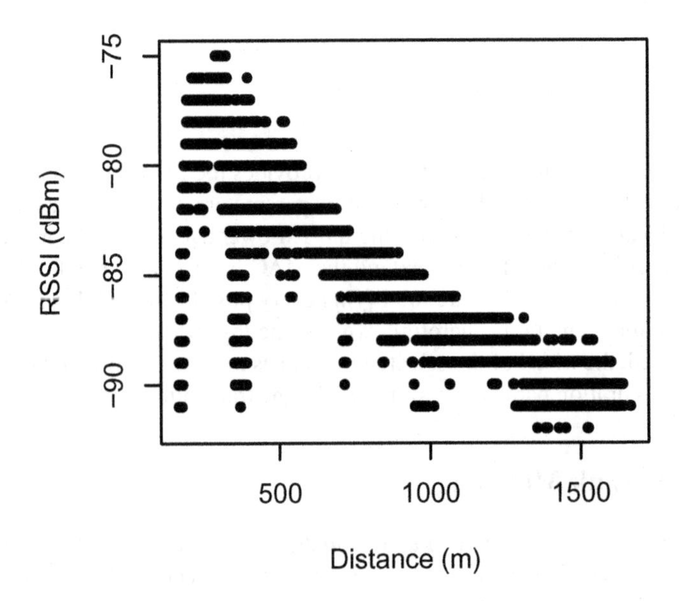

Fig. 7 Scatter plots of Distance vs. RSSI

Using the experimental data, two types of models are proposed for estimating the RSSI. The first type of model being proposed is an empirical model constructed using non-parametric estimation. The main advantage of this approach is that we can introduce as many dimensions as necessary without knowing the real distribution of the data. One of the disadvantages, however, is that the convergence to true values takes longer and it is proportional to the number of dimensions used for the estimation.

The second type of model being proposed is a parametric model based on a theoretical propagation model. The propagation model that was used is known as the log-normal shadowing model [15]. For this model two parameters need to be estimated from the data: the path loss exponent and the variance. The main advantage

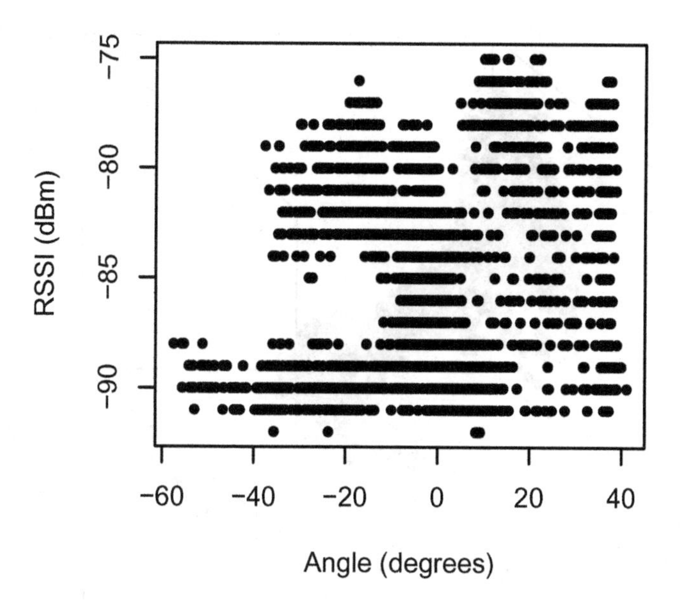

Fig. 8 Scatter plots of Distance vs. RSSI

of this model is that convergence is much faster. One disadvantage with this type of model is that we can only use the dimensions used by the model for the estimation, which in this case is only the distance. A second disadvantage with this type of model is that we are assuming that the distribution of the experimental data is equal to the distribution proposed by the model.

4.1 Empirical Model

For the empirical model, two predictor variables are used to predict the RSSI: the distance and the banking angle of the airplane, also known as the roll. The Cartesian plane formed by these two dimensions is divided in a grid with squares of 10m increments in the distance dimension and 1 degree increments in the banking angle dimension. The distance dimension ranges approximately from 160m to 1700m. The banking angle dimension ranges approximately from -60 degrees to 40 degrees.

For each square of the grid, the RSSI is modeled using the empirical distribution function, which is commonly used as a non-parametric method for estimating the cumulative distribution function (cdf) of an unknown distribution. Let $r_1, r_2, \ldots r_n$ be RSSI data points sampled from the common unknown distribution function. The cdf of the empirical distribution is defined by:

$$F(r) = \frac{1}{n} \sum_{i=1}^{n} I(r_i \leq r), \tag{1}$$

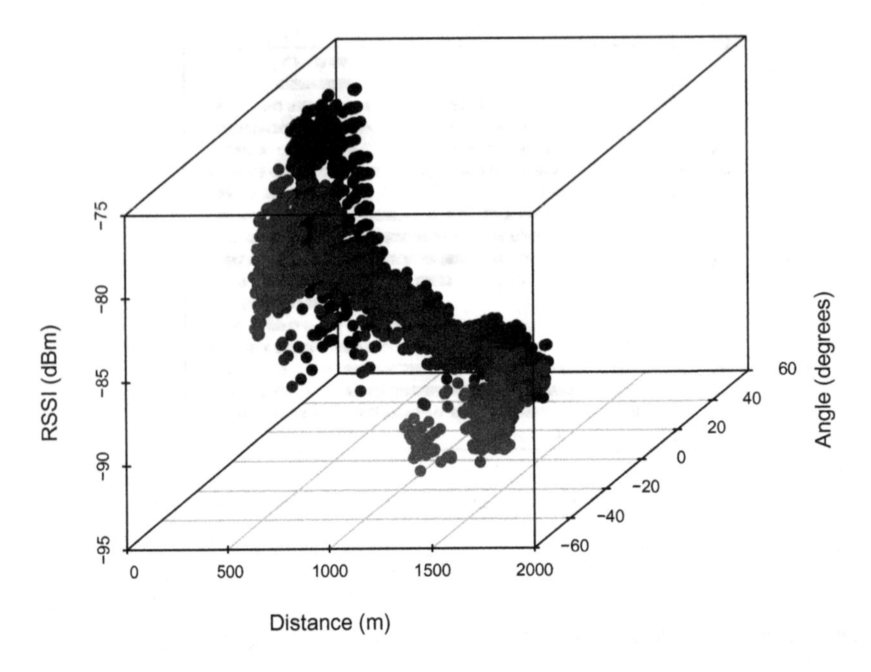

Fig. 9 Scatter plot of Distance and Angle vs. RSSI

where r is the RSSI for which the cumulative probability is being computed and $I(r_i \leq r)$ is the indicator function, defined by:

$$I(r_i \leq r) = \begin{cases} 1 & \text{if } r_i \leq r \\ 0 & \text{otherwise.} \end{cases} \quad (2)$$

Having modeled the RSSI for each square in this way, we can now generate pseudo-random numbers drawn from this distribution by generating a $p \sim U(0,1)$, and then computing $F^{-1}(p)$. So, to estimate the RSSI for a given distance and banking angle, we find the corresponding square, and then use the empirical cdf for that square to randomly generate the RSSI value.

4.2 Theoretical Model

The theoretical model used is known as the log-normal shadowing model [15], which is the same probabilistic model used by NS-2. This model assumes that the average received signal power decreases logarithmically with distance, and that the path loss is randomly distributed log-normally (normal in dB) about that mean. The equation describing the path loss at a given distance is:

$$PL(d) = \overline{PL}(d_0) + 10 \cdot \beta \cdot \log\left(\frac{d}{d_0}\right) + X_\sigma, \quad (3)$$

where $\overline{PL}(d_0)$ is the average path loss at the close-in reference distance which is based on measurements close to the transmitter or on a free space assumption from the transmitter at distance d_0, β is the path loss exponent which indicates the rate at which the path loss increases with distance, and X_σ is a zero-mean Gaussian distributed random variable (in dB) with standard deviation σ (also in dB).

Using the free space assumption, the average path loss at the close-in reference distance can be estimated using the following equation:

$$PL(d) = -20 \cdot \log\left(\frac{\lambda}{4\pi d}\right), \tag{4}$$

where λ is the wave length of the carrier signal.

To be able to use the log-normal shadowing model, we need to estimate two parameters from the experimental data: the path loss exponent, and the standard deviation of the random variable. These two parameters can be estimated by doing a linear regression of the path loss using as predictor the logarithm of the distance. The path loss in the experimental data for any given point can be computed using the following formula:

$$PL = P_t - P_r, \tag{5}$$

where Pt is the transmitted power and Pr is the received power, both in dBm. The slope of the regression line can be used as an estimate of the path loss exponent by dividing the slope by 10, and the variance obtained from the regression can be used as an estimate of the variance of the random variable. For example, using the data from the same flight shown in Figs. 7 and 8, the path loss is 0.127 and the deviation is 0.25. Fig. 10 illustrates the relationship between RSSI and distance, for the values generated using a log normal shadowing model with the same parameters.

4.3 Model Validation

Validating the models means making sure that data generated by the models resembles the distribution of the experimental data. Using a goodness-of-fit test would only be appropriate for validating the theoretical model, because the empirical model is non-parametric.

It would be ideal to use a validation method that would allow us to compare which model fits better. With this purpose in mind, we used the following validation process. Let $\{(d_1,r_1,p_1),(d_2,r_2,p_2)\dots(d_n,r_n,p_n)\}$ be the testing data, composed of distance (d_i), roll angle (r_i) and RSSI (p_i) tuples. For each of the models to validate, we will generate points $\{(d_1,r_1,\hat{p}_1),(d_2,r_2,\hat{p}_2)\dots(d_n,r_n,\hat{p}_n)\}$, where $\hat{p}_i$ is the estimated RSSI from the model being validated.

Dividing the testing and estimated data in the same fashion as it is divided in the empirical model (a grid divided by distance and angle), we can compute the average for each of the squares in the grid and then compute the relative error between the average of the testing data and the average for the estimated data for the same square in the grid.

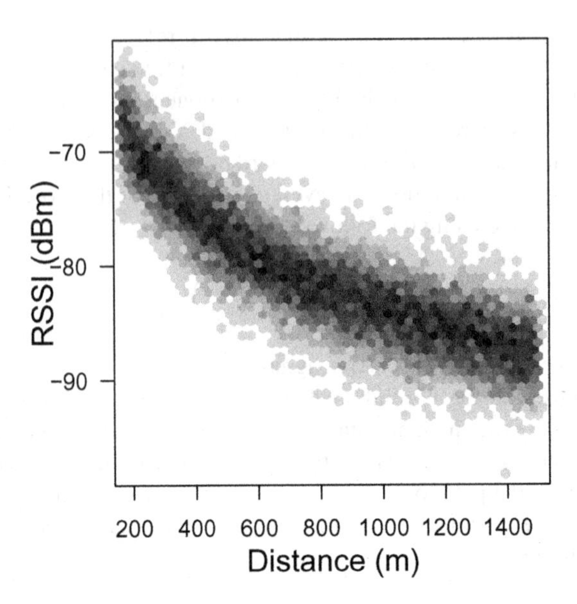

Fig. 10 Scatter plot of Distance vs. RSSI for RSSI points generated using the log-normal shadowing model

The errors for each of the squares in the grid can later be combined using a simple average, providing a single estimate of the error for the estimated model acquired for a given test set. If the models are then tested using multiple test sets, we can get a better estimate of the error, by averaging the error obtained for each of the test sets.

Ideally, the best way to test the models would be to use data from a different flight, but given that different flights may take place under different conditions, the estimates of the error for each of the models might be misleading. On the other hand, using for evaluation the same data that was used for estimating the models would not be appropriate as it would lead to over-fitting. A better way to empirically validate the models is to divide the experimental data in a training set and a testing set, using k-fold cross-validation [18] to estimate the accuracy of the model.

We used test flights from the PENGWUN datasets to validate and compare the accuracy of both empirical and theoretical models when predicting the RSSI of signal transmitted by an airborne node and measured from a node in the ground. We selected flights for which we have information about the characteristics of the transmitter, in particular the frequency and transmitter power.

Using the specified transmitter power and transmitting frequency, and using the measured RSSI at each data point, we computed the path loss exponent (PL) needed for the shadowing propagation model. With the calculated PL value in hand, we were now able to compute the theoretical estimated value of RSSI between the nodes of the dataset.

We estimated the accuracy of the theoretical model using k-fold cross-validation with 90% of the data used for training the model and the remaining 10% for validation. Likewise, we used the same data to compute and feed the empirical model with the distance and antenna angles between sender and receiver, and proceeded to perform k-fold cross validation using the same data partitions.

Table 2 shows the results of the validation of both the theoretical and empirical models. The results show that in all cases, the empirical model was able to successfully predict the RSSI of each link with an error rate that is a fraction of the error rate of the theoretical model. These results show that despite its disadvantages, the empirical model provides better estimates of the RSSI for airborne wireless links than the theoretical log-normal shadowing propagation model.

Table 2 Relative error rates between estimated and observed RSSI averages.

Flight No.	Source Node	Target Node	Empirical Model Error	Theoretical Model Error
1	UAV	GN1	9%	57%
1	UAV	GN2	18%	47%
2	UAV	GN1	7%	53%
2	UAV	GN2	17%	48%
3	UAV	GN1	11%	56%
3	UAV	GN2	16%	57%

5 Conclusion

In this chapter we propose a hybrid emulation approach for airborne network environments, and describe the designed and implementation of the mLab-Pengwun emulation framework. mLab-Pengwun integrates an experimental flight test database with EMANE [13] to provide a high-fidelity link emulation capability for known flight patterns.

The proposed hybrid approach for airborne emulation environments allows for the design, test and evaluation of airborne protocols in laboratory, under comparable conditions expected for field tests. The separation of external link effects from the protocol design allows for an evaluation of the protocol properties under different realistic link conditions, without the need for repeated flight tests. Furthermore, it allows for field-validation tests of the proposed protocols, with the proper isolation of environmental and external effects that are generally not considered in theoretical link models.

The use of both theoretical propagation models and statistical data-driven models derived from experimental flight data makes mLab a well-fit emulation environment for airborne wireless networks, where multiple degrees of freedom and the complexity of the nodes make it very difficult to create reliable theoretical models that would suffice for the emulation. While there are multiple approaches in the literature for data-driven modeling, in this work we have chosen a simple parametric, and a non-parametric data model for links, with good results.

The proposed bi-dimensional statistical data-driven propagation model has shown to provide better estimates for RSSI values of airborne links than the theoretical propagation models that were considered for comparison in this work. The selection of parameters used for data-driven models (both parametric and non-parametric) used in this work were based on the flight experiments available in PENGWUN. While, based on our limited results, they cannot be claimed to be the best models in general, they have given us the best accuracy in our experimental tests.

Acknowledgements. This work was developed in part under sponsorship of the U.S. Air Force Research Laboratory, under contract number FA8750-10-2-0145.

References

1. The ns-3 network simulator (2009), `http://www.nsnam.org`
2. Ahrenholz, J., Danilov, C., Henderson, T.R., Kim, J.H.: Core: A real-time network emulator. In: IEEE Military Communications Conference, pp. 1–7 (2008)
3. Bajaj, L., Takai, M., Ahuja, R., Tang, K., Bagrodia, R., Gerla, M.: Glomosim: A scalable network simulation environment. University of California, Los Angeles (UCLA). Computer Science Department Technical Report 990027, 213 (1999)
4. Chao, W., Macker, J., Weston, J.: NRL mobile network emulator (2003)
5. Erceg, V., Greenstein, L.J., Tjandra, S.Y., Parkoff, S.R., Gupta, A., Kulic, B., Julius, A.A., Bianchi, R.: An empirically based path loss model for wireless channels in suburban environments. IEEE Journal on Selected Areas in Communications 17(7), 1205–1211 (1999)
6. Hortelano, J., Cano, J.C., Calafate, C.T., Manzoni, P.: Testing applications in manet environments through emulation. EURASIP Journal on Wireless Communications and Networking, 47 (2009)
7. Ivanic, N., Rivera, B., Adamson, B.: Mobile ad hoc network emulation environment. In: IEEE Military Communications Conference, MILCOM 2009, pp. 1–6 (2009)
8. Jiang, W., Zhang, C.: A portable real-time emulator for testing multi-radio manets. In: 20th International Parallel and Distributed Processing Symposium, IPDPS 2006, pp. 7–10. IEEE (2006)
9. Kiess, W., Mauve, M.: A survey on real-world implementations of mobile ad-hoc networks. Ad Hoc Networks 5(3), 324–339 (2007)
10. Konishi, K., Maeda, K., Sato, K., Yamasaki, A., Yamaguchi, H., Yasumoto, K., Higashino, T.: MobiREAL simulator-evaluating MANET applications in real environments. In: 13th IEEE International Symposium on Modeling, Analysis, and Simulation of Computer and Telecommunication Systems, pp. 499–502. IEEE (2005)
11. Kotz, D., Newport, C., Gray, R., Liu, J., Yuan, Y., Elliott, C.: Experimental evaluation of wireless simulation assumptions. In: Proceedings of the 7th ACM International Symposium on Modeling, Analysis and Simulation of Wireless and Mobile Systems, pp. 78–82. ACM (2004)
12. Kurkowski, S., Camp, T., Colagrosso, M.: Manet simulation studies: the incredibles. SIGMOBILE Mob. Comput. Commun. Rev. 9, 50–61 (2005)
13. Labs, C.: Emane user training workbook: 0.6.2. Tech. rep., CenGen Inc.,, Columbia, Maryland (2010)
14. Pawlikowski, K., Jeong, H.D.J., Lee, J.S.R.: On credibility of simulation studies of telecommunication networks. IEEE Communications Magazine 40(1), 132–139 (2002)

15. Rappaport, T.S.: Wireless communications: principles and practice. Prentice Hall PTR, New Jersey (1996)
16. Setiwan, G., Iskander, S., Chen, Q., Kanhere, S., Lan, K.: The effect of radio models on vehicular network simulations. In: Proceedings of the 14th World Congress on Intelligent Transport Systems (2007)
17. Shrivastava, L., Jain, R.: Performance analysis of path loss propagation models in wireless ad-hoc network. International Journal of Research and Reviews in Computer Science 2(3) (2011)
18. Wasserman, L.: All of statistics. Springer, New York (2004)
19. Zhou, G., He, T., Krishnamurthy, S., Stankovic, J.: Impact of radio irregularity on wireless sensor networks. In: Proceedings of the 2nd International Conference on Mobile Systems, Applications, and Services, pp. 125–138. ACM (2004)

Novel Theory and Simulations of Anticipatory Behaviour in Artificial Life Domain

Pavel Nahodil and Jaroslav Vitků

Abstract. Recently, anticipation and anticipatory learning systems have gained increasing attention in the field of artificial intelligence. Anticipation observed in animals combined with multi-agent systems and artificial life gave birth to the anticipatory behaviour. This is a broad multidisciplinary topic. In this work, we will first introduce the topic of anticipation and will describe which scientific field it belongs to. The state of the art on the field of anticipation shows main works and theories that contributed to our novel approach. The parts important for presented research are further detailed probed in terms of algorithms and mechanisms. Designed multi-level anticipatory behaviour approach is based on the current understanding of anticipation from both the artificial intelligence and the biology point of view. Original thought is that we have to use not one but multiple levels of unconscious and conscious anticipation in a creature design. The aim of this chapter is not only to extensively present all the achieved results but also to demonstrate the thinking behind. Primary industrial applications of this 8-factor anticipation framework design are intelligent robotics and smart grids in power energy.

1 Introduction

This chapter describes the current situation of the studied problem. The area of our work is dealing with the theory of anticipatory behaviour and its applications usable for Artificial Life (ALife). This area is still considered as one of the so far unresolved topics of Artificial Intelligence.

Nature evolves in a continuous anticipatory fashion targeted at survival. Sometimes we humans are aware of anticipation, as when we plan. Often, we are not

Pavel Nahodil · Jaroslav Vitků
Department of Cybernetics, Faculty of Electrical Engineering,
Czech Technical University in Prague,
Technická 2, 166 27 Prague, Czech Republic
e-mail: {nahodil,vitkujar}@fel.cvut.cz

A. Byrski et al. (Eds.): Advances in Intelligent Modelling and Simulation, SCI 416, pp. 131–164.
springerlink.com

aware of it, as when processes embedded in our body and mind take place before we realize their purpose. We can take an example from any sport or game which requires precise and fast body movement. For example, in tennis the return of a professional serve can be successful only through anticipatory mechanisms. Even very fast but conscious reaction takes too long to process. With anticipation we start the action even before the event that would normally trigger this action occurs. Creativity in art and design are fired by anticipation. Before the archer draws his bow his mind has already hit the target. Motivation mechanisms in learning, the arts, and all types of research, are dominated by the underlying principle that a future state controls the present action, aimed at some goal. The entire subject of prevention entails anticipatory mechanisms. We could continue in naming all the areas of life where we can find a trace of anticipatory principles. It is true that an overwhelming part of every being's everyday behaviour is based on the tacit employment of predictive models.

2 Anticipation — Definition and State of the Art

There are several definitions and descriptions of anticipation, some of them are just broadening the initial definition of Robert Rosen [1]. These definitions are not in contradiction, they describe anticipation from different points of view. Over the last few decades, research in anticipation advanced rapidly but not only in the ALife domain. Experimental psychology research gradually started to accept the notion of anticipations beginning with Tolman's suggestion of "expectancies" [2] due to his observation of latent learning in rats (learning of environmental structure despite the absence of reinforcement). More recently an outcome devaluation procedure [3] has been employed that provides definite evidence for anticipatory behaviour in animals. The most recent works that inspired me in this chapter were the works of Martin V. Butz [4], Daniel Dubois [5] and Carlos Martinho [6].

We would like to point out that a significant work was done on the field of anticipation and there were several accomplishments published. As one example for all we will name a conference held each two years and dedicated to anticipation named Computing Anticipatory Systems (CASYS). This conference organized and chaired by Daniel Dubois has been held since 1998 and has become an excellent opportunity for researches in this area to exchange opinion. We consider it to be an honour that our article was accepted and published on this conference in 2009 [7].

2.1 The Basics of Anticipation

A basic definition of anticipatory systems was published in 1985 by the biocyberneticist Robert Rosen in his book Anticipatory systems [1]. He defined an anticipatory system as follows: *"A system containing a predictive model of itself and/or its environment, which allows it to change state at an instant in accord with the model's predictions pertaining to a latter instant"*. Rosen in his book was inspired by his observation of live organisms, namely the ones with higher intelligence.

Especially by their ability to predict the future and make adaptations based on them. This ability of live beings was already discovered before. Rosen however utilising this knowledge, created a theory which was abstracted for various systems in the following way. Rosen in his work exposed a recurring basic pattern of causality and laws, arising initially in physics and generalized over the years stating that: "*in any law governing a natural system, it is forbidden, to allow present changes of state to depend upon future features*" ([1], page 9). This law is widely followed in technical sciences such as physics or control theory. Past states are allowed (in systems with memory) but not the future states. This may seem like a denial of causality and thus it appears to be an attack on the ultimate basis on which science itself rests, while as a matter of fact it is not the case. If we consider the behaviour of a system which contains a predictive model and which can utilize the predictions of its model to modify its present behaviour. If we further suppose that the model can approximate by its predictions the future events with a high degree of accuracy then this system will behave as if it was a true anticipatory system (i.e. a system of behaviour that depends on future states). So we do not have the present state available only its estimate, and this estimate is not based on the information about the future state but on information from past and current states. This system will not violate our notions of causality, but since we explicitly forbid present changes of states to depend on future states, we will be driven to understand the behaviour of such a system in a purely reactive mode (i.e. one in which present change of state depends only on present and past states). Since we claimed that the information we can derive about future can be based only on present and past information we respected the causality.

Let's describe this in a more formal way to clarify the thoughts, see Fig. 1. Let us suppose that we are given a system S, which is the system of interest. For the sake of simplicity let us consider that S is a non-anticipatory dynamic continuous system. We will associate another dynamic system M with system S, where M is a model of S. We require that S is parameterized in real time and that M is parameterized by a time variable that goes quicker than that. In this way, the behaviour of M predicts the behaviour of S. By looking at the state of M at time t, we get information about the state that S will be in at some time later than t. We shall now allow M and S to interact with each other. We shall suppose that the system M is equipped with a set of effectors E, which allow it to operate either on S itself, or on the environmental inputs of S, and change the dynamical properties of S. For M to be a consistent

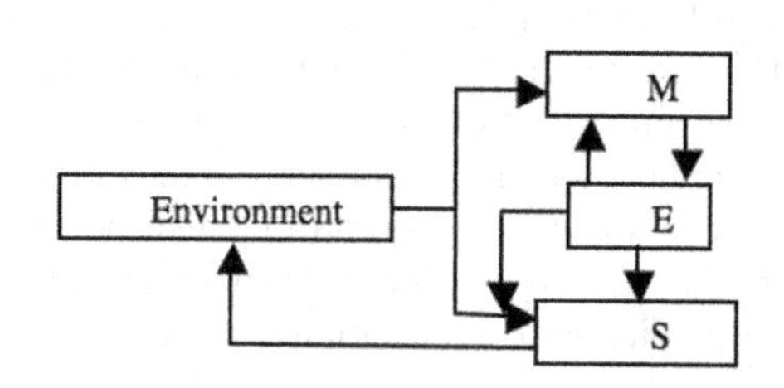

Fig. 1 Rosen's Definition of an Anticipatory System. S *is the system of interest;* M *is the model of* S*, equipped with a set of effectors* E *that changes the dynamical properties of* S *or its environmental inputs. For consistency, these changes are also reflected in* M.

model, the actions operated on S should also be operated on S. Fig. 1 represents such a system. If we put this system into a single box, that box will appear to us to be an adaptive system in which prospective future behaviours determine present changes of state. We will call this system an *anticipatory system*.

It may seem that anticipation is a matter just in biological systems and what more that it is present only in simple animals. On the contrary anticipation plays important role in all living and also non living systems. Our work is focused on artificial life hence mostly concerned about the living systems. One of the researchers that noticed anticipatory behaviour even in non living systems is Daniel Dubois. The demonstration of the fundamental property of anticipation in electromagnetism is made on the well-established and well experimentally verified Maxwell Equations. It is shown that very famous physicists like Feynman, Wheeler and Dirac thought about anticipatory solutions to resolve big problems in theoretical physics. At one hand, many physical processes deal with electromagnetism, and at the other hand, many biological systems deal also with electromagnetism, like, for example, the nervous system, the brain, the heart, etc... in living systems. Robert Rosen argued that anticipation distinguishes the living systems from the non-living ones. Dubois shows that physical systems deal with strong anticipation because the anticipation is fundamentally embedded in these physical systems. Rosen's anticipatory system deals with weak anticipation, because the anticipation is based on a model of the system and thus is a model-based prediction and not a system-based prediction [8].

2.2 Current Types of Anticipation

One of the contributions Kohout's work brings is the attempt to unify and to complete the categories of anticipation. In order to build this in later chapters, it is necessary to briefly describe the current state. This chapter was composed based on the work of Martin Butz [4]. According Butz anticipations are an important and interesting concept. They appear to play a major role in the coordination and realisation of adaptive behaviour. Looking ahead and acting according to predictions, expectations, and aims seems helpful in many circumstances. For example, we say that we are in anticipation, we are looking forward to events, we act goal-oriented, we prepare or get ready for expected events, etc. Despite these important approaches, it is still hardly understood why anticipatory mechanisms are necessary, beneficial, or even mandatory in our world. It might be true that over all constructible learning problems any learning mechanism will perform as good, or as bad, as any other one, the psychological findings suggest that in natural environments and natural problems learning and acting in an anticipatory fashion increases the chance of survival. Thus, in the quest of designing competent artificial animals, the so called animats, the incorporation of anticipatory mechanisms seems mandatory.

Without a conceptual understanding of what anticipatory behaviour is referring to, scientific progress towards more elaborate and competent anticipatory behaviour systems is hard to achieve. The term anticipation is often understood as a synonym for prediction or expectation - the simple act of predicting the future or expecting a future event or imagining a future state or event. Anticipation really is about the impact of a prediction or expectation on current behaviour. Thus, anticipation means more than a simple look ahead into the future. The important characteristic of anticipation that is often overlooked or misunderstood is the impact of the look into the future on actual behaviour. We do not only predict the future or expect a future event but we alter our behaviour — or our behavioural biases and predispositions — according to this prediction or expectation. Here we are moving in definition from anticipation towards anticipatory behaviour. This is the very core of ALife research, the behaviour is the main area of interest. Butz defines the anticipatory behaviour as follows: A process, or behaviour, that does not only depend on past and present but also on predictions, expectations, or beliefs about the future. In fact, any "intelligent" process can be understood as exhibiting some sort of anticipatory behaviour in that the process, by its mere existence, predicts that it will work well in the future. This implicit anticipatory behaviour can be distinguished from explicit anticipatory behaviour in which current explicit future knowledge is incorporated in some behavioural process. This defines two very intuitive categories of anticipation.

Implicitly anticipatory animat-type is the one in which no predictions whatsoever are made about the future that might influence the animat's behavioural decision making. Sensors input, possibly combined with internal state information, is directly mapped onto an action decision. The predictive model of the animat is empty or does not influence behavioural decision making in any way. One of the reasons for this might be memory limitations. Moreover, there is no action comparison, estimation of action benefit, or any other type of prediction that might influence the behavioural decision. In nature, even if a life form behaves purely reactively, it still has implicit anticipatory information in its genetic code in that the behavioural programs in the code are (implicitly) anticipated to work in the offspring.

If an animat considers predictions of the possible payoff of different actions to decide on which action to execute, it may be termed *payoff anticipatory*. In these animats, predictions estimate the benefit of each possible action and bias action decision making accordingly. No state predictions influence action decision making. There is no explicit predictive model however the learned reinforcement values estimate action payoff. Thus, although the animat does not explicitly learn a representation with which it knows the actual sensed consequences of an action, it can compare available action choices based on the payoff predictions and thus act payoff anticipatory.

While in payoff anticipations predictions are restricted to payoff, in *sensory anticipations* predictions are unrestricted. However, sensory anticipations do not influence the behaviour of an animat directly but sensory processing is influenced. The prediction of future states and thus the prediction of future stimuli influence stimulus processing. As will be shown later, comparison of the expected value with the actual value can be used to focus attention as well as to produce emotions. Expected sensory input might be processed faster than unexpected input or unexpected input with certain properties (for example possible threat) might be reacted to faster.

Maybe the most interesting group of anticipations is the one in which animat behaviour is influenced by explicit future state representations. As in sensory anticipations, a predictive model must be available to the animat or it must be constructed by the animat. In difference to sensory anticipations, however, *state anticipations* directly influence current behavioural decision making. This means that the predicted future state(s) directly influences the actual action selection.

2.3 Strong and Weak Anticipation

This chapter was based on the work of Daniel Dubois [5],[8]. In his work he deals with some mathematical developments to model anticipatory capabilities in discrete and continuous systems. He also noticed that even non-living systems without any possibility of construction of a model (like electromagnetism and relativity transformations) exhibits some anticipatory behaviour. Dubois puts a tentative definition of anticipation: *"An anticipatory system is a system for which the present behaviour is based on past and/or present events but also on future events built from these past, present and future events. Any anticipatory system can obey, as any physical systems, the Maupertuis least action principle."*

In view of explicitly mathematically defining systems with anticipation, Dubois introduced the concept of incursion, an inclusive or implicit recursion. An incursive system is a recursive system that takes into account future states for evolving. Some nonlinear incursive systems show several potential future states, that he called hyperincursion. A hyperincursive anticipatory system generates multiple potential states at each time step and corresponds to one-to-many relations. A selection parameter must be defined to select a particular state amongst these multiple potential states. Here we can apply criteria to select the best states from the potential states. These multiple potential states collapse to one state (among these states) which becomes the actual state the anticipation of a system can be based on a model of its environment.

In this case, the notion of exo-anticipation is introduced, with the following definition: *"An exo-anticipation is an anticipation made by a system about external systems. In this case, anticipation is more related to predictions or expectations"*. This defines a weak anticipation.

The anticipation of a system can be based on itself, rather than its environment. In this case, the notion of endo-anticipation is introduced, with the following definition: *An endo-anticipation is an anticipation built by a system or embedded in a system about its own behaviour.* This is not a predictive anticipation anymore but a built anticipation. In this case, this is a *strong anticipation.*

2.4 Anticipatory Classifier System

One of the most successful approaches using Markov chain theory is anticipatory modification of *Learning Classifier System* (LCS) invented in 1975 by John Holland [10]. All LCSs have in common that they are rule-based systems able to automatically build the rule set they work on [9]. LCSs are based on two fundamental mechanisms — *Genetic Algorithms* (GAs) and *Reinforced Learning* (RL). The

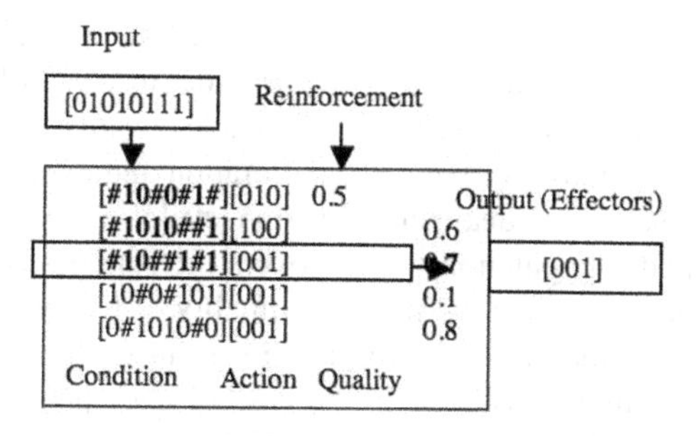

Fig. 2 Learning Classifier System Example

anticipatory modification of these is called *Anticipatory Classifier System* (ACS). ACS consists of a set of rules called *classifiers* combined with adaptive mechanisms in charge of evolving the population of rules (see Fig. 2). Classical *Reinforced Learning* (RL) algorithms such as *Q-learning* rely on an explicit enumeration of all the states of the system. But, since they represent the state as a collection of a set of sensations called *attributes*, ACSs do not need this explicit enumeration thanks to a generalisation property that will be described later on. This generalisation property has been recognized as the distinguishing feature of ACSs with respect to the classical RL framework. An LCS is composed of a population of classifiers. Each classifier is a triple $\langle c, a, p \rangle$ containing a $[Condition]$ part, an $[Action]$ part, and an estimation of the expected accumulated reward that the agent can get if it fires this classifier. The c and a represent the condition and action of the agent, and p the current estimate of the long term reward that the agent can expect from this (s,a) pair. Formally, the $[Condition]$ part of classifiers is a list of tests. There are as many tests as attributes in the problem description, each test being applied to a specific attribute. In the most common case where the test specifies a value that an attribute must take for the $[Condition]$ to match, the test is represented just by this value. There exists a particular test, denoted as "#" and called "don't care", which means that the $[Condition]$ of the classifier will match whatever the value of the corresponding attribute. At a more global level, the $[Condition]$ part of a classifier matches if all its tests hold in the current situation. In such a case, the classifier can be fired. After describing the representation manipulated by LCSs, we must present their mechanisms. The general goal is to design an RL system, thus there will be at its heart an action selection mechanism relying on the value of all actions in different situations. Furthermore, these systems are endowed with a generalisation capability which relies on classifier population evolution mechanisms in order to reach a satisfactory level of generality. We present both categories of mechanisms in the next sections and we will show afterward that families of systems can be distinguished by the way they deal with interactions between these mechanisms. The set of classifiers whose $[Condition]$ part matches the current situation is called the "match-set" and denoted $[M]$. Furthermore, we denote by $[A]$, the "action-set", the set of classifiers in $[M]$ which advocate the action a that is actually chosen. Given the generalisation property of classifiers, the $[Condition]$ part of several classifiers can match at the same time, while they do not necessarily specify the same action. Thus, LCSs must contain an action selection mechanism which chooses the action executed given the list of classifiers in $[M]$. In order to benefit from RL properties, this mechanism must use the expected accumulated reward of each classifier, but it must also include some trade-off between exploration and exploitation. Ensuring that each classifier reaches the ideal generalisation level is a crucial concern in LCSs. The system must find a population which covers the state space as compactly as possible, without being detrimental to the optimality of behaviour. The mechanisms responsible for this property differ from one system to the other, but they all rely on adding and deleting classifiers. In the case of anticipation-based systems, more deterministic generalisation and specialisation heuristics are being used.

Although they share a number of common characteristics ACSs deviate from the classical framework on one fundamental point. Instead of $[Condition] \rightarrow [Action]$ classifiers, they manipulate $[Condition]$ $[Action]$ $\rightarrow$ $[Effect]$ classifiers. The $[Effect]$ part represents the expected effect (next state) of the $[Action]$ part in all situations that match the $[Condition]$ part of the classifier. Such a set of classifiers constitutes what is called in the RL literature a model of transitions. Since they learn a model of transitions, ACSs are an instance of model-based RL architecture. As a result, ACSs can be seen as combining two crucial properties of RL systems. First property is that they learn a model of transitions, which endows them with anticipation and planning capabilities and speeds up the learning process. The second is that they are endowed with a generalisation property, which lets them build much more compact models. The first design of ACS (*Anticipatory Classifier System*) was introduced by Stolzmann [11]. ACS was later extended by Butz to become ACS2 [12]. The ACS use classical solutions to deal with the exploration versus exploitation trade-off. The agent first chooses actions bringing more information about the transitions that have not been tried enough. Then, if the best actions are equivalent with respect to the first criterion, it chooses actions bringing more external reward, as any RL system does. Finally, if the best actions are equivalent with respect to the first and second criteria, it chooses actions that have not been tried for the longest time, so as to handle non-stationary environments as efficiently as possible. In order to obtain a model of transitions as general, accurate and compact as possible, ACSs generally rely on the combination of two heuristics. A specialisation heuristic is applied to inaccurate classifiers and a generalisation heuristic is applied to overspecialized classifiers. When appropriate, the combination of both heuristics results in the convergence of the population to a maximally general and accurate set of classifiers. For the specialisation process, all ACSs rely on the same idea. When a general classifier oscillates between correct and incorrect predictions, it is too general and must be specialized. Its $[Condition]$ part must be modified so as to match only in situations where its prediction is correct. ACS randomly chooses a # test and changes it into a specialized test. The generalisation process is more complex. Usually in ACS a GA is used to replace specific classifiers with more general ones.

2.5 Emotivector

Emotivector was proposed and described by Carloss Martinho in his dissertation thesis [6]. The emotivector architecture is based on four main ideas: (a) to be based on the software agent architecture, (b) to not alter or interrupt the flow of the agent architecture, (c) to be transparently addable or removable from the agent architecture, (d) to be usable in both symbolic and sub-symbolic processing models. The architecture design builds above the Russell and Norvig architecture [13], which is an approach used in most of the designs nowadays where an agent perceives its environment through its sensors (e.g. sns_i) and acts upon that environment through effectors (e.g. eff_i). This basic design is used also in our work. It is typically composed of three phases, executed as a sequence or running in parallel:

- *Sensing*, that is providing the agent with percepts translated by the sensors from the environment signals according their capabilities.
- *Processing*, that is mapping the percepts and constructs into a set of effector actions and updating the current constructs.
- And as the last step is *Acting*. That is modifying the environment through the agent effector actions, within their limitations.

Graphical representation of Russell-Norvig architecture together with block of modification by Martinho is shown on Figures 3(a) and 3(b). The approach that Martinho used was enriching this architecture by a semi-autonomous module he called salience module. This will perform context-free monitoring of the percepts flowing from the sensors to the processing module as well as of the action commands flowing from the processing module to the agent effectors. In more detail the information flowing from the sensors to the processing module of the agent is observed by the salience module that computes its a-priori salience. Each sensor (sns_i) is associated with an emotivector (emo_i) that computes a context-free a-priori salience for the signal and sends it alone with the signal to the processing module.

Isla and Blumberg [14] define salience as the "degree to which an observation violates expectations" $s(x) = (1 - c(x))/c(x)$. As noted by Martinho there seems to be no need for context to estimate this a-priori salience. Salience could be performed using only the changes in percept values over time. The salience module is context-free and leaves to the processing module the responsibility of putting the salience in the context of the agent and its environment. Of course, the processing module can use this recommendation or ignore it, according to its processing resource policies. So we can conclude that to detect when the mismatch between our expectation and the percept value is significant and, when it does, tag the percept as salient, we don't need context or interpretation. When the salience information reaches the processing module, sensations are appraised in the context of the agent and its environment, and emotions may be generated and expressed accordingly by our agent. Please note that the evaluation is a context free; it just measures the mismatch between expectation and value, in relation to a desired value. The code of the information flowing through a sensor is usually consistent, in the sense that it is the repeated measurement of a specific aspect of the environment on a same scale

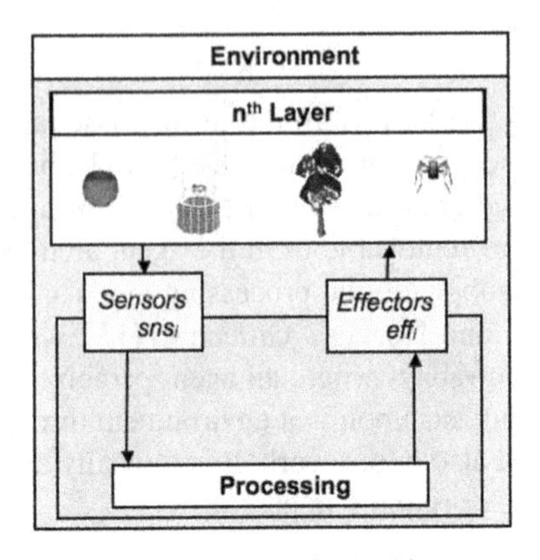

(a) Russell and Norvig Architecture

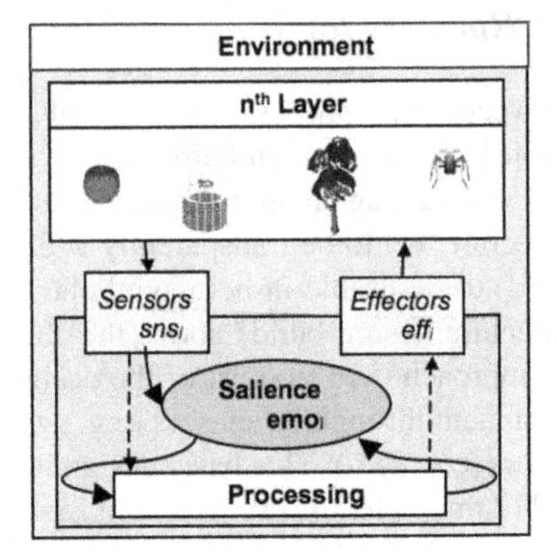

(b) Martinho's Modification with Emotivector

Fig. 3 ACS architectures

over time. We define our universe of perceptions as an n-dimensional vector space where n one-dimensional vectors (or a n-dimensional vector) define a perception in time. Each one-dimensional vector is thus the perception of a specific aspect of the environment at a certain moment in time. Note that we do not associate any a-priori semantics with the one-dimensional vector. Additionally, to ensure that our mechanism can be used in a variety of situations, every aspect of the world is reduced to a value in the normalized range $\langle 0,1\rangle$. The normalisation function may be customized according to the characteristics of each dimension of perception. We would like to model the fact that a same difference between two measurements is more relevant near the agent than far away from it. Depending on the situation we can use specific modulation function to stress out changes in particular interval of values like in the example where closer changes are more relevant that changes far from agent.

The definition of emotivector provided in Martinho's work is as follows. *"Emotivector is a one-dimensional vector with a memory and mechanisms using this memory using an anticipatory affective model to assert the salience of a new value. The anticipatory affective model generates an affective signal from the mismatch between sensed and predicted values, providing some qualitative information regarding the salience of the new value."* The emotivector is used to generate the low-level context-free attention and also emotion.

Architecture and computational details was described and presented in Martinho's work and we have presented our use and modification in [19], [20], [21]. The general principle of the emotivector is the following. Using the signal history of a sensor, the emotivector computes the next expected signal value of the sensor. Then, by comparing the expectation with the actual sensor value the emotivector is evaluated for attention potential. Afterwards, a sensation is generated. The combination of both attentional and emotional salience is then fed to the processing module to be used to support resource management. The Martinho's model of attention presented in his thesis is inspired by Posnera's exogenous and endogenous systems [15] and MÃijllera's and Rabbit hypothesis [16]. This inspiration is reflected in the two components that are used to compute the emotivector salience. The exogenous component, inspired in bottom-up, automatic reflex control of attention, and emphasising unexpected values of a signal. The endogenous system, inspired in top-down, voluntary control of attention, and emphasising the closeness of a signal value to actively searched values.

3 The Proposed Approaches and Their Simulation

Theory and new design of anticipation described in this paper represents the original results and common long-term work in anticipation, achieved at the Department of Cybernetics at the CTU in Prague. This chapter is devoted to the details of our common contribution (working methods) to the field of anticipation in *artificial life domain* during the last six years. The core of this part presents main results published by Karel Kohout in his dissertation, supervised by Pavel Nahodil. One of main contributions as we see it was to propose a single architecture called 8-factor anticipation. This term and the whole architecture are original terms and suggestions that were introduced in Kohout's dissertation work [20]. Each "factor" of *8-factor anticipation architecture* will be now described in this part.

What is not that obvious and sometimes even missed, anticipation is not matter of one mechanism in a living organism. Anticipation happens on many different levels in one creature. The works studying anticipation seems to overlook this fact so far, focusing on the anticipatory principles, mechanisms and their optimisation. There was undeniably great progress in past years in theory and applications of the anticipation. What we miss in the deployment of anticipation in Artificial Life domain is to follow the nature's example and use anticipation principles in more design blocks. Several researchers categorized the anticipation already. Even though we embrace the categorisation of anticipation Martin Butz did we was not fully satisfied with it. We missed the connection between the types of anticipation and the consciousness in his work. In addition to the existing types Mr. Kohout added the consciousness and thus created 8 types of anticipation [19], [20]. This idea is the basis of our theoretical contribution to the field. Our thinking here is that each algorithm is better in a different way and by combining them and properly selecting the right one we can improve the results.

All the types are schematically shown on the Fig. 4. We can say that the complexity grows in the picture from left to right and from bottom to top.

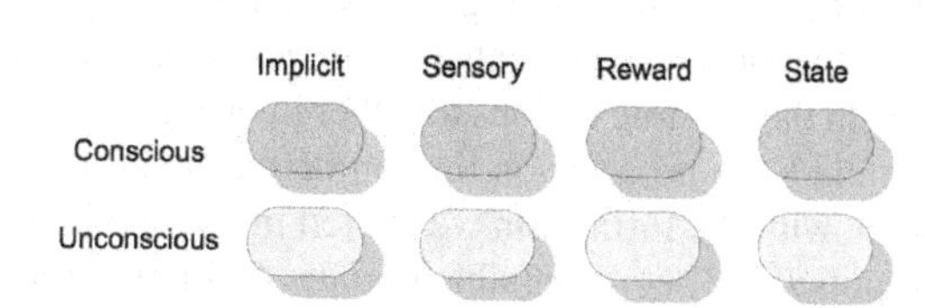

Fig. 4 The 8-Factor Anticipation

3.1 Unconscious Implicit Anticipation

Unconscious implicit anticipation (UIA) concludes the behaviour that was imprinted in the creature by nature or creator (in our case) and that is not voluntary. Under this we can understand the very basic reactions with anticipation imprinted in the design. Reactions and reactive behaviours itself is not anticipatory and in fact is very often used and understood as exact opposite of anticipation. So what exactly are reactions with anticipation? We cannot say that the reaction is associated with prediction of next sensed value, state or reward because these are subject of the other anticipation types. There classical view of implicit anticipation would be satisfied with the fact that it is the prerequisites given to the system wither by the long evaluation or by the creative mind of architecture designer. In order to describe it we need to define a formalism to approach this is systematic manner. At this very basic level we have only the set of inputs and set of possible outputs O. By this we implicitly assume discrete values which we typically have in a virtual environment. Please also note that we are not speaking about agent sensory inputs or actions yet. The reason is that I'm trying to generalize this description so it can be used for agents' internal blocks and not only for agent as whole. The reaction base is typically in form of projection $I \Rightarrow O$. The inference mechanism is very simple: if any of the input matches the output is executed. There are couples of possibilities from binary rulebase to ACS. On the contrary the anticipatory approach would be to expect another input after the executed action $I \Rightarrow O \times I$. It still might seem as nothing new one can say that everything was already presented. We must realize here that on unconscious implicit anticipation there is no mechanism to modify this rulebase other than evolution. As was said above it reflects only the non-learned behaviour. The interesting question is if the same rule as here can be then created in some other probably conscious level. The answer is yes the same rule can be inferred by the consciousness of the artificial creature but its execution takes longer path so it is less likely to be executed. We do not need to solve creation of new or forgetting of obsolete rules here because the rulebase is fixed and it is subject of only minor evolutionary changes.

3.2 Conscious Implicit Anticipation

The combination of *Conscious Implicit Anticipation* (CIA) may seem illogical because as we said above implicit anticipation is something imprinted in the creature by design. How can this be consciously controlled is the right question and the moment. Here still everything depends on the design but the results are available to the higher levels and also higher levels data such as desired state (converted to the desired value in the current step) are available as inputs. This means that here we can chain the existing actions together in order to create a new non-atomic action, which would have no decision time in between and focus the attention. We will continue here with the formalism we started in the previous part. We still have only the set of inputs I and set of possible outputs O. In previous chapter we ended with anticipatory projection from input to output and expected new input $I \Rightarrow O \times I$. We explore this further here with two modifications described below. First suggestion to this is to add to the expectation also expected next action. This is expected to improve the reaction time. In our formalism, we are now projecting the current input and the output to current output, expected output and expected input $I \times O \Rightarrow O^2 \times I$. Please note that in the agent terminology we moved from the term output to action.

Imagine the predator evasion scenario and imagine two prey agents. One of them equipped with standard prediction scheme ($I \Rightarrow O \times I$) and second with the suggested modified one ($I \times O \Rightarrow O^2 \times I$). Both prey-agents are in the vicinity of predator which will through the sensors result in (input I) the action of both is to flee (output O). Even if the reaction process is fast it still takes some time to search through the rule base for a match. Our question is what will happen in case agent will have the chance to take another action before it recalls the appropriate action from the rulebase? Since there will be no action selected yet it must wait till it the next step. On the contrary if this moment comes to the modified agent it can straight away execute the *"prepared action"*. This is graphically demonstrated on the Fig. 5.

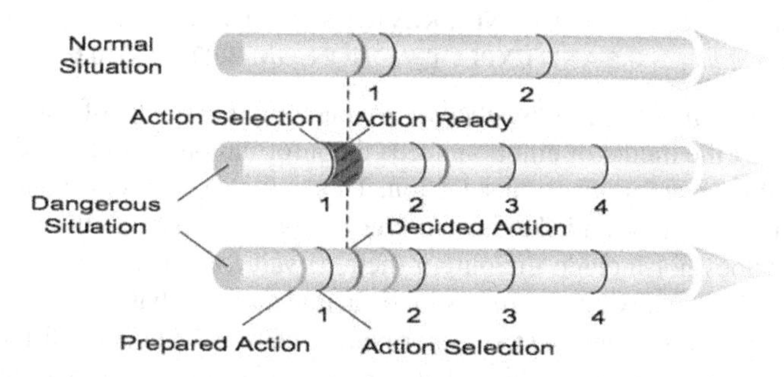

Fig. 5 Action Selection without the Action Anticipation (top) and with Action Anticipation (bottom)

At this point because it is conscious part we can introduce Kohout's second suggestion. We do not have the reward yet but we can have a rate of change for the input value. For output values, because they are typically in ALife a discreet values not expressed by numbers, statistical measure such as probability or likelihood can be measured. This is another parameter that can add value to the decision process and help to choose the right action in the correct moment. This describes the typical scenario but in fact any combination in term of discrete and continuous in the input or output can occur. So we are adding two new values the r_i and r_o which we will call rateability (the combination of word *rate* and *probability*). This enriches the projection $I \times O \Rightarrow O^2 \times I \times \mathfrak{R}^2$. For example we have a proximity sensor for exploring creature that provides one input called *distance*. Let's have a rule to change direction when the distance is lower than half a meter to avoid collision. We have two actions available "move" and "turn in one direction by a given angle". Let's also assume that the previous action was to move straight. This simple example shows that even on a very basic level the amount of information available can vary.

1. The classical reactive approach

   IF x THEN y, where $x \in I,\ y \in O$
   example: IF $distance < 0.5$ THEN $turn(90)$

2. The classical anticipatory approach

   IF x THEN y EXPECT z, where $x, z \in I,\ y \in O$
   example: IF $distance < 0.5$ THEN $turn(90)$ EXPECT $distance > 0.5$

3. First suggested improvement — action anticipation

   IF x AND PREVIOUS_ACTION a THEN y EXPECT z AND EXPECT_ACTION b where $x, z \in I,\ y, a, b \in O$
   Example: IF $distance < 0.5$ AND PREVIOUS_ACTION $move$ THEN $turn(90)$ EXPECT $distance > 0.5$ AND EXPECT_ACTION $move$

4. Second suggested improvement — rateability evaluation

   IF x AND PREVIOUS_ACTION a THEN y EXPECT z AND EXPECT_ACTION b WITH $\langle r_i, r_o \rangle$ where $x, z \in I,\ y, a, b \in O,\ r_i, r_o \in \mathfrak{R}$
   Example: IF $distance < 0.5$ AND PREVIOUS_ACTION $move$ THEN $turn(90)$ EXPECT $distance > 0.5$ AND EXPECT_ACTION $move$ WITH $\langle 0.1, 0.6 \rangle$

The approach is *emotivector* described above and only the model of attention since the second part the model of emotion needs the information about reward too. This determines this level for attention selection. This model does not have the *rateability* factor, but this can be added to the *emotivector theory*. The first difference is that emotivector does not include the output (action) value estimation and evaluation. Since actions are usually not expresses in the actual numbers but as worded abstraction, it would be very complicated to normalize them and calculate differences. It is not even required, the only thing that is required is to have an expected output stored (in other words) the prepared action. As mentioned this estimation will be based on previous action therefore $\hat{a}_t = a_t$. The other difference is the *rateability* evaluation. For input value it is simple, the speed of change (velocity) for discrete values is calculated as a difference $v = \Delta x / \Delta t$ in one step we calculate the velocity $v = \Delta x = x_{t-1} - x_t = r_i$. However we will argue that there is a better measure of how the object is interesting and that is the salience computed by the emotivector so we will use it instead of simple change speed $r_i = salience_i$. For output rateability will be counted as frequency of occurrence of the output across the whole actions $r_O = n_O / N$.

The model of attention is implemented as follows. Using its history at time $t-1$, the emotivector estimates a value for next time $t(\hat{x}_i)$ and predicts that its value will change by $\Delta\hat{x}_t = \hat{x}_t - x_{t-1}$. At time t, a new value is sensed (x_t), and a variation $\Delta x_t = x_t - x_{t-1}$ is actually verified. The newly sensed value triggers the computation of the emotivector components. The exogenous component at time t (EXO_t), is based on the estimation error and reflects the principle that the least expected is more likely to attract the attention. EXO_t is computed as follows $EXO_t = |x_t - \hat{x}_t|$. If the emotivector has no associated desired value, the exogenous component will be the only factor contributing for the emotivector salience. However, if there is a desired value (d_t), then the endogenous component of the emotivector is also triggered by the newly sensed value. Whenever a desired value is present within the emotivector at time t, the endogenous component (END_t), is computed. It is a function of the distance of the sensed value to the desired value Δs_t and of the estimated distance of the expected value to the desired value Δs_t. END_t is computed as follows $END_t = \Delta \hat{s}_t - \Delta s_t$ where $\Delta s_t = |x_t - d_t|$ and $\Delta \hat{s}_t = |\hat{x}_t - d_t|$. Together, the exogenous and endogenous components define an a-priory salience for the emotivector. This salience can computed by adding the absolute value of both components as $salience_t = EXO_t + |END_t|$. Of course, other emotivectors are being evaluated at the same time, each one with its own salience computed based on the described process.

3.3 Unconscious Sensory Anticipation

Moving on to the *sensory anticipation on the unconscious level* (USeA) concludes all the sensory input gathering, pre-processing and data filtering. Basically here we can meet all the functions that cannot be voluntarily influenced. In broader sense by this we can simulate the situation where the input magnitude is so huge that it cannot be processed all by the conscious processes. This information is collected, processed, stored or disregarded based on the attention and other factors. We will not go that far to implement this in full scope and we will stay just with the input gathering and pre-processing. In anticipation we talk all the time about some estimated future value but less we speak about the means how to get this value. In most implementation very basic approaches such as the no-change rule are used. However statistics provides a wide variety of very powerful and complex methods to estimate the future values based on arbitrary long history data. In our opinion these have place exactly in this part. For anticipation purposes they are just tools that present us with the estimated value that we can in other levels use and further process.

The second function that we sometimes require to get closer to the animal world is filtering the information in order to reduce the information value so more less informative data can be processed at the same time. Let me demonstrate on an example what we mean by this. We can take again our robot. The robot has some hardware limitations in terms of data size it can process. So it is in our best interest to give it more accurate data about the objects it has focus on (we already know where the focus is from the previous level) and other objects data can be reduced to some approximation. Let still continue with the distance measure but this time the distance is measured in four directions. In the direction we are close to the wall we would be interested in the number, how close are we. In other directions the information if the distance is short, medium or long would suffice. Of course one can object that in narrow spaces we would need detailed information in all the directions in worst case scenario. Yes that is correct and that is why it can take then more than one time unit to process the data and the robot would slow down.

Here we meet first possible conflict of the levels if we want to combine them together. For the focus attention mentioned in the previous chapter we would need the exact data to measure the changes and decide about the object we want to focus on but here we hid this data in some intervals. We have two resolutions to this situation, leave it as it is and accept the fact that objects can be in the focus only if the cross borders of the intervals and if they are close. This variant does not seem to bring value in certain sense it would impede the previous two factors significantly. Even if it is more probable to have closer object in focus than the further one we still would want our agent to be able to focus even to object in long distance if they are interesting enough so we need another solution. The second solution is to bypass this filtering for attention focus. That would solve our problem with focus, but neglect the reason why we used the filtering so this does not seem to be optimal as well. The solution is at hand, the sensors alone have the data and these are being pre-processed by the unconscious layers, so these can have access to the full information and present only results to fit with the above scheme of a limited data size that can be processed at one time.

For the implementation we continued using Martinho's emotivector and the suggested simple predictor. In this simple predictor the prediction of the next value is the weighted sum of two parameters, the previous prediction $\hat{x}_{t-1}$ and the sensed value x_t. Both compete for influence in the computation of the new prediction x_t. In a certain way, the weight w_i accounts for the certainty of the system in its previous prediction. When there is no desired value, the exogenous component EXO_t is used for the value of $w_t = |x_t - \hat{x}_{t-1}| = EXO_t$ and the prediction is then $\hat{x}_t = \hat{x}_{t-1}(1 - w_t) + x_t w_t$. When there is a desired value in the emotivector the learning rate is set to the intensity of the current sensation associated with the END_t. As such, the change of w_t (Δw_t) at each step is computed as $\Delta w_t = \xi_t(x_t - \hat{x}_{t-1})$ where $\xi_t = END_t$.

3.4 Conscious Sensory Anticipation

As in every chapter it is discussed first what we understand under this category and support it by examples. We are on the *conscious sensory anticipation* (CSeA) level at the moment, so we have access to the sensor data and from the previous level even to the estimate of future data.

What else would we need at the sensory level? There are still many things that would be very helpful for the artificial creature to derive that would require some higher level processing than just having several expected values for each sensor. A dog hunting a rabbit does not need to sense the hare continuously. If the hare, for example, disappears behind a bush the dog predicts the future location of the hare by anticipating where it is going to turn up next and continues its hunt in this direction. This behaviour described below needs little bit more than a pure sensory input. It requires recognition of the objects (rabbit, bush) and making projections of a sensory data that cannot be directly measured at the moment by sensors. Also some knowledge about the rabbit and the environment has come into play. This means that we would need some already stored data to be recalled from the memory and associated with the recognized objects. As we can see, the situation is complicated and requires mechanisms that we did not described so far. It is obvious that model of some kind would be very helpful at this level. The solution to this is that we are on conscious part of the architecture and the consciousness has access to the memory, plans, active knowledge etc... This knowledge is shared across all 8 anticipatory levels. This means that we need to introduce a shared part for the conscious levels that they can either utilize or contribute to. It is schematically shown on Fig. 6. For the lack of better expression we call it *memory*. This is only a logical design in the physical design of the whole architecture several components will use their own memory that is not shared or use this shared memory. With the use of memory the sensory anticipation can be used now in our example to predict even more complex events that just sensory input. It can then abstract objects and predicts future sensory input for these objects. We are still on the sensory anticipatory level so we cannot derive yet any other observations than future sensory inputs. Filling the memory, building models and beliefs, planning and other cognitive tasks, will be subject of further levels.

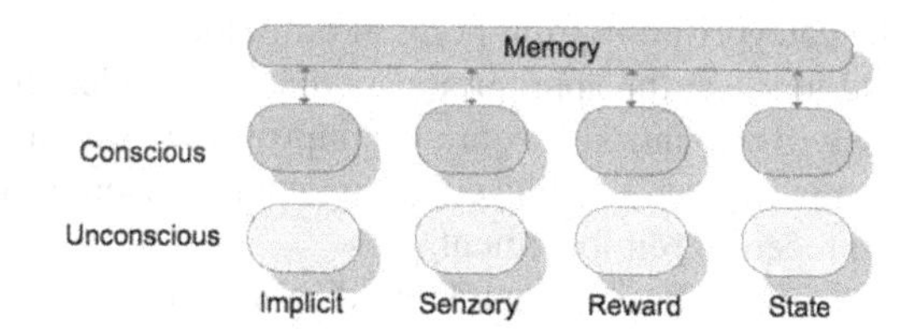

Fig. 6 Shared Media for Conscious Factors (original idea of Kohout K.)

Now we will get more in details about how to implement this. One of the suitable approaches was suggested by Isla and Blumberg in their work [14] the Probabilistic Occupancy Map (POM). This is simple yet efficient approach to track objects even when they are not visible (hidden behind another objects) and estimate the probable position. It is based on separating the environment to hexagons (also called nodes) and assigning each of them for each object the probability of its presence in it. This probability is diffused using simple isotropic diffusion

$$p^{t+1}(n) = (1-\lambda)p^t(n) + \frac{\lambda}{8} \sum_{n' \in neighbours(n)} p^t(n') \tag{1}$$

where λ is a diffusion constant in the range [0,1] and $p^t(n)$ is the probability of the node n at time t to reflect the last motion pattern of the observed object. The diffusion constant can be modified

$$\lambda_i = \lambda_c + max\left(0, \frac{vl_i}{l_i^2}\right) \tag{2}$$

where v is the velocity vector, l_i is the position offset between the current node and the node's *i-th* neighbour, and λ_i is the diffusion rate along the *i-th* connection. λ_c is a constant diffusion rate, ensuring that some probability is diffused to every neighbour, even if that neighbour does not lie in the direction of the velocity vector. One of the main contributions of this work is the fact that the levels can support each other. In the USeA we implemented for the sensors several estimators of the future values. These can be used instead of using the method of altering the diffusion constant to aid the algorithm. The formula for diffusion constant helps only to propagate the probability in the map in the right direction and to decrease it for more distant nodes. But we can use estimators to aid this process and provide estimated position based on the history of measured values.

3.5 Unconscious Reward Anticipation

We are now finally approaching the area that almost all current anticipation behaviour designs operates with (sometimes with combination with lover levels in the sense of description above) the *unconscious reward anticipation* (URA). The reason is at hand, the reward or better said reinforcement to include also punishment is a powerful way to learning. *Reinforcements together with the expectations (anticipation) also serve to generate emotions.* The contribution to this area is to argue about the categorisation from point of consciousness to fit this into presented framework and to select the appropriate approach to implement it. As in every level of design we will also introduce our own improvements to the design. The generation of emotion is in general achieved through comparison of the expected reinforcement with the received reinforcement.

This is a basic principle used in most of the current works. The absolute value of the difference can drive the strength of the emotion. Multiple emotions such as happiness, surprise, disappointment, frustration, sadness etc... can be generated. This means that without the algorithm to verify the expectation against the real reinforcement, also the set of emotions needs to be defined or in more general case the set of rules for emotions and their generation. The first *Kohout's Contribution* to this level is in *definition of three main emotion elements* to consider the reinforcer (generated by received reward or punishment), expectation difference (generated by comparing the expected and received reward) and the surprise (evaluates the expected versus real value). We can normalize each of these to an interval $[0,1]$ and then draw an *emotion cube* with these values on each of the axis Fig. 7.

Adding names of all the emotions to the cube would make it difficult to read. The Table 1 below gives the mapping of the emotions and contains our own definition of the combination of the emotions. The cells of the table maps to the cube as the intensity of each emotion grow. The intensity of the final emotional state is created by superposition of the intensity of the three parts of emotion. Their intensity is given for reward and punishment by the amount of the reward/punishment received and for the surprise by the distance of the expected value from the observed value. As we mentioned already above, a complex creature pursues more goals at once and hence the creature also has multiple reinforcement (reward/punishment) expectations. The final emotional state is generated by a superposition of all the current emotions.

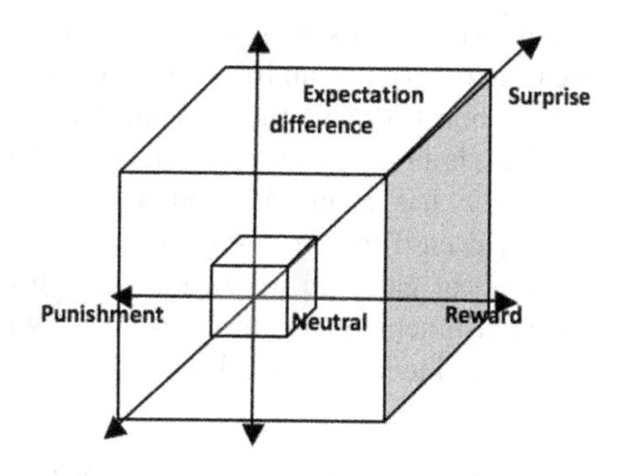

Fig. 7 The Emotion Cube. Reward/punishment on x axis, expectation difference on y axis and surprise on the z axis

Table 1 Designed Emotion Mapping on the Nine Sensation Model

	More reward received	Received as expected	More punishment received
Reward expected	Joy + Surprise (Pride)	Joy	Sadness + Surprise (Suffering)
Negligible	Joy + Surprise (Happiness)	Neutral	Sadness + Surprise (Disappointment)
Punishment expected	Joy + Surprise (Relief)	Sadness	Sadness + Surprise

The described approach to emotion uses very simple mathematical approach. The only complexity is the superposition and evaluation of the emotion to be generated from the emotional state. There is a value x observed by the agent's sensors and its expected value $\hat{x}$. This value can be connected with reinforcement. We define r as the reinforcement that was actually received and $\hat{r}$ the as the reinforcement expected. The difference between expected and received reinforcement will be denoted as Δr. The surprise factor is Δs .

$$\Delta r = \frac{\hat{r} - r}{max\left(|r|, |\hat{r}|\right)} \qquad \Delta s = \frac{\hat{x} - x}{max\left(|x|, |\hat{x}|\right)} \tag{3}$$

All these values need to be normalized in order to be comparable and projectable to the cube. We use both the actual observed value and the associated reward in order to be able to capture situations where there is small increase in reward/punishment, but still a significant difference between expected value and the actual value leading to higher value of surprise.

The evaluation of final emotional state is done as mentioned above by superposition of the partial emotions. The resulting vector depicts the sum of all the emotions and its position in the cube then dictates the final emotional state. There is another aspect to emotion that we want to capture in our work as well. Usually in simulations the emotion is in effect until it is changed by another emotion. In a simulated world that has many agents and objects this is a good approximation. However what if there is no reinforcement for a longer period of time, the emotion will not definitely stay the whole time with the same intensity (i.e. the emotion intensity decreases over time). This is another addition we made to the emotion approach implemented. Each emotion that has occurred is counted in, but it loses intensity with each time step. This way even if there is a bigger reward received followed by minor punishment, the emotional state will still favour the outweighing contribution of the reward, but if the punishment comes few steps later, where the joy of the reward should wear off, if can influence the emotional state.

3.6 Conscious Reward Anticipation

On the contrary to the previous chapter, on the conscious level there is the advantage of the consciousness shared media so the possibilities are much wider. At this *Conscious Reward Anticipation* level (CRA) we are finally reaching full capabilities of the current architectures, plus Kohout's design has two additional levels above this one thus leaving space for further advancements.

At this stage we are looking for framework that works with reward, and is able of working with the observations and gained knowledge including creating, modifying and deletion. We need to keep in mind that these might serve for other conscious levels to work with and hence they need to be compatible or abstract enough so all levels can understand them. This also means that the system should be open enough in terms of inputs it requires and outputs it provides co we can easily integrate it in the complex architecture. ACS seems to be an ideal algorithm for the conscious reward anticipation due to its generalisation and specialisation properties. We selected one of the basic ones that gave the idea to others i.e. the work of Stolzmann [11].

3.7 Unconscious State Anticipation

This is the last of unconscious levels at the same time the most sophisticated one and most complex one. This *Unconscious State Anticipation* level (USA) has a similar problem to the CIA — *conscious implicit anticipation.* The combination itself seems at the first sight confusing. However it is important part of the architecture and has its meaning. All the state creations manipulations, and estimation of next states that are not brought to consciousness right away or at all have place here. One example is the internal state of the creature. It is monitored through internal sensors, it is regulated and working without external actions needed but some unusual states should be reported to the conscious levels. The motivation was taken from the nature as always. As long as all the internal variables are within the certain boundaries there is no need to alert consciousness Once some of them drops or exceeds the threshold and external action is required to get it back within the safe range (energy is for example low — i.e. hunger). The consciousness controlled action is needed as the situation needs to be evaluated and proper actions executed in order to address the situation. So the artificial creature internal state can be monitored and partially controlled from this level. Second consideration is if models of other agents or environment can be created on the unconscious level. We can think of one possibility — latent learning. It is surely subject for discussion if the latent learning is triggered consciously — learning something "just in case" we will need it sometime. But we would here say that it is not. Our argument will be based on the available storage space and processing speed. It seems very unlikely to create models and states of the whole environments and store them for a long term and then for every new goal going through them if they can be used or not. More likely the models are created subconsciously in the short term memory and when the proper reinforcer appears, and the model is proven useful then it is kept in the conscious long term memory. This scenario seems to be more efficient and reasonable from the resource optimisation point of view unfortunately we are not aware of a nature experiment to support Kohout's theory.

What exactly is anticipatory about keeping the internal state and latent learning? We will start from latent learning because the answer is straightforward, creation of knowledge and maps about the situations we are going through and storing them anticipates that they might be useful in the future. So latent learning is anticipatory in its very nature. With the internal state it is not that easy as the principle seems more or less reactive (value drops to certain level — alert is triggered). But the rate internal values decrease is not same under different circumstance (more energy is consumed when running than when exploring etc...). So anticipatory monitors using the information about current actions and observed internal values behaviour would help to optimize the system and bring it again from reactive to anticipatory.

For the implementation of this level, the ACS framework described already in previous section can be used. Instead of creating an environmental map in its explicit representation (even if that is also possible). I've decided to capture the latent knowledge in terms of the ACS rule base. The advantage is, that 8 levels architecture has support for ACS already build in and as such can work with it. Another such advantage is that degradation over time of such knowledge representation is then trivial and means removal of random classifier from the rule base. The disadvantage is that the next level than would not be able to use for example planning directly on such knowledge representation. We have decided to stay with the ACS latent knowledge representation for this level.

3.8 Conscious State Anticipation

The most complex and therefore the most interesting factor of 8 level architecture is *conscious state anticipation* (CSA). Basically all the classical AI approaches can find their place here starting from state space search through different methods of planning (for example based on Markov decision chains) up to the reasoning about others and self. These tasks typically require more time to process. This can be imagined as a state of the agent where there is no urgent internal need (food, sleep, etc..) and also the agent external goals are satisfied. In that case the agent can select the action to be to build, review, updated or evaluate the model of its own state or of others. This type of meta reasoning case still have anticipatory basis. The research was focused on the interconnection of several levels together connected by memory as described above.

Working on the previous two levels where it was used the ACS algorithm we noticed and pointed out several weak spots of the approach. Thanks to the probabilistic approach and low degree of specialisation of the newly created classifiers the behaviour even after long learning cycle is still random. While this greatly promotes the environment exploration it lacks the deterministic use of the gained knowledge. In Kohout's design [20], [21] the agent creates parallel to the rulebase of the ACS a map of the environment in the memory. This map is however tight to the ACS very closely as is it created by the ACS exploration phase and is composed of the applicable actions successfully executed. In a dynamic environment the applicable actions can change, thus this map needs to be able to adapt to these changes. The second modification is a priority queue of the goals. This queue has multiple levels and priorities as we have a planning (deliberative approach), exploration (reactive approach) and internal state (hysteretic approach) competing for actions. Please note that this is a simplified situation. In Kohout's architecture there will be up to 8 competing layers of the priority queue.

The last modification is the actual planning approach. The planning is done on the map of the environment created through the discovery of the environment. This means that the agent is not able to plan action that has not yet been applied as there will be no knowledge in the agent map and rule base about such action. This fact helps to balance the exploration and the planning phase. The state space is then created by a position and the applicable actions. Such state space can be searched for goal state by different algorithms from depth or breadth state space search, through A* up to approaches based on Markov decision chains such as dynamic programming. We chose the A* algorithm.

4 Simulations and Results

The simulations in the virtual environment are the most common method how to test theories and compare effectiveness of results with others, also in the ALife domain. The Kohout's experiments were conducted in the REPAST simulation tool [17]. The presented simulation scenarios are intentionally made as simple as possible to clearly demonstrate and articulate the functionality.

4.1 Unconscious Implicit Anticipation

The setup of the experiment is placing a robot (agent) into a world with several objects. There are walls obstructing way and the beverage which the agent has to reach. The goal for the agent architecture is to reach the goal in effective manner. There are three layouts of the environment, one without obstacles, and two containing different types of obstacles. In each simulation there are always two agents, each trying to reach different goal location. This experiment was designed to show in practical sense the meaning and differences in the implicit anticipation and we understand it. The setup of this experiment counts two instances of an agent with similar sensors, effectors and action selection mechanism. Each agent has a "proximity sensor" of the 8-neighbourhood. It gives the agent information about presence of objects. In order to ensure agents follow goals, one agent is thirsty and goes to water and second agent is hungry and searches for food. Both agents for the sake of simplicity know the location of the object they are searching for. The task is to get there. To complete the task the agents have a set of nine possible actions and those is movement in all possible 8 directions and "do nothing" action. In each scenario the agent is given a set of rules that maps the inputs to the outputs. In the first experiment the agents moves randomly, the rule base has just one rule that tells the agent to stop when the food is found. There is little anticipation in this scenario. In the second experiment reactive behaviour was implemented with the implicit anticipation of "right angles" which means that when meeting the wall rotation by 90 degrees will help to avoid it. The rulebase here has the rule from previous agent plus 8 other rules that gives appropriate action to the met walls. In the third experiment, Karel Kohout enriched the reaction base with different mechanism for navigating in space with obstacles which is called "wall following" [21]. This means that we have even larger reaction base as more situations of wall presence in the base is needed to successfully navigate along the wall.

All these agents were tested in three different scenarios where first was without obstacles, second contained only straight obstacles and the third contained also curved obstacles. In the first scenario it takes long until the agent randomly stumbles on the food regardless the obstacles (please note here that some single cell organisms use this method of navigation). In the second experiment the 3rd scenario is not achievable as the rules do not allow the agent to cope with the obstacle. In the last experiment we can see that it is quantitatively better than the previous one, not only it can reach the goal in fewer steps, it also can complete the third scenario. The Table 2 shows the number of steps necessary to complete the scenario.

Table 2 Results of the Experiment with Unconscious Implicit Anticipation

Experiment	Scenario 1	Scenario 2	Scenario 3
(a)	Rand	Rand	Rand
(b)	26	38	∞
(c)	26	35	38

4.2 *Conscious Implicit Anticipation*

The aim is to test and prove the emotivector attention focus features. For this purpose we designed scenario including several types of agents. The predator agent shown as *"wolf"* is observing the environment and its task is to pick a target of interest based on their salience, there are three agents to be observed. Two *"piglets"* agents both with similar characteristics, except the move pattern, while one of them uses a random move method to navigate through the environment, the second one moves in a constant cyclic pattern. The last agent depicted as *"flower"* is a static agent. It was confirmed by this experiment that the moving agents are more interesting for the observing agent than the static ones, which was expected based on the fact that emotivector is sensitive to the observed value change in time. This reveals the strong and weak sides. For the attention focus only the changes in the environment are relevant on this level. This is acceptable on the basic "reactive" level.

One of our suggested improvements to the emotivector approach was introduction of the *rateability*. In the second experiment we demonstrated that it can be beneficial and lead to improvement. We used the same setup as in the first experiment but we added one more agent this time with the enhanced emotivector. In order to maintain the same conditions we kept both agents on the same position. For visualisation purposes we show one of the agents above the other and depicted the additional agent as a *"bear"*. In this experiment we *compared* the performance of *emotivector improved by Kohout's original design* below referred to as *"enhanced emotivector"* against the Martinho's original version of emotivector below referred to as *"standard emotivector"*.

The results were evaluated statistically due to the random movement of one of the agents in this example. Kohout simulated the change in attention over 1000 steps of the simulation, and repeated the same simulation 3 times [19].

The results are summarized in the Table 3 above and also a fifty steps sample is shown on the two figures below. In each of the simulations our enhanced emotivector exhibited better *stability of attention* (in the experiments conducted in average by 2.2%) while still being able to change the attention focus if the other moving object is more interesting.

Table 3 Enhanced Emotivector Test

Experiment	Scenario 1	Scenario 2	Scenario 3
1	12.7%	14.9%	2.2%
2	11.6%	13.3%	1.7%
3	11.0%	14.3%	3.3%

Fig. 8 show the difference in both approaches, while Kohout's approach exhibits stable attention focus areas (shown above), the standard emotivector approach shows unnecessary oscillations between the objects of attention.

The main experiment here is to test the estimators and evaluate their qualities. In the referenced work [6] there are some conclusions about them, but none of them is shown or proven.

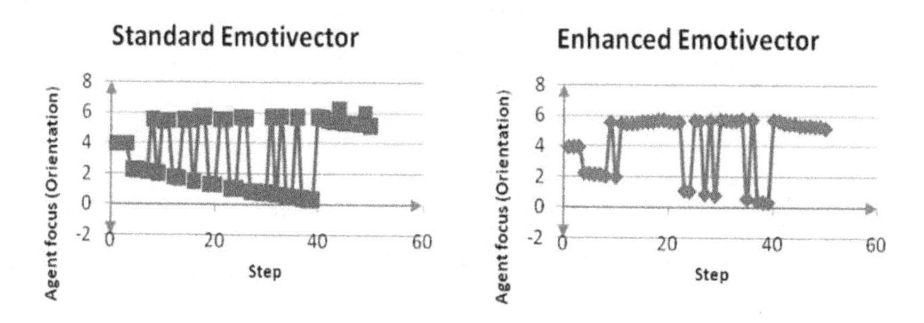

Fig. 8 Standard Emotivector (left) and Enhanced (right) Emotivector of Attention

4.3 *Unconscious Sensory Anticipation*

The predictors tested are referenced by the abbreviations *Simple Predictor* (SP) uses the equations described above, *Limited Simple Predictor* (LSP) uses the same equation, but also keeps history of the input values calculates the mean and the deviation and limits the prediction if outside the statistical range and lastly *Desired Limited Simple Predictor* (DLSP) uses also the desired value. We decided for this experiment to set the desired value to 0.6.

Both the predicted value and also the predictor error of all 3 estimators are shown on the Fig. 9(a) and 9(b). It is clearly visible that the LSP (*Limited Simple Predictor*) and the DLSP (*Desired Limited Simple Predictor*) are nearly identical.

There are several conclusions that can be drawn from the results above. The *Simple Predictor* has poor results. It is not shown in the table above, but the convergence speed was very slow. In this experiment the convergence to value 0.1 ± 0.01 was 87 steps. The other two predictors showed comparably better performance. As can be seen above they are both able to converge in 5 steps. In situation of oscillating values they are not able to adapt and they oscillate too.

4.4 *Conscious Sensory Anticipation*

This level has only one experiment and that is the object persistence scenario. This enables the agent to be able to estimate the position of another agent which is hidden behind obstacle and thus cannot be directly perceived by the sensors. The experiment compares the original work with my suggested modifications. For this experiment we used similar setup to the one used in 4.2. We still kept a stationary observing agent. This is purely not to bring another variable into the experiment. The algorithm works even if the observing agent is moving. The observer shown again as *"wolf"* is trying to follow up the movement of a moving agent shown as *"piglet"* similarly to the previous experiment. This scenario contains also a wall which hides the observed agent (*piglet*) and thus renders is hidden for the observing agent. When the moving agent disappears the probabilistic occupancy map is

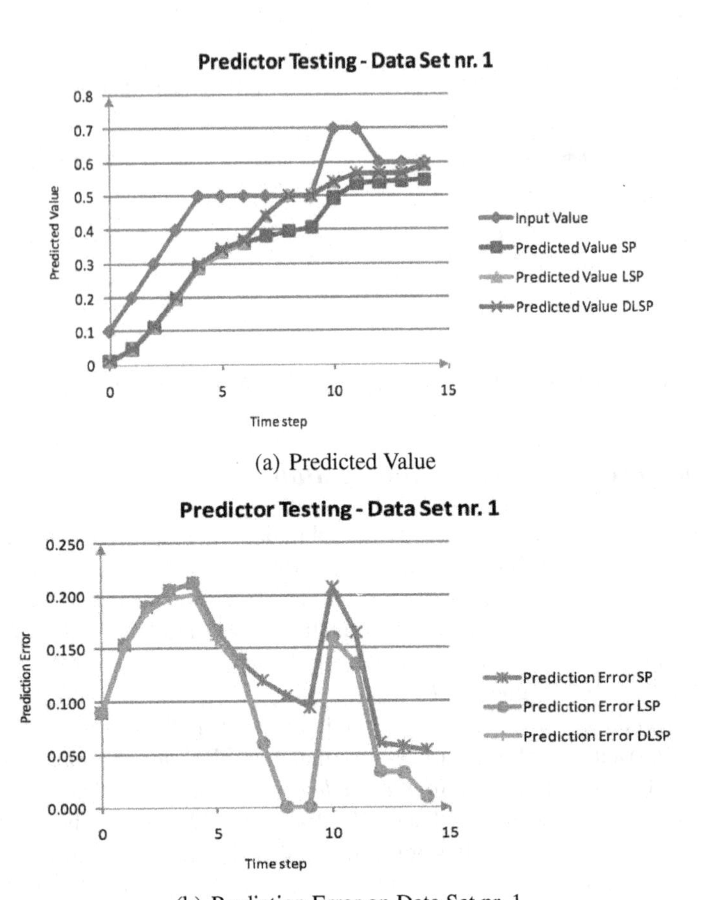

(a) Predicted Value

(b) Prediction Error on Data Set nr. 1

Fig. 9 Results of the Predictor Testing

initialized and probabilities are diffused each step to estimate the position based on the known last position speed vector. In one experiment the probabilities are diffused using the modification of the diffusion constant. In another my suggested modification with also the position estimation was used.

The original approach had problems to keep propagating the probability in the right direction and in some experiments tend to follow up in the last observed direction and speed but after few steps the probabilities were so dispersed that the estimated position stopped being propagated in this direction. In the described approach, the estimator was used to estimate the next position based on the same values but also the observed history values. However this position has only a certain degree of reliability, for this we used the probabilistic occupancy map to reflect the fact that other positions next to this have also some probabilities of occurrence of the observed agent because the agent could have changed the direction of the movement or even stopped.

Fig. 10 shows the exact moment when the agent is about to disappear behind an obstacle (*wall*) and it also shows the generated occupancy map, where the height of a rectangular prism represents the probability of the occupancy of that position by the hidden agent.

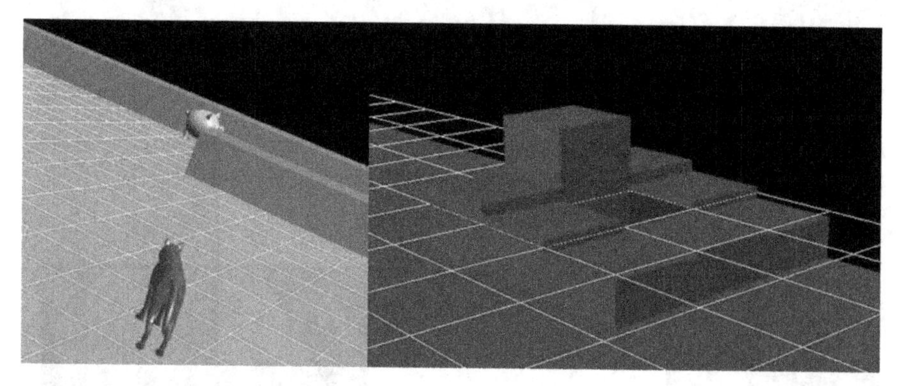

Fig. 10 Object Persistence — 3D Visualisation of the Detail of Agent *(Pig) Starting to be Hidden for the Observer (- Wolf) Behind a Wall (left) and a Probabilistic Occupancy Map created (right)*

4.5 Unconscious Reward Anticipation

This level focuses on emotion generation and superposition. For this the scenario with the predators *"wolf"* and prey *"piglet"*) was still used. The experiments are focused on emotion generation and to confirm that the correct emotions are generated in the correct situation. Emoticons are the widely accepted for of expressing the emotion in a simulation.

We still have an agent *"wolf"* observing another agent *"piglet"* and estimating its position after it disappears behind the wolf. Once the observed agent is visible again and the observing agent can verify his expectation an emotion can be generated, also the intensity of the emotion is evaluated and the final emotional state generated. For this purpose the behaviour of the observing agent was modified and once it is hidden, it can decide to turn back and continue its motion counter clockwise, which will lead to surprise in the observing agent. The Fig. 11 and show two different situations with either confirmed or unconfirmed expectations and the corresponding emotion.

4.6 Conscious Reward Anticipation

Karel Kohout has copied the scenario from work of Kadleček [18] in order to compare my results and to show strengths and weaknesses of my approach. In this scenario the main actor is a Taxi agent, shown as a *"yellow van"* (or dot). This agent's goal is to pickup client agents shown as a *"red woman"* (blue triangle) and take them to their destination, the desired destination is shown as a *"white house"*, the

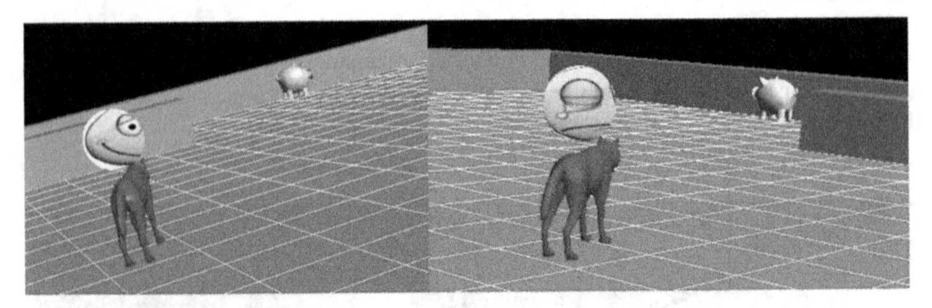

Fig. 11 Emotion Generation — 3D Visualisation of the Agent *(Pig) Reappearing Where Expected (left) and where NOT expected (right) by the Observer (Wolf) and the Positive (left) or Negative (right) Emotion Generation*

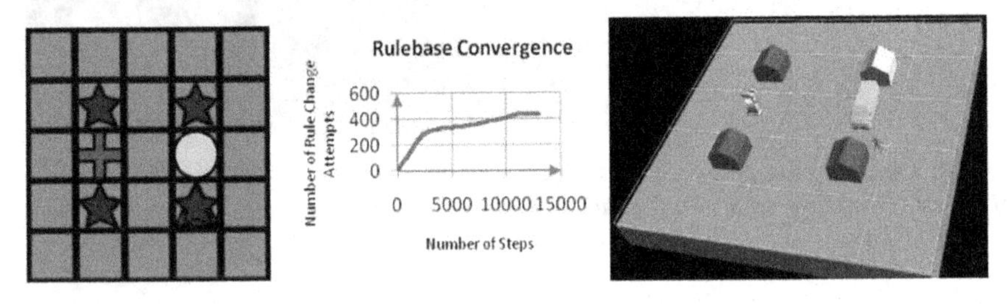

Fig. 12 The Taxi Problem Scenario Layout in 2D View (left) and 3D View (right) and The Convergence Speed of the ACS (middle)

rest of the houses are shown in *"brown"* (stars). The client agent is generated at random intervals (the probability of client appearance is 1/6). At a time there can be only one client agent until the client is delivered to the final destination. The scenario contains also a *"filling station"* (red cross) as the taxi agent consumes energy by moving and transporting client agents. Both the 2D and 3D view of the scenario is shown on Fig. 12.

This experiment shown, that the ACS approach is capable of learning to navigate in multi-goal scenario. However it also shown that the ACS approach alone has many weaknesses. First of all the learning phase greatly depends on the random behaviour. It happens quite often that in the early phase a certain element of behaviour such as *fill up* or customer *pick up* is not learned. Then in a later phase due to better strength of the already learned behaviour these elements have smaller chance of being selected and strength improved. We have analyzed the algorithm, and we believe the root cause is in the new rule generation step. The rules that are generated by this approach are still not specific enough and do not allow to unlearn conditions under which the action has no effect. This can partially be remediated by deleting rules however that step does not help to create more specific rules. The convergence speed of the learning phase shown on Fig. 12 is quite slow. As is shown on the graph below it can be also misleading. Stopping the learning process after 5000 steps would suggest a good convergence, but running the simulation for longer time revealed that there were additional 100 rule base change attempts in the next 7000 steps. The rulebase in the scenario fully converged after 12000 steps. It is due to say that thanks to the high count of #-symbols, it can even happen that a rule that is connected with drop off or pick up action is deleted and then it can never be executed again. This shows another weakness of the ACS approach that we wish to highlight. It can be concluded that ACS alone is not the optimal driving mechanism, and is rather limited. While it can help to build latent knowledge as will be shown below, it should not be used as the only decision making mechanism. But as a part of my architecture it serves its purpose as one of the eight levels.

4.7 Unconscious State Anticipation

This level focuses on latent learning. The experiment we used similar setup to Stolzmann, and even he took the experiment from an *ethology example of rat learning*. In this scenario an agent (in this implementation depicted as *piglet*) is placed in a simple E shaped maze. The agent starts in the middle, and has a choice to go left or to go right. The end boxes of each branch have different colour *black* and *white*. The agent is allowed a free run in the maze pre-learning phase. After a certain period the agent is placed in the left side (*black*) and is presented with reward (*food*). Then the agent is gain placed in the starting point. If the latent learning is correct the agent should be able to run straight to the left black box with anticipation of a reward there. The scenario is shown in Fig. 13 where the agent in the starting position is depicted as red circle.

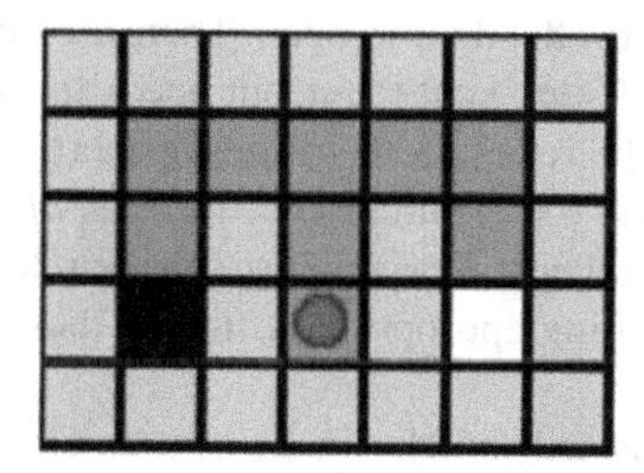 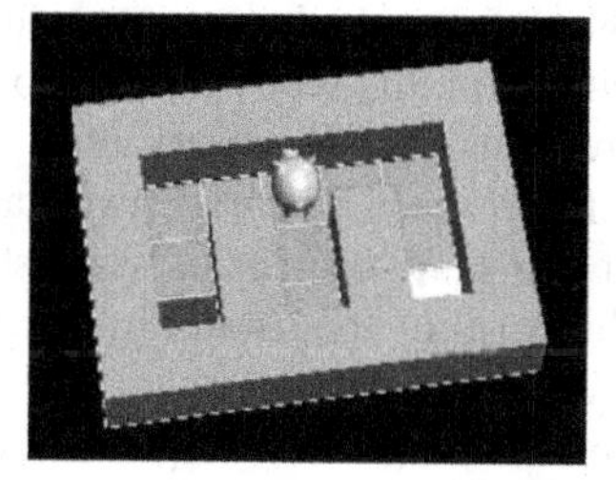

Fig. 13 Latent Learning Scenario *Layout in 2D view (left) and 3D view (right)*

To compare the results with Stozlmann's work (see Table 4), we have conducted the same experiments. This means after the learning we executed ten times a 30 tries to observe, how many times the agent will turn left to reach the reward. The results of these experiments are shown in the Table 5.

There are two conclusions to be made from this table. One is that the latent learning was successfully tested and gave reasonable results. Second is that compared with Stolzmann's result, my results are slightly worse, but since that heavily depends on how the rule base was trained we cannot make a definitive conclusion on the second point.

Table 4 **Latent Learning** — Stolzmann's Results

Trial number	1	2	3	4	5	6	7	8	9	10	Average
Go left action 1	28	26	26	28	22	27	29	24	24	26	26.0
Go left action 2	4	6	6	4	10	5	3	8	8	6	6.0

Table 5 **Latent Learning** — The Statistical Evaluation of the Experiment

Trial number	1	2	3	4	5	6	7	8	9	10	Average
Go left action 1	24	24	25	15	26	24	19	21	20	25	22.3
Go left action 2	6	6	5	15	4	6	11	9	10	5	7.7

4.8 Conscious State Anticipation

For this complex scenario we again chose to compare with work of our colleague David Kadleček. Except the taxi problem already introduced among many other he also used so called Treasure Problem. Again we aim to compare my results and to show strengths and weaknesses of my approach.

In this scenario the main actor is an agent, shown as a *"robot"* (or purple dot). This agent's goal is to reach and open the *treasure chest* (yellow triangle). This chest is not reachable because it is behind *door* (brown cross). To open the door the agent needs to place *heavy stones* (gray star) on the *pressure pads* (black rectangle). The stone needs to remain on the pressure pad for the door to open. There are three such stones available in this scenario. On top of this goal the agent needs to satisfy its own requirements for self preservation. The robot needs to supply energy in terms of food and also water. There are two static food sources shown as *red apples* (red circle) and two static sources of water shown as *blue buckets* (blue circle). Agent can refill his level of food/water by executing appropriate action of these sources (*apple*/*bucket*).

The hybrid ACS-planning algorithm as will be shown below can successfully solve the treasure problem. Unlike the pure ACS approach it also significantly decreases the run time after the initial exploration and convergence of the ACS rule-base. Thanks to the learned environment the planning can then significantly reduce the run time while at the same time being able to satisfy the internal needs.

Fig. 14 below shows the cumulative reward over time (green line). The training takes place for 8772 steps when the treasure is reached. The individual reward spikes were created by multiplying the actual received reward by 5 so it is visible and to scale with the cumulative reward. The smallest spikes show fulfilment of the internal needs of eating and drinking. The middle spikes (in steps 2961 and 4559) show the placement of the stones on the pressure pads. The highest spikes identify when the treasure is found. As previously mentioned once the treasure is once found and the environment is successfully learned the planning algorithm takes precedence

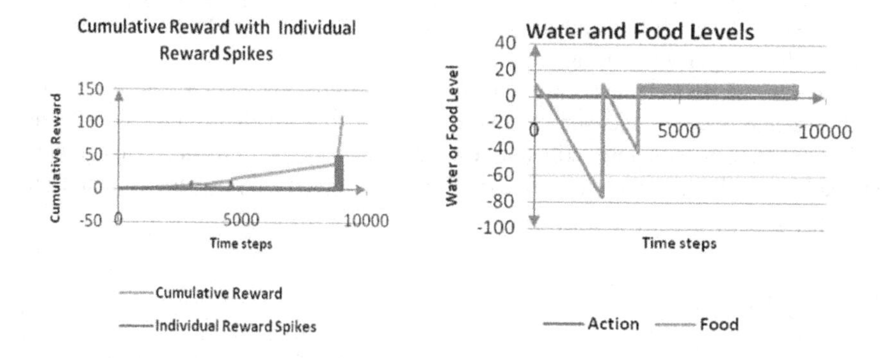

Fig. 14 Treasure Problem The Cumulative Reward over Time with Individual Reward Spikes (left) The Level of Water and Food during the Simulation (right)

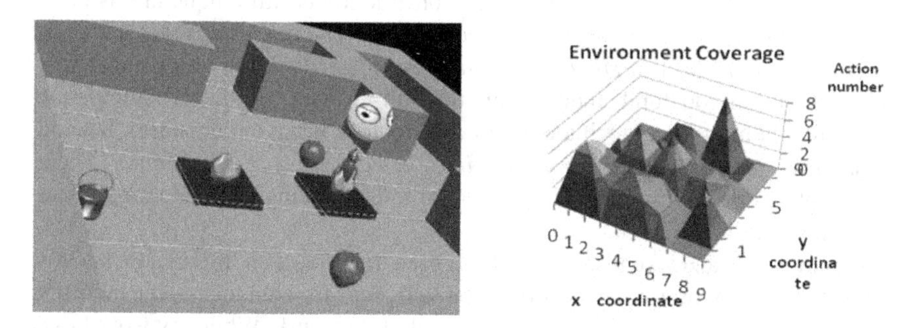

Fig. 15 Treasure Problem — *Emotion Expression* (left) and The Coverage of the States of the Environment (right)

and reaching the goal again becomes very fast. As mentioned it took 8772 steps to reach the goal for the first time after that it took in average 93 steps to reach the goal again.

For the same simulation the level of water and food is shown on the Fig. 14. During the learning phase it is shown that the levels go deep into negative values. What is also interesting that the water level dropped below zero even after the first learning cycle at the step 8823? It is obvious that reaching the treasure (goal) for the first time does not guarantee that the algorithm covered the environment fully. Even if planning is helping significantly to keep the needs for food or water satisfied during exploration the agent can wonder so far from the source that it is not possible to the back in time. The coverage of the environment in this case is shown on the 3D mesh on Fig. 15, where for each of the position we used only the action with the highest number attached. It is simplified but shows clearly the positions of the stones, the pads and the treasure and that they have been successfully mapped.

Emotions can be also generated here by plugging in the appropriate layer of my design. Since the emotion cube was already prepared then it is not a problem to add it as another property of this agent.

Since the scenario counts only with reward and not punishment and the agent in this scenario is not focused on anticipating the reward then there is just one emotional state to generate, still to show how each factor influences each other. Mr. Kohout embedded the emotion cube here [20]. Fig. 15 describes this situation schematically.

5 Conclusion

Anticipation is not plain prediction or estimation of the future. Anticipation in AL-ife sense is much more than just prediction it is utilising the obtained information about the future for the cognitive processes such as decision control and planning. It is also about generating emotions, controlling attention and many other things. Our simulations showed that the described construction is still unique in this field, mainly because of two main ideas have been proven. The first idea is that *anticipation is not a matter of just single mechanism* (similarly to any living being). This is reason, why my doctoral student came up with his idea of 8-factor anticipation [20] (which is multi-level architecture of anticipatory behaving creature). Second idea is *the introduction of consciousness into the categories of anticipation*. There are multiple secondary findings as a product of implementing and evaluating of our long-time scientific research.

The simulation and visualisation methods used may create an impression that this work is focused on improving the artificial intelligence for computer game industry. It is due to say that this has never been the goal of this work. While my work has a value in this industry as well it is not the primary one. The main industries to apply my research in are power distribution (*Smart Grid*), Robotics (HRI) and prevention and protection of health and safety of human beings.

Acknowledgements. This research has been funded by the Dept. of Cybernetics, Faculty of Electrical Engineering, Czech Technical University in Prague and Centre for Applied Cybernetics under Project 1M0567.

References

1. Rosen, R.: Anticipatory Systems - Philosophical, Mathematical and Methodological Foundations, 1st edn., p. 436. Pergamon Press (1985)
2. Tolman, E.: Purposive behavior in animals and men, 1st edn. Century, New York (1932), ISBN: 978-0891975441
3. Rescorla, R.: Associative relations in instrumental learning: The eighteenth Bartlett memorial lecture. Journal of Experimental Psychology 43(1), 1–23 (1991), ISSN: 1747-0218

4. Butz, M., Sigaud, O., Gerard, P.: Anticipatory Behavior in Adaptive Learning Systems: Foundations, Theories and System, 1st edn., p. 303. Springer (2003), ISBN: 3-540-40429-5
5. Dubois, D.M.: Mathematical Foundations of Discrete and Functional Systems with Strong and Weak Anticipations. In: Butz, M.V., Sigaud, O., Gérard, P. (eds.) Anticipatory Behavior in Adaptive Learning Systems. LNCS (LNAI), vol. 2684, pp. 110–132. Springer, Heidelberg (2003), ISBN: 3-540-40429-5
6. Martinho, C.: Emotivector: Affective Anticipatory Mechanism for Synthetic Characters. Lisbon: Instituto Superior Tcnico. PhD Thesis (2007)
7. Kohout, K., Nahodil, P.: Reactively and Anticipatory Behaving Agents for Artificial Life Simulations. In: Dubois, M.D. (ed.) Computing Anticipatory Systems, pp. 84–92. AIP (2010), ISBN: 978-0-7354-0858-6
8. Dubois, D.: Review of Incursive, Hyper incursive and Anticipatory Systems Foundation of Anticipation in Electromagnetism. In: Dubois, M.D. (ed.) Proceedings of 3rd International Conference on Computing Anticipatory Systems (CASYS 1999), pp. 3–30. American Institute of Physics, New York (1999), ISBN: 978-2960017960
9. Sigaud, O., Wilson, S.: Learning classifier systems: a survey. Soft Computing - A Fusion of Foundations, Methodologies and Applications 11(11), 1065–1078 (2007), ISSN: 1432-7643
10. Holland, J.: Adaptation in Natural and Artificial Systems: An Introductory Analysis with Applications to Biology, Control and Artificial Intelligence, 1st edn., p. 228. MIT Press, Cambridge (1995), ISBN: 978-0262581110
11. Stolzmann, W.: An Introduction to Anticipatory Classifier Systems. In: Lanzi, P.L., Stolzmann, W., Wilson, S.W. (eds.) IWLCS 1999. LNCS (LNAI), vol. 1813, pp. 175–194. Springer, Heidelberg (2000)
12. Butz, M.V., Stolzmann, W.: An Algorithmic Description of ACS2. In: Lanzi, P.L., Stolzmann, W., Wilson, S.W. (eds.) IWLCS 2001. LNCS (LNAI), vol. 2321, pp. 211–229. Springer, Heidelberg (2002)
13. Russel, S., Norvig, P.: Artificial Intelligence: A Modern Approach, 2nd edn., p. 1132. Prentice-Hall (2002), ISBN: 978-0137903955
14. Isla, D., Blumberg, B.: Object persistence for synthetic characters. In: Castelfranchi, C., Johnson, W.L. (eds.) Proceedings of the 1st International Joint Conference on Autonomous Agents and Multiagent Systems, pp. 1356–1363. ACM Press, New York (2002), ISBN: 1-58113-480-0
15. Posner, M.: Attention before and during the decade of the brain. In: Meyers, D.E., Kornblum, S.M. (eds.), pp. 343–350. MIT Press (1993), ISBN: 0-262-13284-2
16. Müller, H., Rabbit, P.: Reflexive orienting of visual attention: time course of activation and resistance to interruption. Journal of Experimental Psychology: Human Perception and Performance 15(2), 315–330 (1989), ISSN: 0096-1523
17. Recursive Porous Agent Simulation Toolkit, http://repast.sourceforge.net (cited September 29, 2011)
18. Kadleček, D.: Motivation Driven Reinforcement Learning and Automatic Creation of Behavior Hierarchies. Prague: Dept. of Cybernetics, Czech Technical University in Prague, p. 142. PhD Thesis (2008)
19. Kohout, K., Nahodil, P.: Simulation Environment for Anticipatory Behaving Agents from the Artificial Life Domain. In: Pelachaud, C., Martin, J.-C., André, E., Chollet, G., Karpouzis, K., Pelé, D. (eds.) IVA 2007. LNCS (LNAI), vol. 4722, pp. 387–388. Springer, Heidelberg (2007)

20. Kohout, K.: Multi-level Structure of Anticipatory Behavior in ALife. PhD Thesis. Prague: Dept. of Cybernetics, Czech TU in Prague, p. 162 (2011)
21. Nahodil, P., Kohout, K.: Types of Anticipatory Behaving Agents. In: 24th European Conference on Modeling and Simulation ECMS 2010, Kuala Lumpu, pp. 145–168 (2010)
22. Butz, M.V., Sigaud, O., Gérard, P.: Anticipatory Behavior: Exploiting Knowledge About the Future to Improve Current Behavior. In: Butz, M.V., Sigaud, O., Gérard, P. (eds.) Anticipatory Behavior in Adaptive Learning Systems. LNCS (LNAI), vol. 2684, pp. 1–10. Springer, Heidelberg (2003)

An Integrated Approach to Robust Multi-echelon Inventory Policy Decision

Katja Klingebiel and Cong Li

Abstract. To cope with current turbulent market demands, robust multi-echelon inventory policies are needed for distribution networks in order to lower inventory costs as well as to maintain high responsiveness. This paper analyzes the inventory policies in the context of complex distribution networks and proposes a new integrated approach to robust multi-echelon inventory policy decision, which is composed of three interrelated components: an analytical inventory policy optimisation, a supply chain simulation module and a metaheuristic-based inventory policy optimiser. Based on the existing approximation algorithms designed primarily for two-echelon inventory policy optimisation, an analytical multi-echelon inventory model in combination with an efficient optimisation algorithm has been designed. Through systematic parameter adjustment, an initial generation of optimised multi-echelon inventory policies is calculated. To evaluate optimality and robustness of these multi-echelon inventory policies under market dynamics, they are automatically handed over to a simulation module, which is capable of modeling arbitrary complexity and uncertainties within and outside of a supply chain and simulating them under respective scenarios. Based on the simulation results, i.e. the robustness of the proposed strategies, a metaheuristic-based inventory policy optimiser regenerates improved (more robust) multi-echelon inventory policies, which are once again dynamically evaluated through simulation. This closed feedback loop forms a simulation optimisation process that enables the autonomous evolution of robust

Katja Klingebiel
Chair of Factory Organisation,
Technical University of Dortmund, Leonhard-Euler-Str. 5, 44227 Dortmund, Germany
e-mail: katja.klingebiel@tu-dortmund.de

Cong Li
Department of Supply Chain Engineering,
Fraunhofer Institute for Material Flow and Logistics,
Joseph-von-Fraunhofer Straße 2-4, 44227 Dortmund, Germany
e-mail: cong.li@iml.fraunhofer.de

A. Byrski et al. (Eds.): Advances in Intelligent Modelling and Simulation, SCI 416, pp. 165–197.
springerlink.com

multi-echelon inventory policies. The proposed approach has further been validated by an industrial case study, in which favorable outcomes have been obtained.

1 Motivation and Problem Description

As one of the most important logistical drivers, inventory plays a significant role for the supply chain's performance [14]. Consequently, inventory policy decisions have a huge impact on the efficiency and responsiveness of a supply chain. As collaboration between different supply chain echelons gains increasing attention, it is imperative to consider inventory policies from a network perspective rather than supposing each echelon to be a single isolated player.

As depicted in Fig. 1, a multi-echelon distribution network is composed of warehouses (square shaped symbols) which represent stock points for multiple products (triangle shaped symbols). Every warehouse may be regarded as storing the same variety of products; each stock point corresponds to exactly one kind of product. From the supply chain (multi-echelon) view the overall objective for this distribution network is to optimise the overall value created, i.e., to maximise the service level while minimising overall costs.

Optimised inventory policies obtained through traditional approaches are mostly based on deterministic and stable network conditions. Nevertheless, in today's

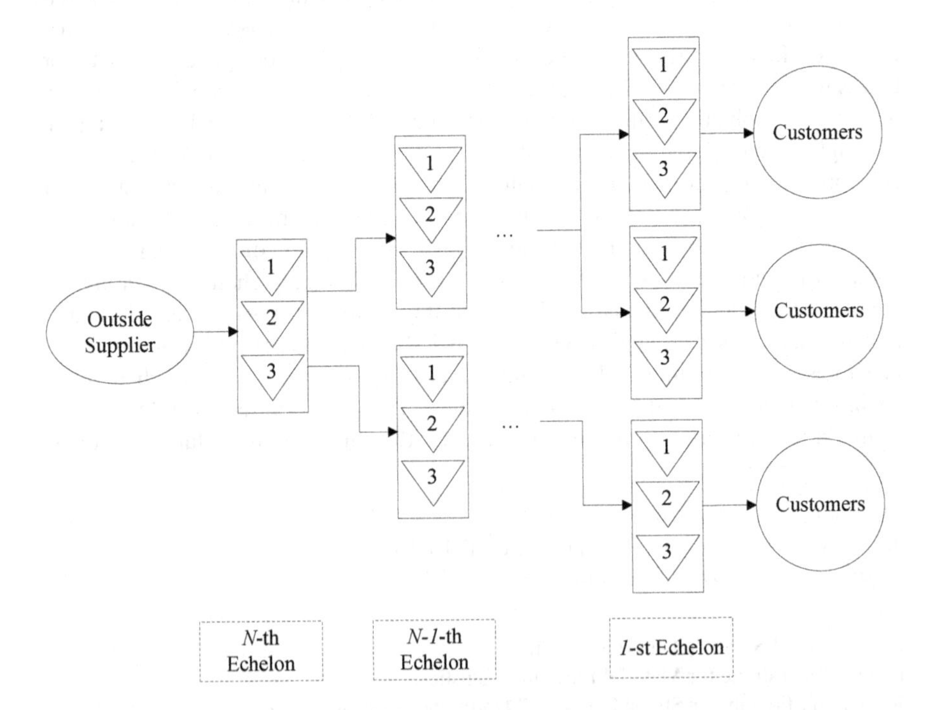

Fig. 1 A Typical Multi-Echelon Distribution Network

dynamic market environment, the level of demand uncertainty has significantly increased. Additionally, product trends like longer lead time, seasonality, larger variety and shorter life cycles have intensified uncertainty [15]. Traditional approaches are no longer capable of delivering the desired results or even greatly deteriorate the performance of the entire supply chain, leading to high stock levels or short sales.

Thus, to cope with current turbulent market demands, robust multi-echelon inventory policies are needed in order to lower inventory costs as well as to maintain high responsiveness. These inventory policies are required to cope with volatility and uncertainties of customer demand and to withstand a certain change as well as defined uncertainties within and outside the distribution network without great loss of performance while in parallel not significantly increasing cost levels. This insensibility to accidental environmental influences is generally defined as robustness [37]. In addition to the assumed optimality (in terms of cost to performance ratio) under stable conditions, robustness inventory policies maintain within a specified cost-performance corridor under dynamic influences.

In this paper, an integrated approach to robust (R, Q) multi-echelon inventory policies for complex distribution networks is presented, of which each stock point is assumed to be continuously reviewed. Thus, when the inventory position, expressed as the physical inventory plus the stock on order minus backorders, is equal to or drops below its reorder point R, a replenishment order is placed towards its supplier with the order quantity Q (see, e.g., [11, 49], etc.).

Two representative performance measures, monetary inventory holding costs (in the following inventory costs) and percentage fill rate, are accordingly adopted to assess the optimality and robustness of specific inventory policies. For these two incommensurable objectives, it is more reasonable to assess each dimension independently than to subjectively transform them into a single objective. Besides, inventory costs and fill rate vary in the opposite directions with the change of inventory policies. In this sense, the concept of Pareto optimal solution or non-dominated solution (see [36, 48], etc.) is well suited for this problem. An inventory policy shall be regarded as Pareto optimal if and only if any improvement in one objective (e.g., inventory costs or fill rate) of the inventory policy will cause deterioration in another objective (e.g., fill rate or inventory costs). Hence, the target of this paper is to present an approach to define the Pareto set of robust multi-echelon inventory policies for distribution networks from which decision makers can select robust inventory policies according to their specific trade-off between inventory costs and fill rates.

The paper is organized as follows: Section 2 reviews the important multi-echelon inventory models and optimisation solutions. Section 3 presents the integrated approach, which is constituted of three interrelated components: an analytical inventory policy optimisation, a supply chain simulation module and a metaheuristic-based inventory policy optimiser. The analytical inventory policy optimisation is presented and discussed in detail in Section 4, which includes model formulation, model calculation and optimisation algorithm. The supply chain simulation module and its implementation tool are described in Section 5. The metaheuristic-based inventory policy optimiser is elaborated in Section 6, which is further divided into the algorithm selection and optimiser design. The proposed integrated approach is then applied to an industrial case in Section 7. Finally, Section 8 concludes the current work and provides directions for further research.

2 Literature Review

An overview of the fundamental ideas about problem assumptions, model designs and solution approaches of inventory policies for single- or multi-echelon logistic networks has been presented by Graves [29], Zipkin [49], Axsäter [11] and Tempelmeier [45]. For the inventory models with stochastic lead time in multi-echelon distribution networks, which is also our research emphasis, Axsäter [9] has provided a quite comprehensive review. Starting from the early famous METRIC model presented by Sherbrooke [41], numerous authors have devoted to this research area, among which pioneering research has been conducted by Deuermeyer and Schwarz [21], Graves [28], Svoronos and Zipkin [43], Axsäter [4, 5, 6, 7] , Kiesmüller and Kok [33]. Apart from the classical multi-echelon model, Dong and Chen [22] developed a network of inventory-queue models for the performance modeling and analysis of an integrated logistic network. Simchi-Levi and Zhao [42] derived recursive equations to characterize the dependencies across different echelons in a supply chain network. Miranda and Garrido [38] dealt inventory decisions simultaneously with network design decisions while Kang and Kim [32] focused on the coordination of inventory and transportation management.

Although great attentions have been paid to the analytical analysis of distribution networks, its application in the optimisation field is still strongly restricted: in large scale distribution networks, modeling complexity and computational requirement can quickly exceed sensible limits. Therefore, various approximation methods and heuristic algorithms have been suggested for real world applications.

Of note in this context is the work of Cohen et al. [17], who developed and implemented a system called Optimiser that determined the inventory policies for each part at each location in IBM's complex network with assumptions of deterministic lead time and ample supply. Caglar et al. [13] proposed a base-stock policy for a two-echelon, multi-item spare part inventory system and presented a heuristic algorithm based on METRIC approximation and single-depot sub problem to minimise the system-wide inventory costs subject to a response time constraint at each field depot. Axsäter [8] used normal approximations both for the customer demand and retailer demand to solve the general two-echelon distribution inventory system. Axsäter [10] considered a different approach to decompose the two-echelon inventory problems. Through providing an artificial unit backorder cost of the warehouse, its optimal inventory policy can be solved first. Al-Rifai and Rossetti [2] formulated an iterative heuristic optimisation algorithm to minimise the total annual inventory investment subject to annual ordering frequency and backorder number constraints. Their approach can be regarded as the further work of Hopp et al. [30], who utilised (R, Q) policies and presented three heuristic algorithms on the basis of simplified representations of the inventory and service expressions to optimise the same inventory problem in a single echelon. Pasandideh et al. [40] developed a parameter-tuned genetic algorithm to solve the effective (R, Q) inventory policies of the same two-echelon inventory problem.

From the above analysis, it may be deduced that multi-echelon inventory policies have been analyzed extensively in recent years. All the above mentioned authors have designed analytical models for inventory policies evaluation or optimisation. However, computational scale, integrity and non-convexity make the corresponding optimisation problem intractable to exact analysis and up till now no general approach is accepted, which might also explain why two-echelon networks are mostly dealt with in various literatures. Moreover, these analytical models, no matter their varying levels of sophistication or abstraction, have all failed to take market dynamics into account—either assuming deterministic demand or approximating a stable probability distribution (e.g., normal distribution). Besides, for inventory policy optimisation, these authors have only adopted a single optimisation objective—either balancing the trade-off between inventory costs and backorder costs or considering service level (e.g., backorder costs or fill rate) as a constraint. The most intuitive multi-objective optimisation problems, which simultaneously tackle inventory costs and fill rate, however, have not been paid attention to. Correspondingly, with these approaches only a single "optimal" inventory policy instead of a series of "optimal" ones is generated, which omits a lot of possibilities for decision makers. In response to such shortcomings, an integrated approach to robust multi-echelon inventory policies is proposed in the next section.

3 Integrated Approach

Even the most delicate analytical model forces abstraction of reality and involves some kinds of simplification or approximation. Models of this kind may not fully represent every aspect of a complex distribution network, typically especially neglecting the outside market dynamics and the inside intricate interactions. Fortunately, these deficiencies can be compensated to a large extent by simulation models, as they are capable of reproducing and testing different decision-making alternatives (e.g., inventory policies) under anticipated supply chain scenarios (e.g., forecasted demand scenarios). This allows ascertaining the level of optimality and robustness of a given strategy in advance [46]. Nevertheless, simulation itself can provide only what-if analysis. Even for a small-sized problem, there exist a large numbers of possible alternatives, making exhaustive simulation impossible.

In the authors' previous work, an integrated approach has been suggested which combines the analytical optimisation with simulation model. Through systematic adjustment of input parameters, a limited set of "optimal" alternatives may be derived from an analytical model. After simulating these multi-echelon inventory policies under realistic environmental conditions (e.g., dynamic and stochastic volatile customer demand), their performance level (e.g., inventory costs, fill rate) can be comprehensively and precisely evaluated and consulted for decision making [35]. The schematic diagram of such an integrated approach is shown in Fig. 2.

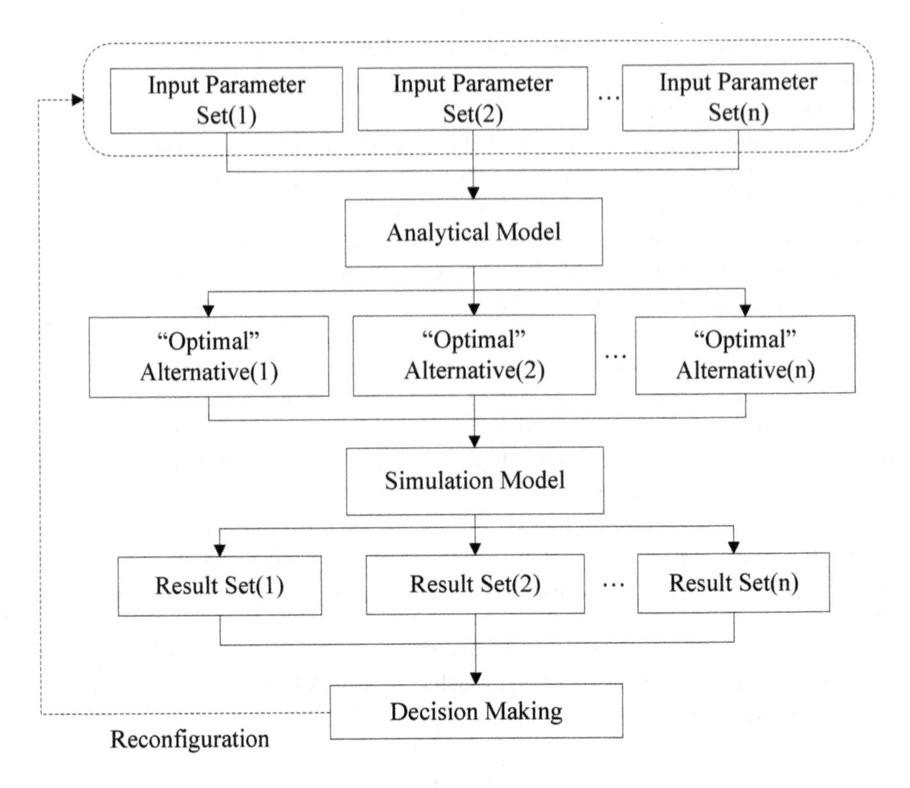

Fig. 2 A Previous Integrated Approach to Robust Multi-Echelon Inventory Policy Decision

However, a problem arises when all the suggested alternatives have not fulfilled the desired expectation. One of the answers is to reconfigure the input parameters of the analytical model based on the simulation result (dotted line in Fig. 2), and then to restart the optimisation process and simulation, so that a closed feedback loop is formed. Nevertheless, such reconfigurations, although feasible, are limited to a certain set of outcomes: Because the analytical model is an abstraction of real world and especially does not integrate aspects of dynamics nor represent the objective of robustness, the "optimal" alternatives obtained from the analytical model can act only as references or starting points. Thus, it is necessary to integrate optimisation and simulation more closely, which leads to the more advanced integrated approach. Specifically, this approach comprises three interrelated components: an analytical inventory policy optimisation, a supply chain simulation module and a metaheuristic-based inventory policy optimiser. The schematic diagram of this integrated approach is illustrated in Fig. 3.

Based on the existing approximation algorithms designed primarily for two-echelon inventory policy optimisation, an analytical multi-echelon inventory model in combination with an efficient optimisation algorithm has been designed to calculate an initial generation of optimised multi-echelon inventory policies. Through systematic adjustment of input parameters, a limited set of "optimal" inventory

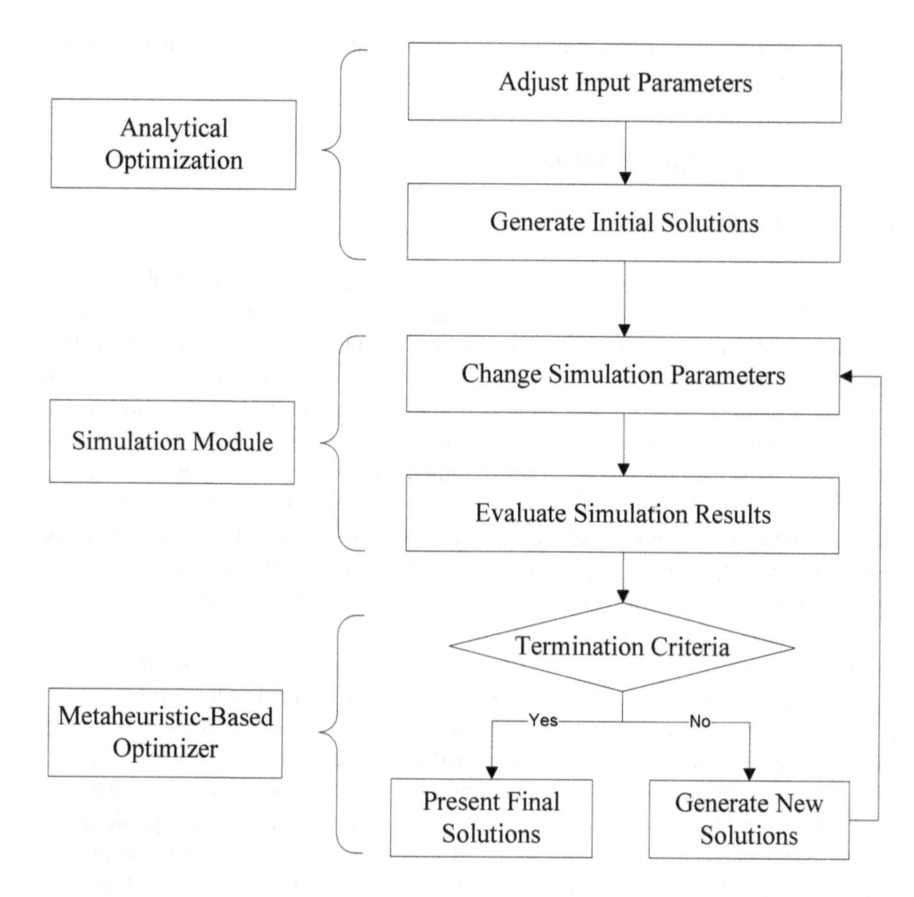

Fig. 3 A New Integrated Approach to Robust Multi-Echelon Inventory Policy Decision

policies may be derived, which serve as the starting points of the integrated approach. These proposed inventory policies are handed over to a simulation module, which is capable of modeling complexity and uncertainties within and outside of a distribution network and simulating them under respective scenarios. Since the comprehensive simulation results fully reflect resulting performance and costs under market dynamics, their levels of robustness can be clearly ascertained. Based on these results, a metaheuristic-based optimiser regenerates improved (more robust) multi-echelon inventory policies, which are once again dynamically evaluated through simulation. This closed feedback loop forms a simulation optimisation process that enables the autonomous evolution of robust multi-echelon inventory policies. The iteration continues until given termination criteria are finally satisfied. Since each inventory policy has been carefully tested under real dynamics, the resulting multi-echelon inventory policies may be specified in their robustness against dynamics and uncertainties within and outside the distribution network.

In the next three sections, these three components of the whole integrated approach are elaborated respectively.

4 Analytical Optimisation

4.1 Model Formulation

Corresponding to the general distribution network illustrated in Fig. 1, the analytical model is represented as a multi-echelon inventory model comprising of multiple warehouses and stock points. The set of all stock points is denoted as M and an arbitrary stock point as $k(k \in M)$. P_k is defined as the set of its immediate predecessors, where stock point $j(j \in P_k)$ is a predecessor of k if and only if stock points j and k store the same item, and the replenishment from k is part of the demand process of j; while S_k represents the set of all its immediate successors, where stock point $l(l \in S_k)$ means replenishment from l is part of the demand process of k . Next, the number of echelons in this network is defined as N and the set of stock points at the n-th echelon as ε_n, of which ε_1 is the set of stock points facing end customers while ε_N is the set of stock points supplied directly by the outside supplier.

Within this general multi-echelon distribution network, the stock point at the most downstream echelon ($k \in \varepsilon_1$) is assumed to face stationary stochastic demand with known average and standard deviation; while the demand process for the stock point k at other upstream echelons ($k \in \varepsilon_n, n = 2, \ldots, N$) is derived as a superposition of the replenishment process from its immediate successors [8, 2]. For the (R_k, Q_k) installation inventory policy, when the inventory position is equal to or drops to below R_k, a replenishment order of size Q_k is placed at its immediate predecessor $j \in P_k$. After placing an order, the actual lead time L_k elapses between placing the order and receiving it. If there is sufficient stock at its processor j to fulfill the entire replenishment order, it will be delivered at once; otherwise, only part of the order can be satisfied. The unfulfilled order has to wait until its predecessor j is replenished. After the arrival of replenishment at j , the outstanding backorders are satisfied according to a first-in, first-out (FIFO) policy. Besides, the outside supplier is supposed to have infinite capacity and no lateral transshipments are permitted between stock points at the same echelon. The list of notations in this paper is defined in Table 1. Notice that the stochastic inventory level I_k here can also take negative values, which interprets backorders as negative inventories. Thus, the on-hand inventory is denoted as $[I_k]^+$ while the backorder as $[I_k]^-$, where $[I_k]^+ = max(0, I_k)$ and $[I_k]^- = max(0, -I_k)$. Then, there is

$$[I_k]^+ = I_k + [I_k]^- \tag{1}$$

4.2 Model Calculation

Corresponding to the two critical performance measures (inventory costs and fill rate) mentioned in Section 1, this analytical model needs to be able to mathematically solve two important variables: average on-hand inventory level and average

Table 1 Parameters that affect the vortex formation probability

μ_k	average demand per time unit at stock point k
σ_k	standard deviation of the demand per time unit at stock point k
μ_l^k	average demand at stock point k from its successor $l \in S_k$ per time unit
σ_l^k	standard deviation of demand at stock point k from its successor $l \in S_k$ per time unit
Y_l^k	number of orders at stock point k from its successor $l \in S_k$ per time unit
I_k	stochastic inventory level at stock point k
$\varphi(x)$	probability density function of standard normal distribution
$\Phi(x)$	cumulative distribution function of standard normal distribution
$\Phi^1(x)$	first order standard normal loss function
$\Phi^2(x)$	second order standard normal loss function
h_k	inventory holding costs per unit and time unit at stock point k
b_k	"artificial" backorder costs per unit and time unit at stock point k

backorder level. A standard approach to the calculation of inventory level is given in Eq. (2), which describes the relationships among inventory level, lead time demand and inventory position [49]:

$$IN(t+L) = IP(t) - D(t,t+L) \tag{2}$$

According to Axsäter [6, 8], the inventory position of any stock point k at steady state could be approximated with a uniform distribution over a range of $(R_k, R_k + Q_k]$. Besides, because the exact demand distribution is intractable in this complex network, a normal approximation is adopted as in Axsäter [8]. For the stock point at the first echelon ($k \in \varepsilon_1$), its average μ_k and standard deviation σ_k is already known, so a normal distribution can be directly fitted. For stock point at other upstream echelons, it can be noted that, in the long run, the demand of one stock point will finally be transferred to its predecessor, i.e. $\mu_l^k = \mu_l$. Nevertheless, due to the effect of batch-order replenishment, it is not that easy to identify the closed-form expression of its standard deviation. To get this expression, the random number of orders at stock point k from its successor l per time unit Y_l^k has to be analysed first, the probability of which can be calculated as:

$$P(Y_l^k = y) = \frac{\sigma_l}{Q_l}\left[\Phi^1\left(\frac{(y+1)Q_l - \mu_l}{\sigma_l}\right) + \Phi^1\left(\frac{(y-1)Q_l - \mu_l}{\sigma_l}\right) - 2\Phi^1\left(\frac{Q_l - \mu_l}{\sigma_l}\right)\right] \tag{3}$$

Thus, the standard deviation of order from its successor $l \in S_k$ is:

$$(\sigma_l^k)^2 = \sum_{y=-\infty}^{\infty} (yQ_l - \mu_l^k)^2 P(Y_l^k = y) \tag{4}$$

Since the different successors are supposed to be independent, the average and standard deviation of demand at stock point k is the sum of all its successors' order, i.e.

$$\mu_k = \sum_{l \in S_k} \mu_l^k \tag{5}$$

$$\sigma_k^2 = \sum_{l \in S_k} (\sigma_l^k)^2 \tag{6}$$

After fitting a normal distribution with parameters μ_k and σ_k^2, the demand distribution at any stock point k is ready for use.

Then, the lead time L_k will be analyzed for each stock point k, which is a function of two components, the constant transportation times L_k^t (including ordering, receiving and handling, etc.) and the random delay at its predecessor due to out of stock L_k^s, i.e.

$$L_k = L_k^t + L_k^s \tag{7}$$

Due to the assumption of ample supply, the lead time of stock point at the most upstream echelon ($k \in \varepsilon_N$) consists of only constant transportation time, i.e. $L_k = L_k^t$, $\forall\, k \in \varepsilon_N$. For a stock point at other downstream echelons ($k \in \varepsilon_n$, $n = 1, \ldots, N-1$), however, its lead time is directly influenced by its predecessor $j \in P_k$. The component of lead time due to stock out at its predecessor has to be determined. Here the METRIC approximation [41] is applied which replaces the stochastic lead time by its mean. To achieve this, the backorder level of stock point is needed, which is:

$$E([I_j]^-) = \frac{L_j \sigma_j^2}{Q_j} \left[\Phi^2 \left(\frac{R_j - L_j \mu_j}{\sqrt{L} \sigma_j} \right) - \Phi^2 \left(\frac{R_j + Q_j - L_j \mu_j}{\sqrt{L_j} \sigma_j} \right) \right] \tag{8}$$

According to the Little's formula [49], the average delay due to out of stock at its predecessor j is:

$$L_k^s = \frac{E([I_j]-)}{\mu_j} \tag{9}$$

After fitting a normal distribution with parameters $L_k \mu_k$ and $L_k \sigma_k^2$ to the lead time demand, the average backorder level is now possible to be solved in analogue with Eq. (8), i.e.:

$$E([I_k]^-) = \frac{L_k \sigma_k^2}{Q_k} \left[\Phi^2 \left(\frac{R_k - L_k \mu_k}{\sqrt{L} \sigma_k} \right) - \Phi^2 \left(\frac{R_k + Q_k - L_k \mu_k}{\sqrt{L_k} \sigma_k} \right) \right] \tag{10}$$

Meanwhile, by applying Eq. (1), the average on-hand inventory level is:

$$E([I_k]^+) = E(I_k) + E([I_k]^-) = R_k + \frac{Q_k}{2} - L_k \mu_k + E([I_k]^-) \tag{11}$$

4.3 Optimisation Algorithm

The target of inventory policy optimisation is to identify the best reorder points that balance the trade-off between economical consideration (inventory costs) and service level (fill rate or inventory backorder costs). Since the decision variables (i.e. reorder points) are not independent, it is impossible to apply blind optimisation. Hence, a decomposed concept is introduced so that the inventory policies can be optimised item-by-item and echelon-by-echelon. The corresponding optimisation problem for each stock point is:

$$Min\, TC_k(R_k) = h_k E([I_k]^+) + b_k E([I_k]^-) \tag{12}$$

where $R_k \geq -Q_k$. Notice that the cost function TC_k is convex in decision variable R_k [10]. The search procedure presented here is a partial enumeration method [1] that exploits the convexity character, implying that the local minimum is also the global minimum. The outline of the search algorithm is given as below:

Step 0: Initialisation
Set $R_k = -Q_k, R_k^{min} = R_k$, the lower bound $R_k^{lower} = R_k$, the upper bound $R_k^{upper} = M$, where M is a sufficiently large integer. The search step $\Delta R_k = \lceil L_k \mu_k \rceil$, where $\lceil x \rceil$ is the smallest integer larger than or equal to x.
Step 1: Local search
For $R_k = R_k^{lower}$ to $R_k = R_k^{upper}$, step ΔR
$TC_k^{min} = TC_k(R_k)$
$R_k = R_k + \Delta R_k$
If $TC_k(R_k) < TC_k^{min}$
Set $TC_k^{min} = TC_k(R_k), R_k^{min} = R_k$
Else
exit the R_k loop
Next R_k
Step 2: Intensified search with smaller granularity
While $\Delta R_k > 1$
Set $R_k^{lower} = R_k - \Delta R_k$, $R_k^{upper} = R_k + \Delta R_k$, $\Delta R_k = \lceil \Delta R_k / 2 \rceil$
Repeat Step 1
Step 3: Get final result
The optimal reorder point R_k^{min} and the corresponding inventory cost TC_k^{min} have been obtained.

With the above search algorithm for a single stock point, a heuristic optimisation procedure is designed to determine the multi-echelon inventory policies for the entire distribution network as follows:

1. Set $n = 2$.
2. For each stock point at that echelon, i.e. $k \in \varepsilon_n$, use Eq. (3) and (4) to determine the replenishment process from any one of its successor.

3. Fit a normal distribution for the demand through superposing replenishment from its successors, shown in Eq. (5) and (6).
4. Set $n = n + 1$ and repeat Step 2 and 3 until $n = N$.

After the demand processes are determined upwardly along the distribution network, the inventory policies are optimised echelon-by-echelon downwardly.

5. Set $n = N$, lead time of stock point k at this echelon is $L_k = L_k^t$.
6. Calculate the mean values of backorder level and on-hand inventory level using Eqs. (10) and (11).
7. Minimise the inventory cost function (12) with the above mentioned search algorithm, so that the optimal inventory reorder point is obtained.
8. Set $n = n - 1$, apply Eqs. (9) and (7) to determine the lead time of each stock point at this echelon, then repeat Step 6 and 7 until $n = 1$.

As a result, a set of "optimal" multi-echelon inventory policies are gained, which in the next step are tested for robustness.

5 Simulation Module

To make a comprehensive and precise evaluation of the robustness of certain multi-echelon inventory policies, a supply chain simulation module is needed to reproduce the real dynamics in the distribution network. Besides, as a component of the entire integrated approach, this simulation module is closely coupled with the other two components, analytical optimisation and metaheuristic-based optimiser. Thus, apart from some general requirements like efficient data communication and storage, user-friendly interface, three specific requirements on this simulation module, which are essential to the success of the integrated approach, have been identified:

- An event-driven simulation, which includes not only basic event scheduling mechanism but also advanced process-oriented mechanism in order to represent supply chain processes and dynamics.
- Coverage of all the requirements of model description with object-oriented approach (stock points, products, demand, lead times, etc.).
- External interface to the other two components in the integrated approach — analytical optimisation and metaheuristic-based optimiser.

Within the here presented approach, the simulation tool OTD-NET, developed by Fraunhofer Institute for Material Flow and Logistics, has been applied, which introduces a holistic approach for modeling and simulation of complex production and logistics networks and delivers in-depth insights into information and material flows, stock levels, stability of the network, boundary conditions and restrictions [47]. The detailed architecture of OTD-NET is manifested in Fig. 4. Having implemented a novel and object-oriented modeling methodology, OTD-NET allows the user to map all relevant network elements (e.g., distribution channels, buffers, customers, and dealers) as well as many influencing parameters (e.g., inventory policies, transportation plan and time windows) in a selectable level of details. As a

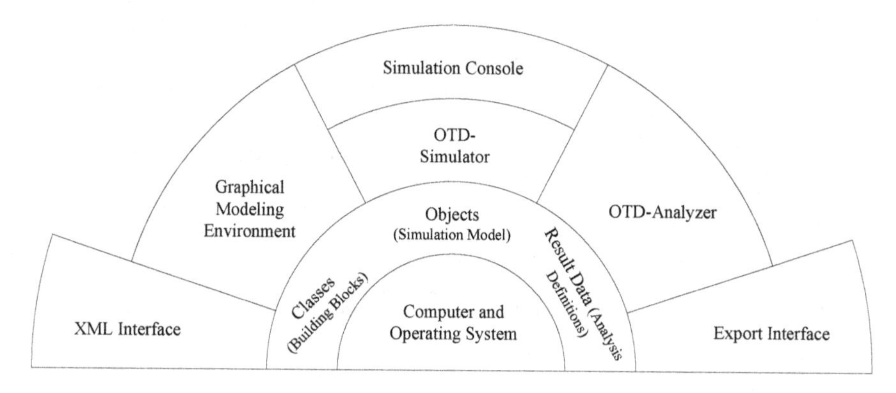

Fig. 4 Detailed Architecture of OTD-NET [34]

discrete-event simulation tool, OTD-NET offers adequate processing and recording of simulation data, which is essential to conduct effective statistical analysis on the simulation results [35]. To facilitate data exchange, OTD-NET provides not only a generic XML-interface, which enables to import data from other independent components; but also auxiliary executable programs, which assist the automatic control of simulation from other independent program modules. Thus, OTD-NET has well fulfilled all the requirements on simulation core, model description and external interface that have been discussed above. Its flexibility and extensibility allow supporting the diversity as well as the complexity of factors inherent in this research problem.

6 Metaheuristic-Based Optimiser

In the integrated approach, the robust level of the proposed multi-echelon inventory policies under market dynamics can be determined through simulation. To promote the evolution of the multi-echelon inventory policies according to their simulation results; however, another independent optimiser (i.e. metaheuristic-based optimiser) is further required (see closed loop in Fig. 3). In the next two subsections, firstly an appropriate type of algorithm is selected to realize the simulation optimisation process and then the optimiser is designed with this algorithm.

6.1 Algorithm Selection

In contrast to a "conventional" deterministic optimisation process, which only cares about the search process (optimisation routine), the simulation optimisation process has to consider an external evaluation process (simulation routine) because its objective values are not directly calculated but assessed through simulation [24].

It seems pretty simple to understand, however, a wide variety of difficulties have arisen when applying a simulation optimisation process in practice. First of all, there

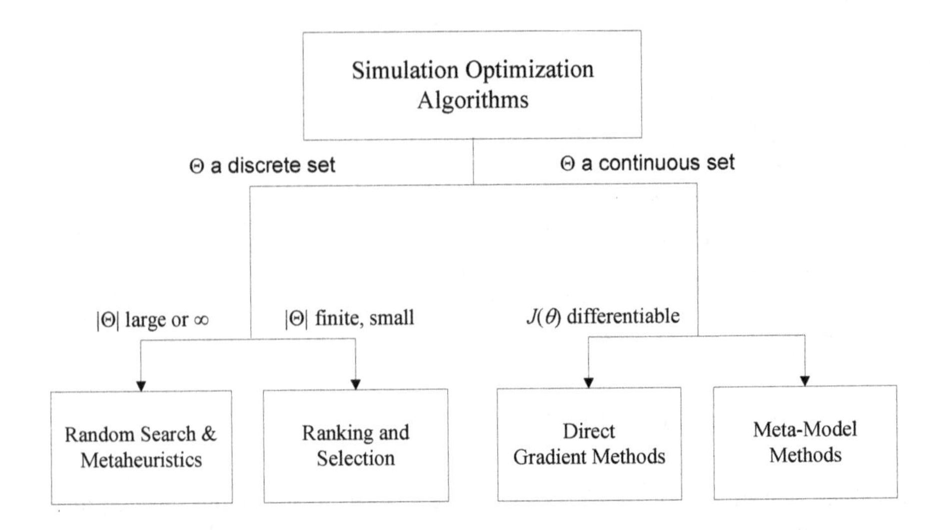

Fig. 5 Classification of Simulation Optimisation Algorithms [12]

typically exist a large numbers of different types of parameters that are subject to optimisation. For this kind of problems, the optimisation itself is not an easy task. Moreover, although simulation can be regarded as a "function" that evaluates the performance of a set of alternatives in a simulation optimisation process [3], little is known about the explicit form of this function. Thus, the effort of accelerating convergence speed or finding an improving direction often ends in vain. Consequently, a great number of alternatives have to be simulated before obtaining favorable ones, which implies great computational effort.

To tackle these difficulties, various kinds of simulation optimisation algorithms have been introduced. In the work of Barton and Meckesheimer [12], these algorithms are classified according to the feature of objective function $J(\theta)$ and the feasible region Θ of the input parameters (illustrated in Fig. 5). In the following, the different simulation optimisation algorithms are analysed so that an appropriate algorithms can be selected for this optimiser to implement the here presented simulation optimisation process.

Proof against Complexity and Intractability

The classical simulation optimisation algorithms such as gradient-based and meta-model based methods usually require users to have a deep understanding of their sophisticated optimisation details and stochastic simulation processes [26]. However, simulation model of the complex distribution network function like a black box. Mostly the inherent relationships between input variables and output results are completely intractable.

Furthermore, as illustrated in Fig. 5, direct gradient methods and meta-model methods are primarily suitable for problems with continuous search space. When there are multiple object functions, which are at the same time discontinuous and non-differentiable, or when the decision variables are qualitative, these methods often fail to find the optimal solutions [44]. Thus, these methods are not qualified for the multi-echelon inventory policy optimisation problem, which involves high-dimensional and integer-value variables. Ranking and search methods could tackle with discrete decision variables, though, these are only appropriate for small-scaled problems, which are not the case here.

Consequently, only random selection and metaheuristics remain. Although feasible, the random selection converges quite slowly (because previous information is not used for iteration) [44]. Hence, it is seldom applied to solve simulation optimisation problems [44]. The metaheuristics, on the contrary, are particularly preferred in those problems that have many local optima and are intractable to define good search directions [39].

Effectiveness vs. Efficiency

Although many traditional optimisation algorithms have exploited some properties of convergence, the respective software implementation is problematic because of their sophisticated assumptions or required prior knowledge. Even if they have been successfully programmed, the full potentials of convergence are rarely realized in practice. This also helps to explain why innovative metaheuristics (e.g., simulated annealing, tabu search, evolutionary algorithm) possess a predominate position in commercial optimisation packages [25]. Based on high-level intelligent procedures, these metaheuristics aim for solution improvement and fast computer implementations [27]. Though not having been theoretically verified, the metaheuristic algorithms, which are designed to seek global optimality (exploration over the entire solution space), seem to have robust features in real applications [24].

It is difficult to simply state which kind of metaheuristic algorithms is the best since each owns its distinct benefits and successful applications. Nevertheless, evolutionary algorithms, which are set-based approaches, have a strong efficiency advantage over those solution-to-solution methods (e.g., simulated annealing and tabu search), i.e. a broader area of the solution space can be explored within a relatively smaller number of objective function evaluations. Since in the context of simulation optimisation object evaluation implies simulation, it is expected that high quality solutions can be found as early as possible [3]. Thus, evolutionary algorithms are considered to be able to provide a good balance of effectiveness and efficiency.

Multi-Objective Optimisation Programming

As discussed in section 1, the set of Pareto optimal inventory policies is preferred to a single optimal policy for decision makers. However, not only conventional simulation optimisation techniques, such as gradient-based ranking and selection, but also most metaheuristic algorithms, such as simulated annealing, tabu search, are not capable of conducting this kind of multi-objective optimisation programming, because they have not incorporated the idea of set-based solutions [23, 39] so that can be extended to such multiple solutions. In practice, multi-objective problems are usually reformulated into single-objective problems before optimisation, which results in only a single "optimal" solution [23].

Evolutionary algorithms (EAs), on the contrary, seem to be especially fit for this task, as they maintain a group of individuals which interact together to search the solution space in parallel [23]. If properly controlled, EAs may provide decision makers a set of Pareto optimal solutions, which are evenly distributed on the Pareto front [48]. Besides, during the optimisation process deep insights into the landscape of these non-dominated solutions may be gained [48].

Intermediate Conclusions

Summarising from the perspective of problem complexity and intractability, general metaheuristics are the first choice for the optimiser which implements the simulation optimisation process. Moreover, with regard to effectiveness and efficiency and especially the last criterion of multi-objective optimisation programming, evolutionary algorithms (EAs) as a special type of metaheuristics are selected as the central algorithm for the realisation of simulation optimisation approach.

6.2 Optimiser Design

To derive multiple Pareto optimal multi-echelon inventory policies, a particular type of evolutionary algorithm called non-dominated sorting genetic algorithm II (NSGA-II) has been adopted. Employing the Pareto ranking as fitness assignment criterion, crowding distance as diversity mechanism and separated storage space as elitism preservation method, NSGA-II is one of the most widely used multi-objective evolutionary algorithms [36].

The general procedures of NSGA-II [20, 48] are illustrated in Fig. 6. Prior to the optimisation process, the fundamental input parameters must be assigned, which include population size POP_SIZE, crossover rate CRO_RATE (probability of crossover), mutation rate MUT_RATE (probability of mutation) and termination criteria (e.g. maximum generations MAX_GEN). Then, the number of generation *genNo* is set to be zero, and two set of individuals are initialized, which are called population $P(genNo)$ (individual set) and archive $A(genNo)$ (elite set) respectively. For the main iteration process, the population and archive of the current generation are firstly combined into a new set, of which the Pareto rank and crowding distance are assigned or calculated for each individual. Based on these two criteria, a new elite set $A(genNo+1)$ is created. Within this new elite set, a mating pool is formed through different selection method (e.g. binary tournament selection) and a new population $P(genNo+1)$ is generated through crossover and mutation from the mating pool. This iterative process will be continuously repeated until given termination criteria are satisfied.

Before describing the detailed procedures of the tailored evolutionary algorithms of the metaheuristic-based optimiser, the data structure of the single individual, has to be properly defined. In this customised evolutionary algorithms, it should not

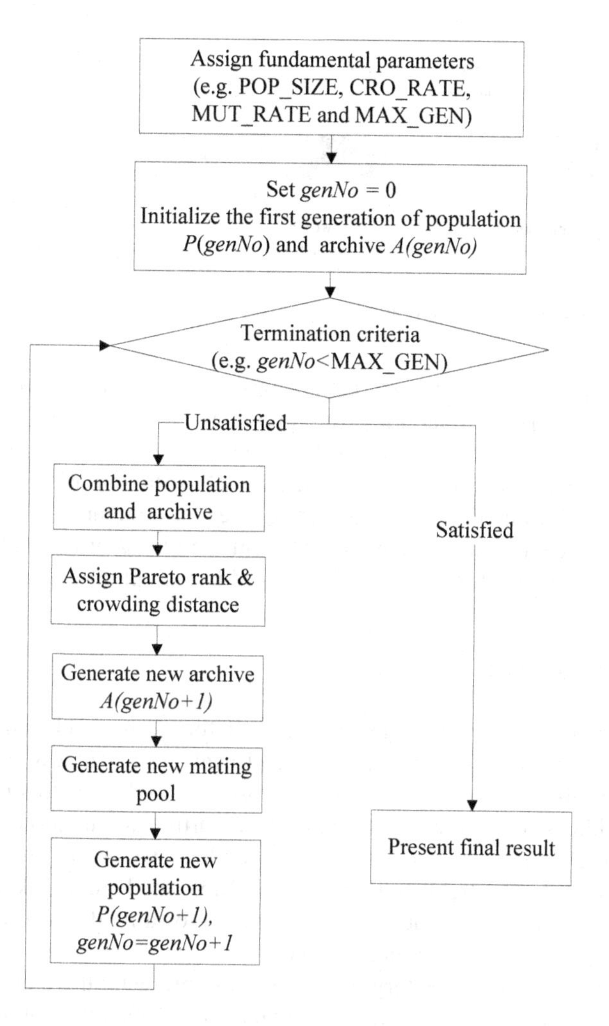

Fig. 6 The General Proccdures of NSGA-II [48]

only be capable of reflecting comprehensive information of a certain solution (multi-echelon inventory policies), but also facilitate the evolutionary process of NSGA-II.

On the one hand, several indispensible components are required for the effective execution of NSGA-II, including chromosome, which is a combination of inventory policies adopted by each stock point, Pareto rank and crowding distance, which are used together to evaluate the fitness of this individual. On the other hand, additional elements can also be integrated into a single individual to manifest the detailed performance measures, although this information is not directly used in NSGA-II (such as on-hand inventory level, out-of-stock level, inventory cost and fill rate of each stock point, and those aggregated over the entire distribution network). As shown in Fig. 7 each square symbol represents a variable. Among them, the grey squares

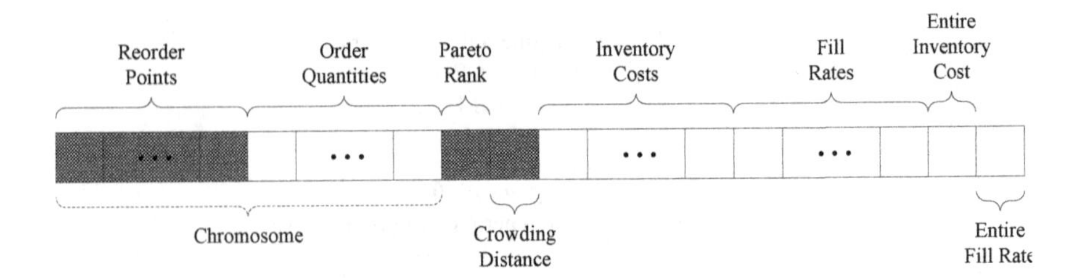

Fig. 7 Data Structure of a Single Individual

stand for compulsory variables while the empty squares stand for non-compulsory variables. Although order quantities are supposed to be a given here, corresponding spaces for decision variables are set aside for future extension. The combination of all decision variables is called chromosome, which is the intrinsic component of a single individual, while the remaining parts of the data structure like Pareto rank or crowding distance are incorporated to represent the fitness of this individual. The sequential procedures of NSGA-II implemented in the optimiser are elaborated as follows:

Individual Evaluation

Traditional evolutionary algorithms usually transform a multi-objective optimisation problem into a single-objective one and evaluate the fitness of individuals according to this objective function. In contrast, NSGA-II tries to simultaneously assess individuals in all dimensions without any artificial alteration or distortion. Correspondingly, there are two criteria of individual evaluation: quality (absolute fitness) and distribution (relative fitness) [48]. While the Pareto ranking method is often employed to measure the quality criterion, crowding distance is a common means of gauging the distribution criterion [48].

In this multi-echelon inventory policy decision problem, there are two incommensurable and conflicting objectives. Inventory cost is measured in monetary unit and is the lower the better, whereas fill rate is evaluated as percentage and is the higher the better. The complete work of Pareto rank assignment involves efficient sorting and successive assignation [20, 31], which can be realized as follows (in the foollowing pseudocodes, underline style is to define a special data format or a class in the actual programming environment, while italic style is to define a certain object):

METHOD: assignParetoRank
INPUT: Set<GenoIndividual> *individualSet*
OUTPUT: List<ParetoFront>

Set an Integer N to be the size of *individualSet*
Sort *individualSet* into a list $\{S_i\}$ so that for any successive individuals S_i and S_j,
$Cost(S_i) < Cost(S_j) \cup (Cost(S_i) = Cost(S_j) \cap Rate(S_i) \geq Rate(S_j))$

Set $E = 1$
Create the first ParetoFront object called PF_1 with S_1, i.e. $PF_1 = PF_1 \cup \{S_1\}$
For $(i = 1; i \leq N; i++)$
    If$(PF_E \prec S_i)$
        Set $E = E + 1$
        Create a ParetoFront object called PF_E with S_i, i.e. $PF_E = \{S_i\}$
    Else
        Search forward the ParetoFront list $\{PF_n\}$ to find the minimum ParetoFront object PF_{min}, that $PF_{min} \nprec S_i$ and add S_i to it, i.e. $PF_{min} = PF_{min} \cup \{S_1\}$
Return $\{PF_E\}$

where the notation $PF_n \prec S_i$ is used to indicate that the individual S_i is dominated by at least one individual in the set PF_n while $PF_n \nprec S_i$ to indicate that there are no individuals in the set PF_n that are able to dominate the individual S_i.

The procedure of crowding distance calculation is simpler, as it only requires the summation of distances in two dimensions: inventory costs and fill rate. Since boundary points of a Pareto front are often beneficial to the exploration of new search areas, respective crowding distances are set to a sufficiently large number M so that the boundary points can be preserved for the next step of archive generation (Fig. 6). With the crowding distance of the i-th individual in a Pareto front defined as $D(S_i)$, the pseudocode is designed as below:

METHOD: calculateCrowdingDistance
INPUT: ParetoFront *paretoFront*
OUTPUT: Void

Set N to be the size of *paretoFront*, and $D(S_1) = D(S_N) = M$, where M is a sufficiently large Integer
For$(i = 2; i < N; i++)$
    Calculate the crowding distance of individual S_i with the following equation:
    $D(S_i) = \frac{Cost(S_{i+1}) - Cost(S_{i-1})}{Cost(S_N) - Cost(S_1)} + \frac{Rate(S_{i+1}) - Rate(S_{i-1})}{Rate(S_N) - Rate(S_1)}$

Archive Generation

Apart from the introduction of special criteria of individual evaluation, another major advantage of NSGA-II is the incorporation of elitism mechanism, which prohibits the best solutions found up till now from being randomly replaced. More specifically, a separate memory called archive is maintained to store the elite solutions during the optimisation. As illustrated in the general procedures of NSGA-II (Fig. 6), once a new generation of population is created, it is combined with the old archive. Afterwards all individuals are evaluated with regard to Pareto rank and crowding distance, so that only the elite individuals are selected to constitute a new generation of archive.

Within the original NSGA-II procedures, individual combination and evaluation are carried out ahead of archive generation (Fig. 6). Under the consideration that a Pareto rank is always considered as a dominant criterion, a certain individual may be abandoned due to its larger Pareto rank. Hence it is not necessary to further calculate its crowding distance. Consequently, it is reasonable to embed the above presented two successive procedures of combination and evaluation into the procedure of archive generation:

METHOD: generateArchive
INPUT: Population *pop*, Archive *arc paretoFront*
OUTPUT: Archive

Combine *pop* and *arc* into a set of GenoIndividual objects called *mergedSet*
Create a set of GenoIndividual objects called *eliteSet*
Call METHOD assignParetoRank with parameter *mergedSet*, which retrieves a list of ParetoFront objects called *paretoFrontList*
Set Integer vars. *currentArcSize* = 0, *nextFrontSize* = 0, and *indexOfNextFront* = 0
While (*currentArcSize* + *nextFrontSize* < POP_SIZE)
    Extract a ParetoFront object called *nextFront* from *paretoFrontList* with parameter *indexOfNextFront*
    Call METHOD calculateCrowdingDistance with parameter *nextFront*
    Add all individuals in *nextFront* to *eliteSet*
    Update the size of current archive, i.e. *currentArcSize*
    Increase the index of *nextFront*, i.e. *indexOfNextFront*
    Update the size of the next Pareto front, i.e. *nextFrontSize*
Extract a ParetoFront object called *nextFront* from *paretoFrontList* with parameter *indexOfNextFront*
Call METHOD calculateCrowdingDistance with parameter *nextFront*
Sort the individuals in *nextFront* according to their crowding distances
While (currentArcSize < POP_SIZE)
    Add the individuals in *nextFront* successively to *eliteSet*
    Increase the size of current archive, i.e. *currentArcSize*
Create an Archive object called *archive* with *eliteSet*
Return *archive*

Mating Pool Generation

After the set of elite multi-echelon inventory policies (i.e. archive) is selected, a new set of multi-echelon inventory policies (i.e. population) can be created through variation operators like crossover and mutation. Nevertheless, according to the standard procedures of NSGA-II, another procedure of mating pool generation is inserted prior to that of population generation (Fig. 6), whose underlying idea is to increase the mating probability of excellent individuals.

Here a useful binary tournament selection is adopted. During each tournament two individuals are picked up randomly with replacement and compared with each other, through which the better one is selected into the mating pool [48]. Similar to the selection process in archive generation, Pareto rank still serves as the primary role while crowding distance as a secondary. The following pseudocode presents its detailed steps.

```
METHOD: generateMatingPool
INPUT: Archive arc
OUTPUT: List<GenoIndividual>

Create a set of GenoIndividual objects called matingPool
Read out all elite individuals, i.e. eliteList, from arc
Set an Integer variable matingPoolSize = 0
While (matingPoolSize < POP_SIZE)
    Select randomly two GenoIndividual objects from matingPool,
    i.e. ind1 and ind2
    Call Method binarySelect with parameters ind1 and ind2,
    which returns a GenoIndividual object called betterInd
    Add betterInd to matingPool
    Update the size of mating pool, i.e. matingPoolSize
Return matingPool
```

To select the better individual from the two candidates, their Pareto ranks are firstly compared. If the two candidates have different ranks, the one with the higher rank is selected. Otherwise, the one with the same rank but larger crowding distance is preferred. The pseudocode of such binary tournament selection is shown as below:

```
METHOD: binarySelect
INPUT: GenoIndividual ind1,GenoIndividual ind2,
OUTPUT: GenoIndividual

Create a GenoIndividual object called betterInd
Extract the Pareto ranks of ind1 and ind2,
which are called rank1 and rank2, respectively
If (rank1 < rank2)
    betterInd=ind1
Else If (rank1 > rank2)
    betterInd=ind2
Else
    Extract the crowding distances of ind1 and ind2,
    which are called dist1 and dist2, respectively
    If (dist1 > dist2)
        betterInd=ind1
    Else If (dist1 < dist2)
        betterInd=ind2
    Else
```

Generate a random real number called *indRand*
If ($indRand \leq 0.5$)
$betterInd = ind_1$
Else
$betterInd = ind_2$
Return *betterInd*

Population Generation

After the mating pool has been formed, a new set of multi-echelon inventory policies (i.e. population) can be generated. Each time two individuals are selected randomly without replacement from the mating pool to perform crossover and mutation. There are numerous kinds of approaches to crossover and mutation, whereas here SBX crossover [18] and polynomial mutation [19] have been applied according to standard NSGA-II. The successive operations of selection, crossover and mutation are repeated until a new population with a defined size of individuals (i.e. POP_SIZE) has been generated:

METHOD: generatePopulation
INPUT: List<GenoIndividual> *matingpool*
OUTPUT: Population

Create a set of GenoIndividual objects called *newIndSet*,
Set an Integer $sizeOfNewIndSet = 0$
While (*newPopSize* < POP_SIZE)
Select randomly the first GenoIndividual object called $parent_1$
Extract an Integer list called $reorderPointList_1$ from $parent_1$
Create an identical clone of $reorderPointList_1$, which is called $clonedReorderPointList_1$
Remove $parent_1$ from *matingpool*
Repeat the same four steps of selection, extraction,
clone and removing for the other parent, i.e. $parent_2$
Generate a random real number *crossoverRand*
If (*crossoverRand* < CRO_RATE)
Call METHOD SBXCrossover with parameters $clonedReorderPointList_2$ and $clonedReorderPointList_2$
Create a GenoIndividual object called $child_1$
Set the reorder point list of $child_1$ with $clonedReorderPointList_1$
Add $child_1$ to *newIndSet*
Do the same three steps of creation, setting and adding for the other child, which is called $child_2$
Update the size of *newIndSet*
Create a Population object called *newPop* with *newIndSet*
Return *newPop*

In this customised evolutionary algorithm, the polynomial mutation operation has been embedded into the SBX crossover operation to reduce the probability of mutation and thus to stabilise the optimisation process. The following pseudocode illustrates the process of SBX crossover.

METHOD: SBXCrossover
INPUT: List<Integer> $reorderPointList_1$, List<Integer> $reorderPointList_2$
OUTPUT: Void

Set the size of $reorderPointList1$ to be N
For $(i = 0; i < N; i++)$
Extract the i-th element of $reorderPointList_1$ and $reorderPointList_2$, which are called $oldValue_1$ and $oldValue_2$, respectively
Generate a random SBX variable called *sbxRand*
Change $oldValue_1$ and $oldValue_2$ through SBX crossover to be $crossedValue_1$ and $crossedValue_2$ with the following equation

$$\begin{aligned} crossedValue_1 &= 0.5 \cdot (oldValue_1 + oldValue_2) \\ &+ 0.5 sbxRand \cdot (oldValue_1 - oldValue_2) \\ crossedValue_2 &= 0.5 \cdot (oldValue_1 + oldValue_2) \\ &+ 0.5 sbxRand \cdot (oldValue_2 - oldValue_1) \end{aligned}$$

Generate a random real number called *mutationRand*
If (*mutationRand* < MUT_RATE)
Set real variables $maxPerturbation_1$ to be $oldValue_1$, $maxPerturbation_2$ to be $oldValue_2$
Call METHOD polynomialMutation with parameters $crossedValue_1$ and $maxPerturbation_1$, which returns a real variable $mutatedValue_1$
Update the i-th element of $reorderPointList_1$ with the up rounded Integer $\lceil mutatedValue_1 \rceil$
Do the same three steps to update that of $reorderPointList_2$ with $\lceil mutatedValue_2 \rceil$

The pseudocode of polynomial mutation is shown as below:

METHOD: polynomialMutation
INPUT: Float $oldvalue$, Float $maxPerturbation$
OUTPUT: Float

Generate a random polynomial variable called *polyRand*
Change *oldValue* through polynomial mutation to be $newValue$ with the following equation

$$newValue = oldvalue + maxPerturbation \cdot polyRand$$

Return $newValue$

With these results the metaheuristic-based optimiser has been completely described. Combing with the other two components of the integrated approach (see Fig. 3) —

analytical inventory policy optimisation and supply chain simulation module — robust multi-echelon inventory policies can be specified. In the next section, an industrial case study is outlined to prove the applicability of the presented approach.

7 Case Study

Within an exemplary distribution network, finished products are manufactured by one single plant centrally and sold in selected 14 European markets. A European distribution center (EDC) and 14 regional distribution center (RDC) hold inventories of the finished products to deliver to end customers. The detailed structure of this exemplary network is illustrated in Fig. 8. The objective is to pursue a high service level while keeping inventory costs as low as possible. As mentioned in Section 1, the focus is laid on the critical performance measures of inventory costs and fill rate. The customer demand is given as one detailed demand forecast scenario which reflects the volatility of the markets. Average and standard deviation of forecasted demand for each regional market may be calculated from this scenario. The order quantity of each distribution center is assumed to be one dayŠs demand. Unit holding cost of EDC is supposed to be 1 monetary unit while that of RDCs to be 2. The resulting basic data is shown in Table 2. Of the 14 European markets, Germany and Great Britain are the first and second largest market respectively, which together contribute to more than three-fifths of the entire demand. Thus, apart from the performance of the entire distribution network, special attention is paid to the European distribution center (EDC) and these two regional distribution centers, i.e. Germany (Code: RDC04) and Great Britain (Code: RDC08).

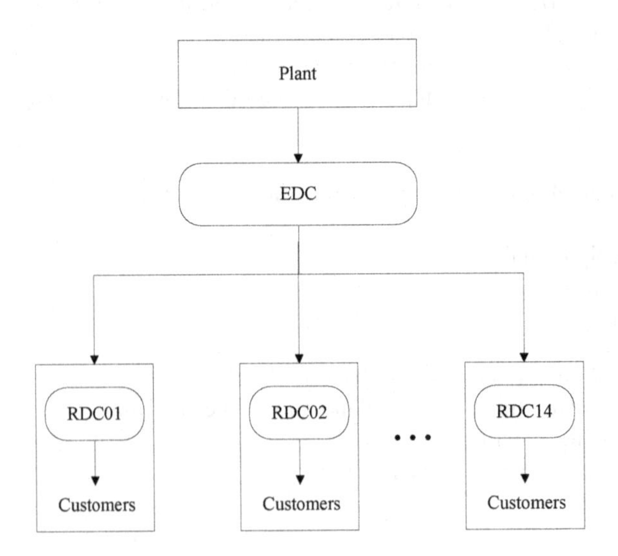

Fig. 8 Structure of the Exemplary Distribution Network

Table 2 Basic Data of EDC and RDCs

Code of Distribution Center	Country	Average Demand	Standard Deviation of Demand	Order Quantity	Unit Holding Costs
EDC		11.49	18.13	12	1
RDC01	Austria	0.66	2.07	1	2
RDC02	Belgium	0.69	2.13	1	2
RDC03	Switzerland	0.31	3.11	1	2
RDC04	Germany	4.87	12.01	5	2
RDC07	France	0.21	0.71	1	2
RDC08	Great Brain	2.56	5.40	3	2
RDC09	Italy	0.27	2.00	1	2
RDC12	Sweden	1.55	4.80	2	2
RDC18	Rheus	0.16	2.59	1	2
RDC19	CZ+SK	0.08	0.75	1	2
RDC20	Poland	0.02	0.42	1	2
RDC21	East	0.03	0.52	1	2
RDC22	Netherlands	0.08	0.76	1	2
RDC26	Hungary	0.01	0.10	1	2

In the following, the proposed integrated approach is applied to this industrial case in order to find robust multi-echelon inventory policies, which is presented according to the procedures of the integrated approach.

Analytical Optimisation

While the unit inventory holding costs are held constant, the "artificial" unit backorder costs can be varied by decision makers in accordance with their preferred service levels. Four exemplary combinations of these two kinds of costs are shown in Table 3. Note that in this exemplary case the unit backorder costs are not varied between the RDCs. With these input parameters, different analytically optimised multi-echelon inventory policies are calculated. These initial solutions are required by the metaheuristic-based optimiser as adequate "starting points" (initial multi-echelon inventory policies). Here, the number of initial multi-echelon inventory policies has been set to 10. The following analysis focuses on the first four solutions (see Table 4). Since the results obtained from analytical optimisation are

Table 3 Combinations of Unit Inventory Costs

Code	Country	Unit Holding Costs	Unit Backorder Costs			
			(1)	(2)	(3)	(4)
EDC		1	3	15	100	5000
RDC04	Germany	2	9	30	200	10000
RDC08	Great Britain	2	9	30	200	10000

Table 4 Analytically Optimised Multi-Echelon Inventory Policies

Code	Country	Analytically Optimized Reorder Point			
		(1)	(2)	(3)	(4)
EDC		16	28	40	58
RDC04	Germany	33	44	61	86
RDC08	Great Britain	23	30	40	54

expected to be compared with that gained from simulation, the same performance measures must be selected, i.e. inventory costs and fill rate. While the former is already known, the latter can be approximately deduced from the average backorder costs. Corresponding to denotations used in section 4, for any stock point k, the mathematical expression of its fill rate f_k can be given by

$$f_k = 1 - \frac{E([I_k]^-)}{\mu_k} \tag{13}$$

The detailed results of the analytical model are presented in Table 5. It is obvious to find that from alternative (1) to alternative (4) the fill rate grows up continuously to near 100%. Yet, this high service level does not come for free: The average inventory costs of the entire distribution network climbs up quickly from 147 to 541. Such findings agree with the expectation that as more severe penalty is imposed on short-of-stock, the resulted multi-echelon inventory policies will lead to higher stock levels and higher fill rate. Although such results comply with expectations, further comparisons can be made to reflect the differences between analytical results and simulation results. To avoid the deviation resulted from measurement scales, the relative difference, instead of absolute difference, is used, which is given by:

$$Difference = \frac{SimulationResult - AnalyticalResult}{AnalyticalResult} \tag{14}$$

As is shown in Table 5, large differences between simulation results and analytical results appear only in the fill rate of RDC08 and entire distribution network of alternative (1). More specifically, almost all of the relative differences of inventory costs are within 10% and those of fill rate are within 20%. Besides, there are two evident characteristics of these differences: the average inventory costs obtained by simulation are mostly higher than those calculated by the analytical model; while the simulated fill rates are clearly lower than the analytical values. These features can be explained by the fact that the real volatile customer demand possesses strong seasonality, interval interruption and abrupt changes. Due to the constraints of the analytical model (e.g., normal demand assumption, METRIC approximation, see Section 4), it is not sufficient to fully grasp the sophistication and intricacy underlying real dynamic distribution networks. Although the analytical model has provided

Table 5 Comparison of Analytical Results and Simulation Results of All 4 Alternatives

Alternative No.	Stock Point	Ave. Inventory Costs			Fill Rate		
		Sim. a)	Ana. b)	Diff. c)	Sim. a)	Ana. b)	Diff. c)
(1)	EDC	11	13	-15,4%	88,2%	81,7%	8,0%
	RDC04	42	44	-4,5%	57,6%	58,2%	-1,0%
	RDC08	22	24	-8,3%	69,9%	52,7%	32,6%
	Network	159	147	8,2%	56,2%	39,5%	42,3%
(2)	EDC	23	23	0,0%	95,6%	96,3%	-0,7%
	RDC04	64	64	0,0%	70,1%	87,7%	-20,1%
	RDC08	38	36	5,6%	81,2%	87,1%	-6,7%
	Network	251	229	9,6%	69,1%	82,7%	-16,5%
(3)	EDC	35	35	0,0%	96,2%	99,6%	-3,4%
	RDC04	98	96	2,1%	83,6%	98,6%	-15,2%
	RDC08	58	56	3,6%	91,1%	98,6%	-7,6%
	Network	375	351	6,8%	80,6%	97,9%	-17,7%
(4)	EDC	53	53	0,0%	97,3%	100,0%	-2,7%
	RDC04	148	146	1,4%	92,0%	100,0%	-8,0%
	RDC08	86	84	2,4%	97,9%	100,0%	-2,1%
	Network	565	541	4,4%	89,9%	100,0%	-10,1%

a) Sim.: simulation result b) Ana.: analytical result c) Diff.: Difference

a relatively good approximation of the real distribution network and is beneficial to the performance prediction and optimisation process, it cannot replace the role of simulation in the accurate evaluation of complex distribution networks with market dynamics. Decision makers must not rely only on analytical models to make final decisions.

Moreover, the analytical model can only offer the average value of performance measures, which might conceal many serious problems. For example, a stock-out situation might last a long time even though the fill rate of the entire distribution network seems to be within boundaries. More detailed information can only be provided by simulation and is beneficial to the performance evaluation. Returning to the previous case, the daily on-hand inventory level of the selected stock points (e.g., EDC, RDC04 & RDC 08) in case of alternatives (1) and (4) are illustrated in Fig. 9 and Fig. 10, respectively. When the line touches the X-axis, stock-out situations occur. From Table 5, for alternative (1) the fill rate level of the EDC may be acceptable (88.2%) (not taking its influence on downstream stock points into account). Yet, the EDC has suffered from stock out for almost two months (54 days). Such stock-out situations occur rarely in alternative (4); nevertheless, much higher safety stock has been built up to tackle with the volatile demand, especially in EDC and RDC08. Thus, simulation can present more comprehensive and informative results of different multi-echelon inventory policies, which are important for reasonable decision making.

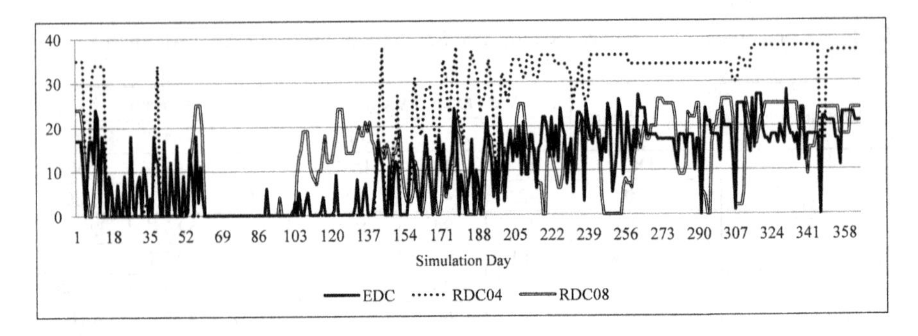

Fig. 9 On-Hand Inventory Level of Alternative (1)

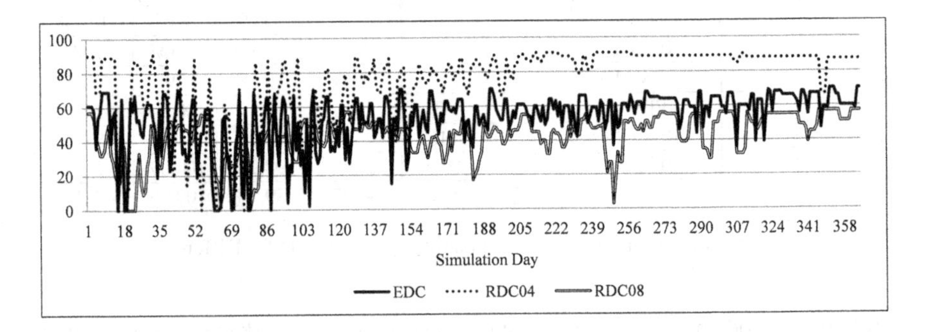

Fig. 10 On-Hand Inventory Level of Alternative (4)

Metaheuristic Optimisation

To further optimise the multi-echelon inventory policies, the customised metaheuristic-based optimiser is applied to evolve alternatives autonomously according to their simulation performances. As mentioned in Section 6.2, the metaheuristic-based optimiser evaluates current multi-echelon inventory policies from two perspectives — Pareto rank and crowding distance. Elite multi-echelon inventory policies are selected and the reproduction process is conducted. This iteration continues until given termination criteria are satisfied.

The simulation results of the ten initial multi-echelon inventory policies (generated by analytical optimisation), which correspond to those different unit backorder costs (see Table 3), are depicted in Fig. 11(a), of which X-axis and Y-axis represent the inventory costs and fill rate for the entire distribution network. Here, inventory costs represent the capital holding costs of all stocks tied up in the distribution network, while the fill rate denotes the service level of all customer orders by the 14 regional distribution centers.

With the 10 initial multi-echelon inventory policies as starting points, the metaheuristic-based optimiser is applied. Fundamental input parameters for the optimiser have been assigned as follows: POP_SIZE = 10, CRO_RATE = 90%, MUT_RATE =

90%, MAX_GEN = 150. For any stock point k, the lower bound of its reorder point R_k^{low} is set to be 0, while the upper bound R_k^{upp} is set with that obtained in the last one of the 10 initial inventory policies, which has been analytically optimised with maximal unit backorder costs. The entire search space includes approximately 5.5×10^{17} points (i.e. inventory policies), and the solutions searched by the metaheuristic optimiser accounts for 2.7×10^{-15} of the entire search space.

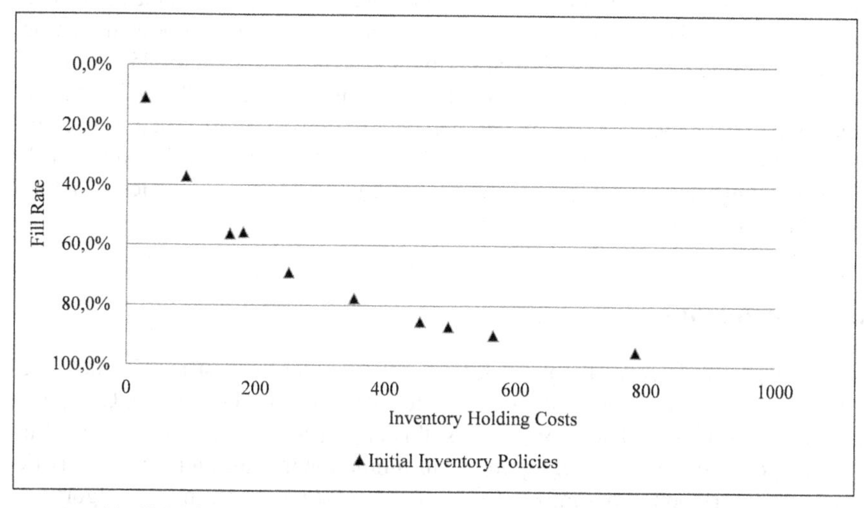

(a) 10 Initial Analytically Optimised Multi-Echelon Inventory Policies

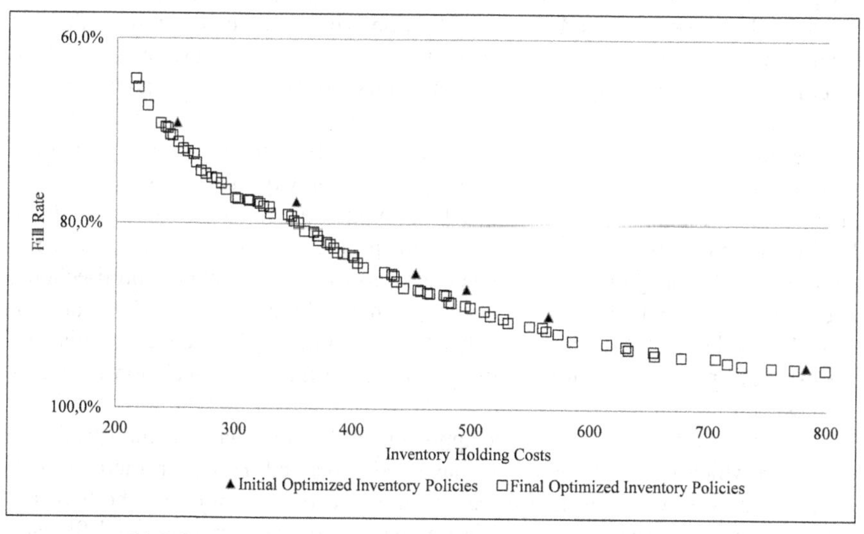

(b) Initial and Final Optimised Multi-Echelon Inventory Policies

Fig. 11 Simulation Results

The simulation results of the initial "optimised" multi-echelon inventory policies (proposed by the analytical optimisation) and the final "optimised" ones (proposed by the metaheuristic-based optimiser) are illustrated in Fig. 11(b). Since the inventory policies that result in too low fill rate are meaningless in real life, only those with fill rates above 60% are drawn in this figure. The simulation results of the optimised multi-echelon inventory policies (white squares) have formed a Pareto optimal front and are completely located below those of the initial ones (black triangles), which shows that these initial multi-echelon inventory policies have been improved simultaneously from both inventory costs and fill rate. For example, one initial optimised multi-echelon inventory policy leads to a fill rate of 95.3% and inventory costs of 783 monetary units, while a higher fill rate of 95.4% is achieved with only 730 monetary units by another final optimised multi-echelon inventory policy. As these simulation results have already reflected the performance of the corresponding multi-echelon inventory policies under market dynamics, a defined robustness is given.

8 Conclusions

In this paper, an integrated approach to the derivation of robust multi-echelon inventory policies for complex distribution networks has been proposed. Composed of three closely interrelated components, this approach integrates an analytical inventory policy optimisation, a supply chain simulation module and a metaheuristic-based inventory policy optimiser. Based on the existing approximation algorithms designed primarily for two-echelon inventory policy optimisation, an analytical multi-echelon inventory model in combination with an efficient optimisation algorithm has been designed. Through systematic parameter adjustment, an initial generation of optimised multi-echelon inventory policies is firstly calculated. To evaluate optimality and robustness of these inventory policies under market dynamics, the suggested inventory policies are automatically handed over to a simulation module, which is capable of modeling arbitrary complexity and uncertainties within and outside of a supply chain and simulating it under respective scenarios. Based on the simulation results, i.e. the robustness of the proposed strategies, a metaheuristic-based inventory policy optimiser regenerates improved (more robust) multi-echelon inventory policies, which are once again automatically and dynamically evaluated through simulation. This closed feedback loop forms a simulation optimisation process that enables the autonomous evolution of robust multi-echelon inventory policies.

Moreover, the proposed integrated approach has been validated in industrial case studies, in which favorable outcomes have been obtained. However, there are still several aspects, to which further research can be devoted. First of all, the two critical performance measures adopted in this paper are inventory costs and fill rate. Although, these KPIs cover the two commonly employed categories of logistic assessment - supply chain costs and supply chain performance [16], other important KPIs can and should be further integrated into this integrated approach so that the optimised multi-echelon inventory policies can be evaluated more comprehensively and systematically. Secondly, the targeted inventory policies of the presented integrated approach are (R, Q) inventory policies, which are frequently employed in continuously reviewed inventory systems. Although this kind of inventory policies have attracted much research interest [49, 11], other types of inventory policies, such as (R, S) or (s, S) policies (e.g., [49, 11]), may be of more practical value. In addition, another form of (R, Q) echelon inventory policies (e.g., [9]) has also been widely analyzed in recent years. Therefore, it is worth extending this integrated approach to other forms of inventory policies. Thirdly, the multi-echelon inventory policies obtained through this integrated approach may indicate that in specific and volatile circumstances the robustness cannot be achieved with one inventory policy. This brings forward the idea of partial policies which are only valid under defined (market) conditions. Hence, a changeable logistics distribution system as sketched by Kuhn et al. [37] can be attained by (self-autonomous) adjustment of policies under changing environments.

Acknowledgements. The Authors gratefully acknowledge the DFG project "Modellbasierte Methoden zur echtzeitnahen Adaption und Steuerung von Distributionssystemen" (MMeAS/KU 619/18-1) for their support. This work was also supported by China Scholarship Council.

References

1. Al-Harkan, I., Hariga, M.: A simulation optimization solution to the inventory continuous review problem with lot size dependent lead-time. Arabian Journal for Science and Engineering 32, 329–338 (2007)
2. Al-Rifai, M.H., Rossetti, M.D.: An efficient heuristic optimization algorithm for a two-echelon (r, q) inventory system. International Journal of Production Economics 109, 195–213 (2007)
3. April, J., Glover, F., Kelly, J.P., Laguna, M.: Practical introduction to simulation optimization. In: Proceedings of the 2003 Winter Simulation Conference, pp. 71–78 (2003)
4. Axsäter, S.: Simple solution procedures for a class of two-echelon inventory problems. Operations Research 38, 64–69 (1990)
5. Axsäter, S.: Exact and approximate evaluation of batch-ordering policies for two-level inventory systems. Operations Research 41, 777–815 (1993)
6. Axsäter, S.: Evaluation of installation stock based (r, q)-policies for two-level inventory systems with poisson demand. Operations Research 46, 135–145 (1998)
7. Axsäter, S.: Exact analysis of continuous review (r, q)-policies in two-echelon inventory systems with compound poisson demand. Operations Research 48, 686–696 (2000)
8. Axsäter, S.: Approximate optimization of a two-level distribution inventory system. International Journal of Production Economics 81-82, 545–553 (2003)
9. Axsäter, S.: Supply chain operation: Serial and distribution inventory systems. In: Handbooks in OR & MS, vol. 11, pp. 525–559. Elsevier (2003)
10. Axsäter, S.: A simple decision rule for decentralized two-echelon inventory control. International Journal of Production Economics 93/94, 53–59 (2005)
11. Axsäter, S.: Inventory control. International Series in Operations Research & Management Science, 2nd edn., vol. 90. Springer (2006)

12. Barton, R.R., Meckesheimer, M.: Metamodel-based simulation optimization. In: Handbooks in OR & MS, vol. 13, pp. 535–574. Elsevier (2006)
13. Caglar, D., Li, C.L., Simchi-Levi, D.: Two-echelon spare parts inventory system subject to a service constraint. Institute of Industrial Engineers 36, 655–666 (2004)
14. Chopra, S., Meindl, P.: Supply chain management: Strategy, planning, and operation, 4th edn. Pearson, Boston (2010)
15. Chopra, S., Sodhi, M.S.: Managing risk to avoid supply chain breakdown. Sloan Management Review 46(1), 53–61 (2004)
16. Cirullies, J., Klingebiel, K., Scarvarda, L.F.: Integration of ecological criteria into the dynamic assessment of order penetration points in logistics networks. In: Proceedings of 25th European Conference on Modelling and Simulation, Krakow, Poland, pp. 608–615 (2011)
17. Cohen, M., Kamesam, P.V., Kleindorfer, P., Lee, H., Tekerian, A.: Optimizer: Ibm's multi-echelon inventory system for managing service logistics. Interfaces 20, 65–82 (1990)
18. Deb, K., Agrawal, R.B.: Simulated binary crossover for continuous search space. Complex Systems 9, 115–148 (1995)
19. Deb, K., Goyal, M.A.: Combined genetic adaptive search (geneas) for engineering design. Computer Science and Informatics 26, 30–45 (1996)
20. Deb, K., Pratap, A., Agarwal, S., Meyarivan, T.: A fast and elitist multiobjective genetic algorithm: Nsga-ii. IEEE Transactions on Evolutionary Computation 6(2), 128–197 (2002)
21. Deuermeyer, B., Schwarz, L.B.: A model for the analysis of system service level in warehouse/retailer distribution systems: The identical retailer case. In: Multi-Level Production/Inventory Control Systems: Theory and Practice, North-Holland, Amsterdam, pp. 163–193 (1981)
22. Dong, M., Chen, F.F.: Performance modeling and analysis of integrated logistic chains: An analytic framework. European Journal of Operational Research 162(1), 83–98 (2005)
23. Fonseca, C.M., Fleming, P.J.: An overview of evolutionary algorithms in multiobjective optimization. Evolutionary Computation 3(1), 1–16 (1995)
24. Fu, M.C.: Simulation optimization. In: Proc. of the 2001 Winter Simulation Conference, pp. 53–61 (2001)
25. Fu, M.C.: Optimization for simulation: Theory vs. practice. INFORMS Journal on Computing 14(3), 192–215 (2002)
26. Fu, M.C., Andradóttir, S., Carson, J.S., Glover, F., Harrell, C.R., Ho, Y.C., Kelly, J.P., Robinson, S.M.: Integrating optimization and simulation: research and practice. In: Proc. of the 2000 Winter Simulation Conference, pp. 610–616 (2000)
27. Glover, F., Kelly, J.P., Laguna, M.: New advances and applications of combining simulation and optimization. In: Proc. of the 28th Conference on Winter Simulation, pp. 144–152 (1996)
28. Graves, S.C.: A multi-echelon inventory model for a repairable item with one-for-one replenishment. Management Science 31, 1247–1256 (1985)
29. Graves, S.C., Rinnooy Kan, A.H.G., Zipkin, P.H.E.: Handbooks in Operations Research and Management Science: Logistics of Production and Inventory. Elsevier, Amsterdam (1993)
30. Hopp, W.J., Spearman, M.L., Zhang, R.Q.: Easily implementable inventory control policies. Operations Research 45, 327–340 (1997)
31. Jensen, M.T.: Reducing the run-time complexity of multiobjective eas: The nsga-ii and other algorithms. IEEE Transactions on Evolutionary Computation 7(5), 503–515 (2003)

32. Kang, J.H., Kim, Y.D.: Coordination of inventory and transportation managements in a two-level supply chain. International Journal of Production Economics 123(1), 137–145 (2010)
33. Kiesmüller, G.P., de Kok, A.G.: A multi-item multi-echelon inventory system with quantity-based order consolidation. Working Paper, Eindhoven Technical University (2005)
34. Klingebiel, K.: Entwurf eines Referenzmodells für Built-to-order-Konzepte in Logistiknetzwerken der Automobilindustrie. Unternehmenslogistik. Praxiswissen, Dortmund (2009)
35. Klingebiel, K., Li, C.: Optimized multi-echelon inventory policies in robust distribution network. In: Proc. of 25th European Conference on Modelling and Simulation, Krakow, Poland, pp. 573–579 (2011)
36. Konak, A., Coit, D.W., Smith, A.E.: Multi-objective optimization using genetic algorithms: A tutorial. Reliability Engineering & System Safety 91(9), 992–1007 (2006)
37. Kuhn, A., Klingebiel, K., Schmidt, A., Luft, N.: Modellgestütztes planen und kollaboratives experimentieren für robuste distributionssysteme. In: Tagungsband des 24. HAB-Forschungsseminars der Hochschulgruppe für Arbeits- und Betriebsorganisation, Stuttgart (2011)
38. Miranda, P.A., Garrido, R.A.: Inventory service-level optimization within distribution network design problem. International Journal of Production Economics 122(1), 276–285 (2009)
39. Ólafsson, S.: Metaheuristics. In: Handbooks in OR & MS, vol. 13, pp. 535–574. Elsevier (2006)
40. Pasandideh, S.H.R., Niaki, S.T.A., Tokhmehchi, N.: A parameter-tuned genetic algorithm to optimize two-echelon continuous review inventory systems. Expert Systems with Applications 38(9), 11, 708–711, 714 (2011)
41. Sherbrooke, C.C.: Metric: a multi-echelon technique for recoverable item control. Operations Research 16, 122–141 (1968)
42. Simchi-Levi, D., Zhao, Y.: Safety stock positioning in supply chains with stochastic lead times. Manufacturing & Service Operations Management 7, 295–318 (2005)
43. Svoronos, A., Zipkin, P.: Estimating the performance of multi-level inventory systems. Operations Research 36(1), 57–72 (1988)
44. Tekin, E., Sabuncuoglu, I.: Simulation optimization: A comprehensive review on theory and applications. IIE Transactions 36(11), 1067–1081 (2004)
45. Tempelmeier, H.: nventory management in supply networks. Problems, models, solutions. Books on Demand, Norderstedt (2006)
46. Terzi, S., Cavalieri, S.: Simulation in the supply chain context: a survey. Computers in Industry 1, 3–16 (2004)
47. Wagenitz, A.: Modellierungsmethode zur Auftragsabwicklung in der Automobilindustrie. Unternehmenslogistik. Verl. Praxiswissen, Dortmund (2007)
48. Yu, X., Gen, M.: Introduction to evolutionary algorithms. Decision Engineering. Springer, London (2010)
49. Zipkin, P.H.: Foundations of Inventory Management. McGraw-Hill, New York (2000)

Survey of Background Normalisation in Affymetrix Arrays and a Case Study

Vilda Purutçuoğlu, Elif Kayış, and Gerhard-Wilhelm Weber

Abstract. The oligonucleotide is a special type of one-channel microarrays which has 25-base pair long and the Affymetrix is the well-known oligonucleotide. In this study the normalisation procedure which enables us to discard the systematic erroneous signals in the measurements is described. Then different gene expression indices which are used to compute the true signals via the background normalisation are described in details. In these descriptions two recently suggested alternative approaches, namely frequentist (FGX) and robust (RGX) gene expression indices are explained besides the well-known approaches in this field. Finally a comparative analysis of the real microarray data with different sizes is presented to evaluate the performance of the underlying methods with RMA which is one of the common techniques in the analysis of microarray studies.

1 Introduction

The microarray technology is a new technique that is used in biological researches to determine expression levels of many genes simultaneously. The structure of microarray includes many single strands of a gene sequence, taken as RNA, on its surface at the specific locations and these strands are called *probes*. Here the RNA gene sequences are attached on the array either by spotting or immobilizing via hydroxylation. In microarray the RNA's are extracted from the interested cells and spread on the array as the complementary of single strand DNA's. If RNA's find its

Vilda Purutçuoğlu
Middle East Technical University, Department of Statistics,
06531, Ankara, Turkey
e-mail: vpurutcu@metu.edu.tr

Elif Kayış · Gerhard-Wilhelm Weber
Middle East Technical University, Institute of Applied Mathematics,
06531, Ankara, Turkey
e-mail: elif.kayis@metu.edu.tr, gweber@metu.edu.tr

A. Byrski et al. (Eds.): Advances in Intelligent Modelling and Simulation, SCI 416, pp. 199–220.
springerlink.com

complementary DNA parts on the array, they bind to them. The underlying process of completion between RNA's and DNA's is called hybridisation [19] which enables us to measure the gene expression levels, i.e. the amount of RNA's coded by the interested genes.

In this study we describe the stages of normalisation techniques which are used to eliminate the erroneous signals of the underlying gene expressions from the true signals. Then we present one of these stages, called the *background normalisation* technique that aims to measure the true signal from the arrays in details. As the novelty, we implement two recently developed background methods, called FGX and RGX, in the analysis of a real dataset. Finally we compare our findings with previous analysis of the data. But before these computations and mathematical details, we initially represent several biological definitions such as the structures of DNA and RNA, and experimental details that are commonly used in this type of analyses [19, 3, 4].

2 Biological Background

The DNA refers to the code of the alives and has a double helix structure like a ladder whose sides have the same elements. These elements are backbone of sugar, phosphate molecules, and nucleotide bases, namely, Adenin (A), Timin (T), Guanin (G), and Cytosine (C). Among them A and T bases, G and C bases are complementary bases of each other. In the binding of these pairs the two sides are bound by weakly hydrogen bonds [16].

In a cell the proteins are produced by the message delivered from DNA via RNA's. The RNA is also a complementary string for DNA whose structure, i.e. complementarity, is the basis of the microarray experiments in the sense that the information about the activities of genes in a cell is delivered to RNA by creating a complementary RNA copy of a sequence of DNA. This procedure is called the *transcription*. Later this information is translated into proteins. Accordingly the microarray experiments use the amount of the transcribed RNA in a cell or tissue to measure the expression levels of genes. In an experiment the RNA's are typically extracted from two different kinds of cells (or tissues). One is called the *control* and the other is named as the interested *treatment* cell [19].

3 Experimental Background

The experimental set-up of a microarray consists of the following steps.

3.1 Gathering Information from RNA

In order to measure the amount of transcribed RNA in a microarray experiment, some organic materials in the cytoplasm is extracted from both control and treatment cells. In this liquid, there may be other contamination of ingredients such as proteins. The ratio of RNA's to proteins in the contaminated sample is measured by spectrophotometry in such a way that a sample with ratio close to two is accepted as clean [16, 19]. This cleaning process, also called the *purification*, is repeated until the ratio of RNA over proteins is reached. Then the remaining mixture for both treatment and control of cells can be concentrated using precipitation. During this process, the RNA is kept in low temperature to prevent its degradation.

3.2 Preparation of RNA Mixture and Labelling

In the microarray experiments, some kinds of fixed points are determined both to control the lowest compatible amount of signals from genes and to check the lowest amount of gene quantities in the mixture. Hereby the RNA's from certain unrelated genes are added to the RNA mixture of samples and the observed levels are measured on the microarray. This process is called the *spiking*.

On the other hand, in some experiments, cDNA, i.e. copy DNA, molecules are taken instead of RNA's since the former are more stable than the latter. Basically the cDNA molecules are inverse copies of RNA's and are produced by an inverse machine, called the *enzyme*.

In a microarray experiment, a *poly-T primer* is also added to the mixtures of treatment and control samples as it enables us to trigger the hybridisation. Then the mixtures are heated up to prevent RNA's from self-hybridisation. Then they are quickly cooled so that RNA's cannot hybridize again [16, 19]. After the hybridisation of RNA's (or cDNA molecules), it may be difficult to measure how many of them are hybridised directly from microarrays. Thus in order to track the signals from genes, the RNA (or cDNA) molecules are signed by dye molecules.

In the following part we present the preprocessing, also known as the *normalisation*, steps of the microarray experiments which is recommended before the data analysis.

4 Normalisation

In the microarray data we can observe two sources of variations, namely, *random* and *systematic* erroneous fluctuations. The normalisation is a technique to decrease the systematic errors in a microarray dataset [19], resulting in removing the artificial bias in the data and producing more reliable measurements for the analysis. Whereas the detection of erroneous signals can be also dependent on the scale of the measurements. In microarray studies we typically prefer the logarithmic scale as it enables us to better observe the fold changes in the gene expressions. Moreover it can satisfy the robustness against the outlier genes since it limits the possible range of intensities in a narrower bound. This feature is also helpful for observing the linearity between the changes in the intensity of spots, i.e. probes, and the changes in the amount of transcribed RNA's under medium intensities [19, 15]. Whereas in certain normalisation methods, the estimates of the true signals are found on the original scale under the assumption of gamma or exponential distributions [11, 12].

Apart from the adjustment of the appropriate scaling of data, there are some other control mechanisms to check the possible problematic signals on the array. One of them is called the *control spots* which are used to capture the irrelevant signals between the unrelated and interesting genes. Hereby expecting very low signals in these probes, any high estimated intensities can be seen as the indication of systematic bias in the measurements [17, 19]. On the other hand the damage of the array may also cause faulty or no signals in the measurements. Therefore such outliers probes need to be detected before the analysis and should be either excluded from the data or replaced with the imputed values [19, 11].

In a microarray study there are four major stages of normalisation which are listed in details as below. Among these steps only the dye normalisation is implemented in the two-channel, i.e. two-condition, microarray, whereas, the remainings are performed in both one-channel and two-channel microarray studies.

4.1 Spatial Normalisation

After scanning the arrays, we can observe some probes whose intensities indicate big differences with respect to the remaining probes of the array. These differences are sometimes caused by spotting biologically related genes, control genes, or both. But the underlying erroneous intensities do not lead to a bias unless they are originated by technical variances such as an uneven wash of chip, i.e. arrays, or scanning defaults. In order to avoid such kind of bias sources, these spatial effects are need to be normalised. There are three main methods which can be used to make correction in the spatial effects.

The first method assumes that the spatial effect is multiplicative and each array, i.e. treatment and control, is affected by the same multiplicative function C. Here if the location of each probe is represented by (x,y) coordinates, $S(x,y)$ denotes the observed intensities of both arrays in the probe which locates on the (x,y) coordinate. Moreover $g(x,y)$ is the true intensities of both arrays and $C(x,y)$ presents the multiplicative function of that probe. Thereby the following relationship between intensities can be described via S and g components such that $g(x,y)C(x,y) = S(x,y)$. Then by taking ratio of the observed intensities from both treatment and control arrays, we can find the ratio of true intensities since it is assumed that the spatial effect should be similar in each array. As the second method, the substraction of a mean term from the edge of arrays is implemented [21]. Although this approach can be useful for discarding some sorts of spatial effects, it may lead to bias on the edges. Finally the third method, called the robust smoothing approach, can remove the spatial effects which are not discarded via the multiplicative method. Here we implement a locally weighted polynomial regression (loess) where a smooth spatial trend is removed from the data in such a way that the resulting transformed data are free from the first and second-order spatial effects [21, 4].

4.2 Background Normalisation

In the normalisation of microarrays, although there is no complementary DNA on a probe, the labeled cDNA's or RNA's can sometimes hybridize to the array surface when there is no probe on the chips or alternatively they can hybridize partially to the non-complementary genes, resulting in erroneous signals. In order to measure these faulty signals, each gene is defined by a *probe pair* whose elements are named as the *perfect matches* (PM) and *mismatches* (MM) probe, and each gene is composed of 11 to 20 probe pairs in oligonucleotide Affymetrix arrays. The *oligonucleotide* is a type of single channel microarray which has 25-base pairs long and Affymetrix GeneChip is the most common oligonucleotide microarray. In this type of arrays the PM probes are used as the complementary probes of RNA's, hereby are though to measure the true signals. Whereas the MM probes are designed as a base change in the 13 entry of the associated perfect matches so that the noisy signals can be detected.

There are several methods to eliminate or normalize the underlying background signals and get the true intensities from the arrays. These methods are known as the *gene expression indices*. In the following part we initially present some of well-known background methods in details. Then we are mainly interested in two recently suggested gene expression indices, namely FGX and RGX, as the background normalisation approaches and as the novelty, we apply them in real datasets with different sizes. We compare our results with the findings from the RMA method which is one of the common background normalisation approaches in the oligonucleotide analysis.

4.2.1 MAS 5.0 Method

The MAS 5.0 (Microarray Suite Software) background normalisation method [6] formalizes the true signal intensity T for gene i on the probe j as

$$log(T_{ij}) = log(PM_{ij} - S_{ij})$$

where PM$_{ij}$ and S_{ij} refer to the signal intensity of perfectly matched and background signal, also called the *stray signal*, respectively, while $i = 1,\ldots,n$ and $j = 1,\ldots,m$. In MAS 5.0 the stray signal is modelled as $S_{ij} = MM_{ij}$ if $PM_{ij} \geq MM_{ij}$ and is computed by

$$\log(S_{ij}) = log(PM_{ij}) - SB_i^+ \tag{1}$$

while $PM_{ij} \leq MM_{ij}$. Here $SB_i^+ = T_{bi}[\log(PM_{ij}) - \log(MM_{ij})]$ is a robust estimate of the log intensity for gene i where T_{bi} presents the one-step Tukey biweight estimator of location [1]. For brevity, T_{bi} downweights the observed intensity in the jth probe, i.e. x_j, according to the distance from its median. Hereby each x_j is calculated as $x_1 = \log(PM_{i1}) - \log(MM_{i1}),\ldots,x_m = \log(PM_{im}) - \log(MM_{im})$ in the sense that a distance measure, u_j, is taken by dividing each x_j by median absolute deviation. If $0 \leq |u_j| \leq 1$, the weighted function for that data point x_j is calculated by

$w(u_j) = (1-u_j^2)^2$. If $|u_j| \geq 1$, the weight of x_j is set to zero. Accordingly T_{bi} is described as follows.

$$T_{bi} = \frac{w(u_j)x_j}{w(u_j)}. \tag{2}$$

In the MAS 5.0 model, SB_i^+ value must be greater than zero for reasonable true signals. Thereby a negative value of SB_i^+ refers to only presence of stray signal intensity for gene i without reasonable true signals. As a result, in order to get a positive SB_i^+ value, the median of x_j's is used as the threshold value γ.

On the other hand if $SB_i^+ > \gamma$, SB_i^+ is taken as its original value in Eq. (1). Whereas if $SB_i^+ < \gamma$, it is adjusted such that $SB_i^+ = \gamma/(1+0.1(\gamma - SB_i^+))$.

In general the MAS 5.0 method uses only stray signals for the MM values and has a smart way of adjustment in observed signal when $SB_i^+ < \gamma$ [7], whereas, this calibration may also lead to biased gene expression levels in inference of the true signals.

4.2.2 MBEI (dChip) Method

The MBEI (Model Based Gene Expression Index) method [10] suggests a multiplicative model for the observed signal for each pair of probe. In this approach it is assumed that there is a linear relation between the observed intensity and the model expression index θ_k for a gene in the ith probe of the kth array. This linear relation differs in each probe and has an increasing property at high level of intensities, i.e at high gene expression levels. Moreover in this approach it is accepted that PM values have an additional increasing rate with gene expression index. Then the whole model of MM and PM values for probe i on the kth array is expressed as

$$MM_{kj} = v_j + \theta_k \alpha_j + \varepsilon_{kj}^m,$$
$$PM_{kj} = v_j + \theta_k \phi_j + \varepsilon_{kj}^p$$

where v_j is the baseline response caused by the non-specific hybridisation, α_j denotes the rate of increase for the MM response, and ϕ_j stands for the additional rates of PM values. Finally, ε_{kj}^m and ε_{kj}^p present random errors of MM and PM values, respectively. Hereby the true signal intensity value of the probe j is modelled as follows.

$$Y_{kj} = PM_{kj} - MM_{kj} = \theta_k \phi_j + \varepsilon_{kj}$$

Regarding the idea of the MAS 5.0 method, the MBEI method is computed on the original scale. Moreover the random error ε_{kj}'s are assumed to be normally distributed with mean zero and variance σ^2, and the estimation of parameters ϕ_j's is found by the least square estimator method [10]. On the other hand from comparative studies, it is shown that this method is stable for the experiments which include at least ten arrays and unstable for experiments that have large number of arrays with single outliers. But it can easily identify unusual probes and arrays as it models the intensity under the probe and array specific effects. Finally it is found that this method can outperform MAS 5.0 in terms of accuracies of estimated parameters [10].

4.2.3 RMA Method

The RMA (Robust Microarray Analysis) [8] represents a model in which the PM intensity value of a gene i, probe j, and array k is composed of the background signal intensity b_{kij} and the true signal intensity s_{kij}. This method uses the non-specific hybridisation measurements for the evaluation of the MM probe while MAS 5.0 and MBEI methods apply stray signal measurements for the same intensity values. Thereby the true signal from PM is estimated as follows.

$$S^{*}_{kij} = E(S_{kij}|S_{kij} + b_{kij}) = E(S_{kij}|PM_{kij})$$

in which S_{kij} is the exponentially distributed true signal and b_{kij} refers to the normally distributed non-specific hybridisation with mean $E(b_{kij}) = \beta_k$. In order to calculate the expression value of a gene i, firstly, the quantile normalisation is applied to the expected true signal intensity so that we can get the same probe intensity distribution for each array. Secondly the normalised value S^{**}_{kij} is transformed to base 2 log scale for computational accuracy, i.e. $\log_2(S^{**}_{kij})$, so that a linear additive model can be described via

$$\log_2(S^{**}_{kij}) = \mu_{ki} + \alpha_{kj} + \varepsilon_{kij}$$

where α_{kj} gives the jth probe effect, μ_{ki} denotes the expression value of a gene i at the kth array on the log scale, and ε_{kij} is the normally distributed random error with mean zero for gene i.

Among previous methods, the RMA index produces smaller deviations, especially, for genes which give low intensities. Moreover it produces more consistent fold-change estimates, even if the concentration levels are different. It also has a high sensitivity to determine differentially expressed genes via the ROC (Receiving operating characteristic) curves [8]. On the other hand, as its limitation, it is not an efficient method for large number of arrays like the MBEI method since the inference is based on the least square method.

4.2.4 GC-RMA Method

The GC-RMA (Robust Microarray Analysis Based on GC Content) method [20], an extended version of RMA, defines MM probe intensities by assuming that the MM values do not give only background signal, but also possess some information about the true signal S. In fact the GC-RMA method is the first method which says that MM probes have relationship with the true signals. In this method PM's are calculated by summation of optical noise (O), non-specific hybridisation noise N, and the true signal intensity S. On the other hand MM's are supposed to include a fraction ϕ of the true signal via the following model.

$$PM = O_{PM} + N_{PM} + S,$$
$$MM = O_{MM} + N_{MM} + \phi S.$$

Here the optical noise O_{PM} and O_{MM} are though to have log-normal distribution. Moreover N values are assumed to be independent functions of probe affinity α where α is the summation of position-dependent base effects.

The GC-RMA uses an empirical statistical inference method in the estimation of model parameters S and ϕ in the sense that it infers the parameters via both maximum likelihood estimation (MLE) and empirical Bayesian (EB) approaches. But from comparative analyses it is found that the GC-RMA with MLE method outperforms MAS 5.0, RMA, and GC-RMA with empirical Bayesian approach in terms of precision, whereas, the GC-RMA with EB outperforms GC-RMA based on MLE, RMA, and MAS 5.0 with respect to the accuracy of estimates. Additionally GC-RMA based on EB can easily detect differentially expressed genes in lower concentrations, on the contrary, in higher concentrations the RMA method performs better [20].

4.2.5 gMOS Method

The gMOS (Gamma Model for Oligonucleotide Signal) gene expression index [12] assumes that PM and MM probe intensities come from a gamma distribution and it is different from the pre-mentioned methods as it is based on a probabilistic model. In this approach it is still accepted that MM probes do not give information about the true signal intensity, whereas, merely consists of non-specific hybridisation.

Accordingly the PM and MM values of the jth probe of a probe set k are modelled as

$$PM_{kj} = S_{kj} + h_{kj} \text{ and } MM_{kj} = h_{kj} \tag{3}$$

considering that PM_{kj} and MM_{kj} are the random values from $Ga(\alpha_k + a_k, b_k)$ and $Ga(a_k, b_k)$, respectively. Here α_k presents the parameter for the true signal intensity and a_k shows the parameter for the background signal intensity for probe set k. Finally in Eq. (3), S_{kj} and h_{kj} refer to the true signal and non-specific hybridisation for the kth gene and jth probe, in order.

4.2.6 mgMOS Method

The mgMOS (Modified gMOS) method [11] differs from the method gMOS by defining a correlation between PM and MM intensities within in a probe set. In this method PM's and MM's are assumed to come from a joint probability density such that

$$P(PM_{ij}, MM_{ij}) = \int p(b_{ij}) P(PM_{ij}, MM_{ij} | a_i, \alpha_i, b_{ij}) db_{ij}$$

where $b_{ij} \sim Ga(c_i, d_i)$. The b_{ij}'s are latent variables representing different binding affinity of probe pairs within the probe set. Hereby this modified model effectively finds the correlated changes in these probe effects caused by similar content of PM and MM probe sequences within the probe pair.

Although the mgMOS produces accurate estimates in both benchmarks and real datasets and is computationally efficient since the likelihood function of data has a closed form [11], it can computes single, rather than multiple, arrays at a time. Furthermore MM probes are only modelled by non-specific hybridisation meaning that it ignores the presence of true signals in these probes.

4.2.7 BGX Method

After the GC-RMA approach, the BGX (Bayesian Gene Expression Index) index [5] is the second method which assumes fraction ϕ of a true signal intensity as an additive effect on the MM intensity. In GC-RMA, although it is supposed that this fraction is zero, BGX considers ϕ from a beta distribution B(1,1) lying between 0 and 1, $0 < \phi < 1$. Hereby BGX models PM and MM values of the gene i on the probe j by using a normal distribution as below.

$$PM_{ij} \sim N(S_{ij} + H_{ij}, \tau^2)$$
$$MM_{ij} \sim N(\phi S_{ij} + H_{ij}, \tau^2)$$

where S_{ij} and H_{ij} represent the specific binding hybridisation and the non-specific hybridisation of the signal intensity, in order. On the other hand τ^2 denotes the constant variance of each probe. In this model S_{ij} and H_{ij} are modelled by truncated normal distribution on the log scale by guarantying the non-negativity via $\log(S_{ij} + 1)$ and $\log(H_{ij} + 1)$ in place of $\log(S_{ij})$ and $\log(H_{ij})$, respectively, such that

$$\log(S_{ij} + 1) \sim TN(\mu_i, \sigma_i^2) \quad (4)$$
$$\log(H_{ij} + 1) \sim TN(\lambda, \eta^2). \quad (5)$$

In Eq. (5), the hyperparameters of the mean terms, i.e. μ_i and λ, are defined from uniform and normal densities, in order. On the other side the associated hyperparameters of variance and precision terms, i.e. $\log(\sigma_i^2)$ and η^2, are generated from normal and gamma distributions whose parameters can be found from the dataset [5]. In inference this method applies the medians of the truncated normal distributions as the measures of gene expressions for the ith gene, θ_i's, via

$$\theta_i = \mu_i - \sigma_i \phi^{-1}(\mu_i / 2\sigma_i)$$

where n stands for the total number of genes in each array while $i = 1, \ldots, n$. Finally from the analysis in benchmark data, it is seen that the BGX model performs better than MAS 5.0 and MBEI approaches in ranking differentially expressed genes. Whereas it is computationally demanding since the inference is based on the Bayesian techniques.

4.2.8 Multi-mgMOS Method

To overcome the computational cost of BGX and the insufficient number of arrays like the absence of the true signal effect in MM values as mgMOS suggests, Liu et al. [11] are proposed the multi-mgMOS method. This method assumes that each PM and MM probe comes from a gamma distribution and the MM probe includes a fraction ϕ of the true signal intensity as the following way.

$$PM_{gjc} \sim Ga(\alpha_{gc} + a_{gc}, b_{gj}), \tag{6}$$

$$MM_{gjc} \sim Ga(\phi\alpha_{gc} + a_{gc}, b_{gj}) \tag{7}$$

in which PM_{gjc} and MM_{gjc} represent PM and MM intensity values of the jth probe pair in the gth probe set on the cth array. b_{gjc} shows a probe-specific parameter and comes from gamma distribution $Ga(c_g, d_g)$ with probe-specific parameters c_g and d_g. The specific parameter b_{gj} is shared across the arrays for each probe pair. Therefore it makes the binding affinity in the mgMOS [11] observable. On the other hand from Eqs. (6) and (7), the true signals for the jth probe pair in the gth probe set, and on the cth array are described by a gamma distribution via $S_{gjc} \sim Ga(\alpha_{gc}, b_{gj})$. In inference of ϕ the method performs the maximum a posterior estimate (MAP) under log normal prior. Then the logarithm of the posterior probability of ϕ is maximised by using PM and MM intensities to estimate other model parameters, namely, α_{gc}, a_{gc}, c_g, and d_g. In the calculation of parameters the distribution of the signal S_{gjc} is formulated as

$$p(S_{gjc}|\alpha_{gc}, c_g, d_g) = \int p(S_{gjc}|\alpha_{gc}, b_{gj})P(b_{gj}|c_g, d_g)db_{gj}.$$

From the analysis it is observed that although multi-mgMOS has lower computational cost than the BGX approach, the estimation of parameters is still costly for large dataset due to its Bayesian settings. On the other hand, with respect to the performance of mgMOS, the multi-mgMOS method gets more efficient results in terms of the accuracy of estimates.

4.2.9 FGX Method

The FGX (Frequentist Gene Expression Index) model [15] is an alternative approach of the BGX method. In this approach, similar to BGX, it is assumed that there is an information about the true signal intensity in the MM values with a fraction ϕ. In other words PM's and MM's are correlated by having a common gene expression signal. But the FGX method uses the maximum likelihood (ML) method, rather than Bayesian setting, to estimate ϕ, true signal intensities, and non-specific hybridisation. Moreover it applies the Fisher information matrix to compute the covariances between model parameters.

Hereby the FGX method describes the logarithm of PM and MM values of the ith gene on the jth probe from normal distribution such that

$$\log(PM_{ij}) \sim N(S_i + \mu_H, \sigma^2) \quad (8)$$

$$\log(MM_{ij}) \sim N(\phi S_i + \mu_H, \sigma^2) \quad (9)$$

in which $i = 1,\ldots,n$ and $j = 1,\ldots,m$ under totally n genes and m probes per gene. Here S_i presents the true signal intensity for the ith gene, ϕ is the fraction of the specific hybridisation for the MM intensities, and μ_H shows the mean of the non-specific hybridisation intensities. From Eqs. (8) and (9), since the averages of log-transformed PM and MM probes are sufficient statistics for mean parameters and the Affymetrix data are analyzed generally on a probe set, rather than an individual probe, the summary statistics $PM_i = (\sum_{j=1}^{m} \log(PM_{ij}))/m$ and $MM_i = (\sum_{j=1}^{m} \log(MM_{ij}))/m$ can be used in inference via

$$PM_i \sim N(S_i + \mu_H, \sigma^2) \ and \ MM_i \sim N(\phi S_i + \mu_H, \sigma^2).$$

In the estimation of model parameters ϕ, S_i, and μ_H, as this method performs the ML approach, the log-likelihood function conditional on $PM = (PM_1,\ldots,PM_n)$ and $MM = (MM_1,\ldots,MM_n)$ is formulated by

$$\begin{aligned} \ell(S,\mu_H,\phi|PM,MM) &= n\ln(m) - n\ln(2\pi) - 2n\ln(\sigma) \\ &\quad - \frac{m}{2\sigma^2}\sum_{i=1}^{n}[(PM_i - S_i - \mu_H)^2 + (MM_i - \phi S_i - \mu_H)^2]. \end{aligned}$$

Then in order to find the maximum likelihood estimators, ℓ is partially derived with respect to each parameter and is equated to zero. Accordingly the associated estimated values of μ_H and S_i which maximize ℓ are found by

$$\hat{\mu_H} = \frac{(\overline{PM}\hat{\phi} - \overline{MM})}{(\hat{\phi} - 1)} \ and \ \hat{S_i} = \frac{(PM_i + \hat{\phi}MM_i - (1+\hat{\phi})\hat{\mu_H})}{(1+\hat{\phi}^2)}$$

in which $\overline{PM} = \frac{\sum_{i=1}^{n} PM_i}{n}$ and $\overline{MM} = \frac{\sum_{i=1}^{n} MM_i}{n}$.

Similarly to obtain the MLE of ϕ, the following partial derivative is set to zero and $\hat{\mu}_H$ and $\hat{S}_i$ are inserted into the expression of the $\hat{\phi}$ via

$$\frac{\partial \ell}{\partial \phi} = -\frac{m}{2\sigma^2}\sum_{i=1}^{n}[-2S_i(MM_i - \hat{\phi}S_i - \mu_H)]. \quad (10)$$

From the solution of Eq. (10), the following expression can be found as the maximum likelihood estimator of ϕ, $\hat{\phi}$.

$$\tilde{\phi} = \frac{(SS_{MM} - SS_{PM}) + \sqrt{(SS_{PM} - SS_{MM})^2 + 4(SS_{PM,MM})^2}}{2SS_{PM,MM}}$$

where $SS_.$ represents the associated correlation between the given intensities such that

$$SS_{PM} = \sum_{i=1}^{n}(PM_i - \overline{PM})^2,$$
$$SS_{MM} = \sum_{i=1}^{n}(MM_i - \overline{MM})^2,$$
$$SS_{PM,MM} = \sum_{i=1}^{n}(PM_i - PM)(MM_i - MM).$$

In this method, since it is assumed that the fraction ϕ is between 0 and 1, $\hat{\phi}$ is taken as $\hat{\phi} = \max\{0, \min\{\tilde{\phi}, 1\}\}$. Furthermore the variance σ^2 is also estimated by the maximum likelihood approach where the likelihood function for σ^2 is conditional on the estimates $\hat{S}_i$, $\hat{\phi}$, and $\hat{\mu}_H$. Hence we can infer σ from the following equation via

$$\hat{\sigma}^2 = \frac{1}{2nm}\sum_{i=1}^{n}\sum_{j=1}^{m}[(PM_{ij}^* - \hat{S}_i - \hat{\mu}_H)^2 + (MM_{ij}^* - \hat{\phi}\hat{S}_i - \hat{\mu}_H)^2].$$

From comparative studies on benchmark datasets it is shown that the FGX method performs better with respect to the other methods in terms of computational efficiency. Moreover it is the only method among other alternatives for capturing too low signal intensities under negligible concentrations. Whereas it can give higher confidence intervals than the other indices as it uses constant variance for all genes and probes.

4.2.10 RGX Method

The RGX (Robust Gene Expression Index) index [14] is an extended version of the FGX method in the sense that it performs the long-tailed symmetric (LTS), rather than normal densities, on the logarithms of PM and MM intensities. On the other hand both methods are based on the likelihood estimation. But in inference of the model parameters, as FGX under LTS does not give explicit solutions, RGX applies the modified maximum likelihood (MML), rather than maximum likelihood, estimators in the calculation.

Accordingly RGX models the logarithms of PM and MM values of the ith gene on the jth probe under LTS densities via

$$\log(PM_{ij}) \sim LTS(S_i + \mu_H, \sigma^2),$$
$$\log(MM_{ij}) \sim LTS(\phi S_i + \mu_H, \sigma^2).$$

Since the majority of Affymetrix analyses are made on the probe set level, the probe values can be summarised by their means like FGX so that $PM_i = \sum_{i=1}^{m} PM_{ij}/m$ and $MM_i = \sum_{i=1}^{n} MM_{ij}/m$ can be represented in the model via

$$PM_i \sim LTS(S_i + \mu_H, \sigma^2/m),$$
$$MM_i \sim LTS(\phi S_i + \mu_H, \sigma^2/m).$$

Then the likelihood function L is defined conditional on $PM = (PM_1, \ldots, PM_n)$ and $MM = (MM_1, \ldots, MM_n)$ for each array as follows.

$$L(\overline{S}, \mu_H, \phi, \sigma | PM, MM) \propto \left(\frac{\sqrt{m}}{\sigma}\right)^n \prod_{i=1}^{n} \left(1 + \frac{z_{PM_i}^2}{k}\right)^{-v} \times \left(\frac{\sqrt{m}}{\sigma}\right)^n \prod_{i=1}^{n} \left(1 + \frac{z_{MM_i}^2}{k}\right)^{-v}$$

where v $(v \geq 2)$ denotes the shape parameter, $\overline{S} = (S_1, \ldots, S_n)$ shows the n-dimensional vector of true signals, and $z_{PM_i} = (PM_i - S_i - \mu_H)/(\sigma/\sqrt{m})$ and $z_{MM_i} = (MM_i - \phi S_i - \mu_H)/(\sigma/\sqrt{m})$ represent the standardised values of PM and MM intensities for $i = 1, \ldots, n$, respectively.

To obtain estimators of parameters μ_H, ϕ, S_i, and σ, the likelihood function L is partially differentiated with respect to μ_H, S_i, and σ. From the derivation we can get the following equations.

$$\frac{\partial \ln L}{\partial \mu_H} = \frac{2v\sqrt{m}}{\sigma k} \sum_{i=1}^{n} [g(z_{PM_i}) + g(z_{MM_i})], \tag{11}$$

$$\frac{\partial \ln L}{\partial \phi} = \frac{2v\sqrt{m}}{\sigma k} \sum_{i}^{n} S_i [g(z_{MM_i})], \tag{12}$$

$$\frac{\partial \ln L}{\partial S_i} = \frac{2v\sqrt{m}}{\sigma k} [g(z_{PM_i}) + \phi g(z_{MM_i})], \tag{13}$$

$$\frac{\partial \ln L}{\partial \sigma} = \frac{-2n}{\sigma} + \frac{2v}{\sigma k} \sum_{i=1}^{n} [g(z_{PM_i}) z_{PM_i} + g(z_{MM_i}) z_{MM_i}] \tag{14}$$

where

$$g(z_{PM_i}) = \frac{z_{PM_i}}{1 + \frac{z_{PM_i}^2}{k}} \text{ and } g(z_{MM_i}) = \frac{z_{MM_i}}{1 + \frac{z_{MM_i}^2}{k}}.$$

By setting Eqs. (11) to (14) to zero, it is observed that there exists no unique solutions for the parameters since $g(z_{PM_i})$ and $g(z_{MM_i})$ are nonlinear functions of z_{PM_i} and z_{MM_i}, respectively. In order to cope with this challenge, some iterative algorithms can be used. The modified maximum likelihood estimation (MMLE) is an alternative approach of MLE when the partial log-likelihood functions have nonlinear form. For solving the nonlinearity, the MMLE technique applies the first order Taylor series expansion of $g(z_{(i)})$'s around $t_{(i)}$'s to linearize $g(z_{PM_i})$ and $g(z_{MM_i})$ where $t_{(i)}$ denotes the ith population quantile of student t-distribution with $(2v-1)$ degrees of freedom. The MMLE technique also uses ordered statistics $\mathrm{PM}_{[i]}$, $\mathrm{MM}_{[i]}$, $S_{[i]}$ in an increasing magnitude instead of PM_i, MM_i, and S_i so that the random samples can have umbrella form weight functions [18]. Hereby the method replaces z_{PM_i} with $z_{PM_{[i]}} = (PM_{[i]} - S_{[i]} - \mu_H)/(\sigma/\sqrt{m})$ and z_{MM_i} with $z_{MM_{[i]}} = (MM_{[i]} - \phi S_{[i]} - \mu_H)/(\sigma/\sqrt{m})$. By means of the underlying linearisation, the MMLE technique approximates to nonlinear functions by $g(z_{PM_i}) \approx \alpha_i + \beta_i z_{PM_{[i]}}$ and $g(z_{MM_{[i]}}) \approx \alpha_i + \beta_i z_{MM_{[i]}}$ in which $\alpha_i = \frac{2t_i^3/k}{(1+t_{(i)}^2/k)^2}$, $\beta_i = \frac{1-t_{(i)}^2/k}{(1+t_{(i)}^2/k)^2}$.

Here β_i refers the weight functions and $\beta_i = \beta_{n-i+1}$ for $i = 1,\ldots,n$ and $\sum_{i=1}^{n} \alpha_i = 0$. Then the estimation of μ_H is derived as

$$\hat{\mu}_H = \left(\sum_{i=1}^{n} \beta_i MM_{[i]} - \hat{\phi} \sum_{i=1}^{n} \beta_i PM_{[i]} \right) / \left((1-\hat{\phi}) \sum_{i=1}^{n} \beta_i \right).$$

Since the information belonging to probe set levels, i.e. PM_i and MM_i, is not sufficient statistics in the estimation of σ, the complete log-likelihood is used in the derivation of $\hat{\sigma}$ similar to the FGX method. Finally by inserting the sufficient $\hat{\sigma}$ into the partial derivative of S_i, $\hat{S}_i$ is defined as

$$\hat{S}_i = \frac{\hat{\sigma}(1+\hat{\phi})\alpha_i + (PM_{[i]} + \hat{\phi} MM_{[i]})\beta_i - \hat{\mu}_H(1+\hat{\phi})\beta_i}{(1+\hat{\phi}^2)\beta_i}.$$

On the other hand in inference of ϕ, we follow a two-stage procedure. At the first stage, an iterative value with initial values of $\hat{\mu}_H$, $\hat{\sigma}$, $\hat{S}_i$, and $\hat{\phi}$ is found and the corresponding α's, β's, and true concomitants are obtained. These α, β, and concomitant values are then used as initial values to estimate $\hat{\mu}_H$, $\hat{\sigma}$, and $\hat{S}_i$ which can be taken as the starting values at the next iteration. This process keeps on until we reach convergent values for $\text{PM}_{[i]}$, $\text{MM}_{[i]}$, $S_{[i]}$, and $\hat{\mu}_H$, $\hat{\sigma}$, $\hat{S}_i$. At the second stage the calculated stable values of $\hat{\mu}_H$, $\hat{\sigma}$, and $\hat{S}_i$ are inserted into the partial derivative of ϕ so that $\hat{\phi}$ can be computed.

On the other side, similar to the FGX method, the variance-covariance structure of the model parameters can be expressed from the inverse of the Fisher information matrix, I which has the following form

$$I = \begin{bmatrix} \frac{\partial^2 \ell}{\partial \mu_H^2} & \frac{\partial^2 \ell}{\partial \mu_H \partial \phi} & \frac{\partial^2 \ell}{\partial \mu_H \partial S_1} & \frac{\partial^2 \ell}{\partial \mu_H \partial S_2} & \cdots & \frac{\partial^2 \ell}{\partial \mu_H \partial S_n} \\ \frac{\partial^2 \ell}{\partial \phi \partial \mu_H} & \frac{\partial^2 \ell}{\partial \phi^2} & \frac{\partial^2 \ell}{\partial \phi \partial S_1} & \frac{\partial^2 \ell}{\partial \phi \partial S_2} & \cdots & \frac{\partial^2 \ell}{\partial \phi \partial S_n} \\ \vdots & \vdots & \vdots & \vdots & \vdots & \vdots \\ \frac{\partial^2 \ell}{\partial S_n \partial \mu_H} & \frac{\partial^2 \ell}{\partial S_n \partial \phi} & 0 & 0 & \cdots & \frac{\partial^2 \ell}{\partial S_n^2} \end{bmatrix}.$$

Similar to FGX, the RGX method also efficiently captures the low signal intensities under negligible concentration compared to other methods. Moreover it preserves the computational gain with high sensitivity and gives more efficient results than FGX in terms of accuracy of the model parameters.

4.3 Dye-Effect Normalisation

In both single and two-channel arrays, the RNA or cDNA molecules are labeled with dye molecules to be identified in the mixtures. Similarly in two-channel arrays, Cy3 and Cy5 dyes are used to label two different conditions. But since the dye molecules show differences in size, reaction to photo bleaching, and produced quantum, the observed signals need to be purified from the dye-effect.

There are two methods suggested to overcome this challenge. The first method, called the dye-swap normalisation, proposes to swap dyes, repeat the experiment, and average the intensity values with respect to the Cy3 and Cy5 channels' outcomes. Although this method is practical, it does not guarantee the purification of intensities from the dye-effect and does not take into account different dye effects of each array. In order to unravel the underlying problem, as the second approach, we can define a set of invariant genes which is irrelevant to the changes in dyes. Then this set is used to normalize the complete data which are log transformed and smoothed by the loess function with a smoothing parameter [21] .

4.4 *Quantile Normalisation*

In microarray experiments, each array represents a replication that is performed under the same condition. Therefore the distribution of probe intensities of each array is expected to be the same. But in practice it is observed that the array's distribution may show difference, in particular, on the tails of the density. To discard the difference across the arrays, resulting in reduction in the variation across the arrays, the quantile normalisation method is applied as the last stage of the overall normalisation procedure. Basically this method represents the probe intensities of each array with a matrix X whose entries are x_{ijk} indicating the ith gene in the jth probe intensity on the kth array while $i = 1,\dots,m$ and $j = 1,\dots,n$. The method arranges the entries of each row in an increasing order and averages the intensities of each probe across the arrays. Finally it rearranges these averaged intensities with respect to the rank of the original X matrix [21].

5 A Case Study

In this study we implement FGX and RGX indices in a real microarray dataset taken from the Turkish Ministry of Agriculture and Rural Affairs, Central Field Crop Research Institute and compare our findings with the outcomes of the RMA approach. The selected data are gathered from the Barley1 GeneChip experiment under different boron toxicity concentrations [13] and can be downloaded from the Gene Expression Omnibus (GEO) at `http://www.ncbi.nlm.nih.gov/geo` under series GSE14521. In the experiment the transcriptome changes in the eight-day-old sensitive barley leaves under 5 mM $B(OH)_3$ and 10 mM $B(OH)_3$ concentrations for 5 days are investigated and compared with the transcriptome changes of the control leaves which are not subjected to the boric acid. At the end of 13 days, the RNA's are isolated from each control and 5 mM $B(OH)_3$ and 10 mM $B(OH)_3$ are subjected to the cells. Then the isolated mixtures of RNA's are purified from ingredients of the contamination until the ratio of RNA's over the elements of the contamination is reached around 2.0. Once the cRNA's are synthetised, they are labeled and hybridised for 16 hours to Barley1 GeneChip. Finally the arrays are washed and scanned according to the standard protocol of Affymetrix [13].

Accordingly each Barley1 GeneChip includes 22840 genes where each of them has a probe pair from 8 to 11 probes. Whereas since in our dataset the majority of genes has 11 probes, we use the associated genes in our analysis, hereby implement 22801 genes in the calculation. On the other hand under both two boron (5mM and 10mM) and control groups, three replications are taken, thereby totally 9 arrays are used in a one-channel microarray experiment.

In the following analysis' part we evaluate the data under small and large number of genes. Hereby in the first analysis we generate a dataset with 1000 genes which are chosen among 22801 genes by taking every 22th gene under both control and 5mM boron groups. We normalize this set by background and quantile normalisation. We assess the transformed data by boxplots. Then we compare the performance of three techniques under fold-change and significance of the genes with respect to the 0.05 cut-off probability value (p-value). But in order to detect their differences under all plausible p-values, we draw their ROC curves, which represent the trade-off between false positive and false negative rates for every possible cut-off point, and compare the findings. On the other hand to evaluate their performance under large data, we use whole 22801 genes under both 5mM and 10 mM boron conditions. Thereby we implement the same stage of analyses for this set and additionally cluster the genes having at least 2-fold-change with respect to the control group by the PAMSAM (Partitioning Around Medoids by using Average Silhouette Weight) method [9, 19]. The results indicate that there is a high correspondence between FGX and RGX outputs as well as the biological knowledge about the system.

5.1 Results of Analyses

In the analysis of the small dataset which is composed of 1000 genes under control and 5mM boron toxicity groups, we compare the performance of RMA, FGX, and RGX methods after the quantile normalisation. As seen in Fig. 1, there is a good correspondence between FGX and RGX outcomes. Moreover with respect to the RMA results, it is seen that the average estimated signals for all genes under each array are lower than RMA plots. Indeed it is an expected finding for both FGX and RGX graphs as found in the analysis of benchmark datasets [15, 14]. Because both indices assume that the MM probes contain a fraction of true signals, hereby can effectively identify the average level of non-specific signal μ_H. In this dataset it is observed that this signal is around 6 on the $\log_2$ scale.

Then for the assessment of fold changes and the detection of significant genes under 0.05 level, we present Table 1. As seen in the table when we check the activity of each gene separately under control and treatment groups, we apply the student t-test and compare our calculated p-values with the cut-off point 0.05. From the results we observe that both FGX and RGX can successfully find more significance genes with respect to the RMA estimates. Whereas they can capture less number of fold changes at least or equal to 2. In fact the detection of fold changes is one of

Table 1 Number of at least 2-fold-change (fc) and significant genes for 0.05 significance level between the difference of control and treatment (10 mM boron toxicity) groups via RMA, FGX, and RGX estimates with 1000 genes.

	RMA	FGX	RGX
At least 2-fc	7	3	3
Significant genes	51	70	73

superior sides of the RMA method [2]. Because RMA is a deterministic model and does not give weights particularly to outlier measurements. On the contrary both FGX and RGX are probabilistic methods and especially, RGX is more concentrated on the tails of densities. Accordingly its advantage over alternatives can be better observed when the data have high fold changes which can be considered more than 2-fold-change. In our data we see that under all models, the estimated signals have fold changes around 2.

Finally as seen in Fig. 2, the ROC curves show that the dataset does not have high percentage of significant genes as the axes are mostly linear. This plot indicates that most genes are inactive and we cannot obtain any new information under treatment group although 10 mM boron injection is thought to be effective in the barley transcriptome and networks of signalling or molecular responses. Whereas since FGX and RGX can be able to detect significant genes better, they can find differentially expressed genes under low cut-offs too. But when we increase the threshold, all graphs indicate similar shape.

From these results we also observe that for this dataset all the outcomes of RGX and FGX are close since the density of the data is not far from normality. So in order to gain from the computational cost between FGX and RGX results due to the fully explicit expressions for the FGX-model-parameter [14], we use the FGX method in the analysis of the same dataset with more genes.

Accordingly in the second analysis with 22801 genes under both 5mM and 10 mM boron toxicity, we compute the number of 2 or higher fold changes and significant genes as seen in Table 2. From the estimates it is observed that there is a correspondence between small and large datasets in the sense that the performance of RMA is better in terms of the investigation of fold-change, whereas, FGX outperforms in the detection of significant genes under 0.05 cut-off points. On the other

Table 2 Number of at least 2-fold-change (fc) and significant genes for 0.05 significance level between the difference of control and treatment (5 mM and 10 mM boron toxicity) groups via RMA and FGX estimates with 22801 genes.

	at 5mM boron		at 10 mM boron	
	RMA	FGX	RMA	FGX
At least 2-fc	175	21	317	84
Significant genes	1181	1715	1202	1465

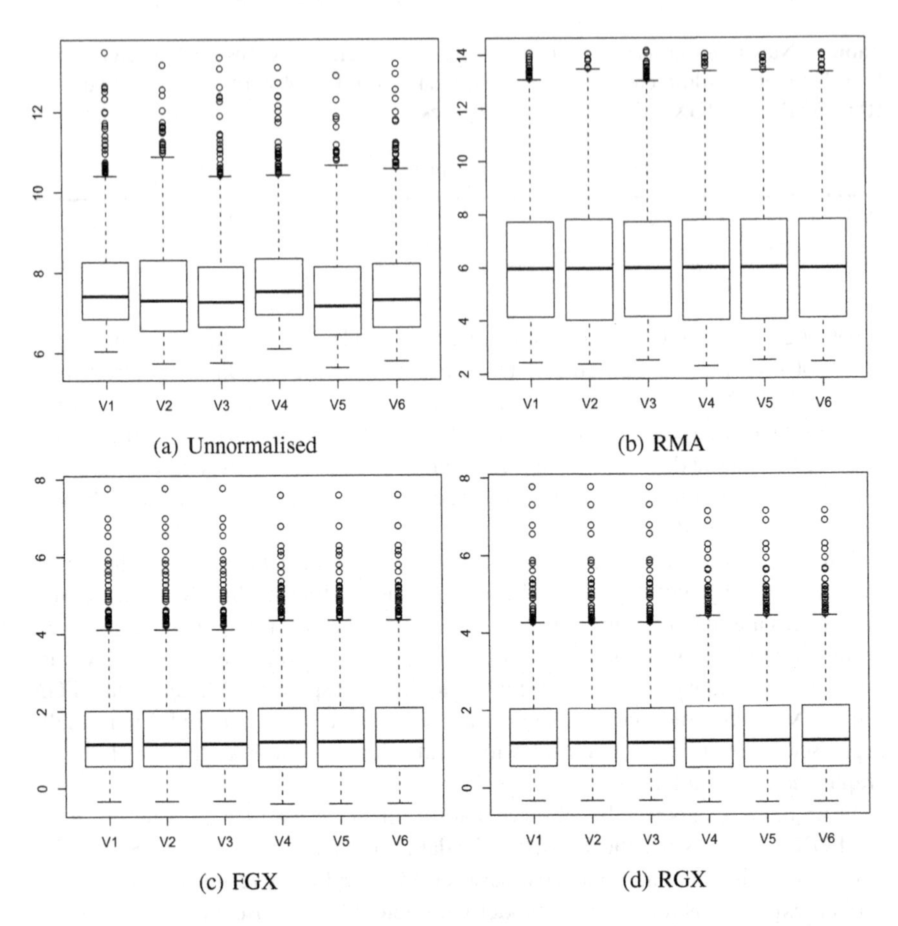

Fig. 1 Boxplots of unnormalised (a), background and quantile normalised (b, c, d) data with 1000 genes whose background normalisation is performed by RMA (b), FGX (c), and RGX (d), respectively, under normal and 10mM boron subjected treatment arrays on the $\log_2$ scale.

side when we plot the ROC curve for both RMA and FGX indices, we get again similar structure as presented in Fig. 2 such that both graphs do not present any active gene under high cut-off points, but FGX performs slightly better than RMA under low cut-off values. Finally, to compare the change in activities under fold-change, we draw the PAMSAM graph of estimated signals 3. In this clustering we use the correlation as the dissimilarity measure and set the number of clusters to 5. From the plots it is found that the fold changes estimated by the RMA method are also found by the FGX estimates under the same clusters. Moreover both indices can infer biologically validated outputs and show functionally close genes in the same clusters [13].

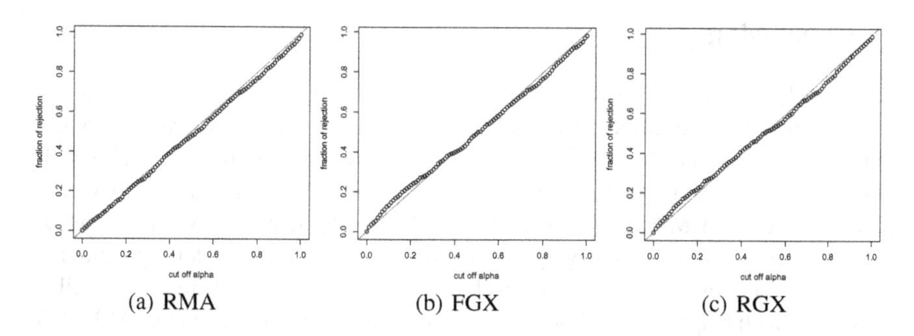

(a) RMA (b) FGX (c) RGX

Fig. 2 ROC curves of differentially expressed genes estimated by (a) RMA, (b) FGX, and (c) RGX methods with 1000 genes and under 10mM boron toxicity.

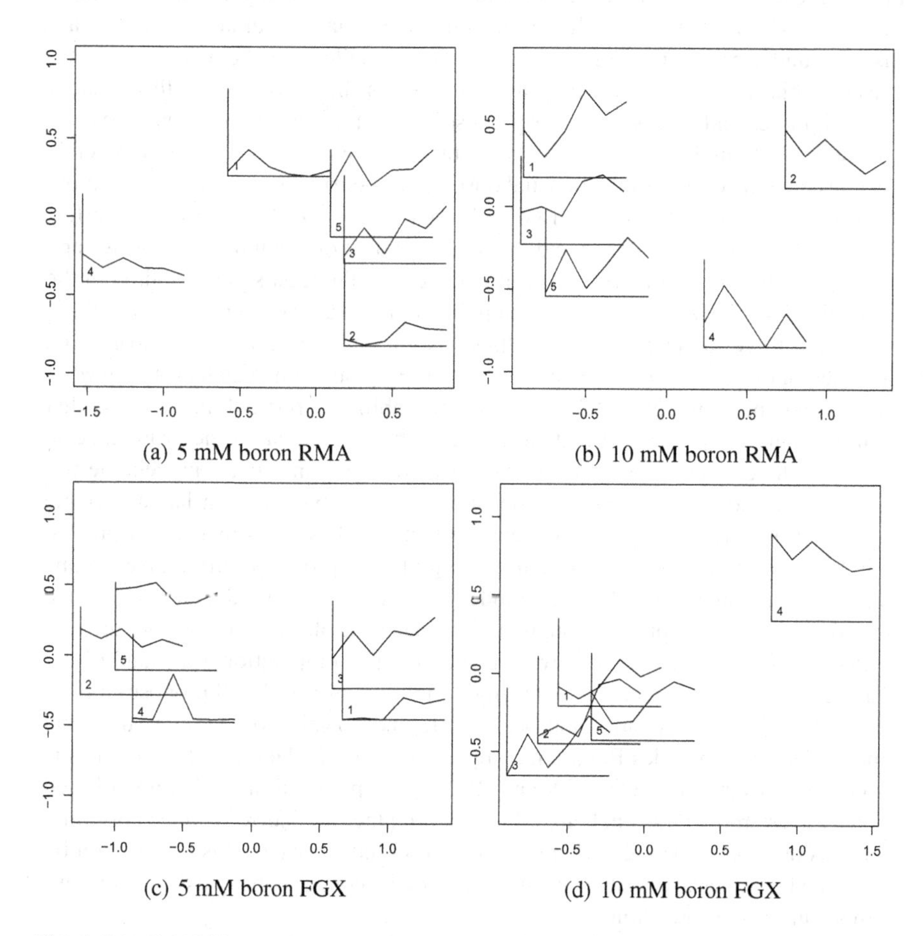

(a) 5 mM boron RMA (b) 10 mM boron RMA

(c) 5 mM boron FGX (d) 10 mM boron FGX

Fig. 3 PAMSAM clustering of at least 2-fold-change genes under 5 mM (a and c) and 10 mM (b and d) boron subjected treatment arrays with 22801 genes estimated by RMA and FGX methods with quantile normalisation.

6 Conclusion

In this study we present the normalisation steps of one-channel microarray data. But our particular interest is the different background normalisation approaches which enable us to estimate the true signal of the genes under highly noisy dataset. In this description, we choose the most recent methods in the field, namely, MAS 5.0, dChip, RMA, GC-RMA, gMOS, mgMOS, multi-mgMOS, and BGX approaches. In the representation of the underlying indices we emphasize their modelling structures for perfect matches and mismatches probes, and discuss their advantages as well as drawbacks with respect to each other. From this comparison, it is seen that MAS 5.0, the oldest method among alternatives, suggests a robust strategy to estimate the true signal by adjusting the observed probe pairs so that their perfect matches values are always greater than their associated mismatches. From the analysis via bench-mark dataset it is found that although the method can estimate accurate signals, the adjustment under perfect matches probes intensities smaller than the mismatches ones can cause bias in inference. dChip can overcome this challenge by initially detecting unusual probes and arrays on the original scale. By this way the estimation can be conducted without the possible effects of outliers. The RMA method comprehensibly considers the effects of the unusual observations during the inference. Therefore it works already quantile normalised and logarithmic scaled data. But the performance of both dChip and RMA indices are dependent on the number of arrays and probes used in the calculation since they implement the least square method in the estimation. GC-RMA solves this problem by applying a mixture of maximum likelihood and Bayesian techniques simultaneously in inference. Moreover apart from the other methods, it accepts the existence of true signal in mismatch probes. gMOS suggests a similar model of dChip but in a probabilistic approach in the sense that both mismatches and perfect matches are distributed gamma on the original scale and mismatches merely possess the background signal. mgMOS unravels the underlying misleading in the description of mismatches by adding a latent variable presenting a correlation terms between perfect matches and mismatches probes. This latent term can denote different binding affinity probes within a given probe set. BGX implements the GC-RMA model by estimating the model parameters in a fully Bayesian framework. Although this approach can allow us to consider the true intensity under each probe and array separately, the computational demand of the estimated model parameters becomes high. Finally multi-mgMOS proposes a mixture of Bayesian ad frequentist approach in inference to gain from the computational time. But since it includes Bayesian components, the calculation of the estimates is still costly for large datasets. FGX and RGX develop a maximum likelihood-based strategy to overcome this challenge. In modelling the first index assumes normality of intensities for probe pair, whereas, the second index relaxes this assumption by long-tailed symmetric distribution which covers the normal density as a special case in this family of distribution.

From the comparative analysis it is observed that even though RMA ignores the intensity of mismatches and merely models the measurement of perfect matches, it performs better than MAS5.0 and dChip in terms of accuracy of the estimates and outperforms GC-RMA besides these two methods in the precision. Also, in particular, for high intensity levels, all these methods give similar results. But FGX and RGX are estimated much faster than their close alternatives, multi-mgMOS and BGX. Therefore in the analysis of a real dataset, we compare the performances of these two recent indices with RMA as it is one of the well-known and strong models among others as stated previously. In the assessment of performance we use different criteria from the number of significant genes to the detection of functionally similar genes. From the analysis we observe that both FGX and RGX methods enable us to obtain more significant genes and produce better ROC curves under distinct cut-off points. Their advantages over RMA can be resulted by the gain in information. Because they can combine the information of the PM probes with the fractional true signals in the MM probes. Moreover they can estimate the non-specific hybridisation component directly since they can measure zero signal under low concentrations. Whereas they are less sensitive to the fold-change at least two with respect to the RMA results since, in particular, RGX focuses on the tails of the density.

As a result we consider that both suggested methods can be seen promising approaches for the background normalisation of oligonucleotide. But their estimates can be improved by using probe specific, rather than probe means, intensities in inference. Furthermore we believe that the estimates can be more accurate if we include gene specific variance in the description of the PM and MM probes.

Acknowledgements. The authors would like to thank to Dr. Tufan Öz and Dr. Remziye Yılmaz for collecting the data and Prof. Ernst Wit for his explanation about different gene expression indices.

References

1. Affymetrix: Statistical Algorithms Description Documents. Santa Clara, CA (2002)
2. Cope, L.M., Irizarry, R.A., Jaffee, H.A., Wu, Z., Speed, T.P.: A benchmark for Affymetrix GeneChip expression measures. Bioinformatics 1(1), 1–10 (2003)
3. Dziuda, D.M.: Data Mining for Genomics and Proteomics Analysis of gene and protein expression data, 1st edn. John Wiley and Sons Ltd. (2010)
4. Gentleman, R., Irizarry, R.A., Carey, V.J., Duboit, S., Huber, W.: Data Mining for Genomics and Proteomics Analysis of gene and protein expression data, 1st edn. John Wiley and Sons Ltd (2005)
5. Hein, A.K., Richardson, S., Causton, H.C., Ambler, G.K., Green, P.J.: Bgx: a fully bayesian gene expression index for affymetrix genechip data. Biostatistics 1(1), 1–37 (2004)
6. Hubbell, E., Liu, W.M., Mei, R.: Robust estimators for expression analysis. Bioinformatics 18(12), 1585–1592 (2002)
7. Irizarry, R.A., Bolstad, B.M., Collins, F., Cope, L.M., Hobbs, B.: Summaries of affymetrix genechip probe level data. Nucleic Acids Research 31(40, e15), 1–8 (2003)
8. Irizarry, R.A., Hobbs, B., Collin, F., Beazer-Barclay, Y.D., Antonellis, K.J., Scherf, U., Speed, T.P.: Exploration, normalization, and summaries of high density oligonucleotide array probe level data. Biostatistics 4(2), 249–264 (2003)

9. Kaufman, L., Rousseeuw, P.J.: Finding Groups in Data. John Wiley and Sons, New York (1990)
10. Li, C., Wong, W.H.: Model-based analysis of oligonucleotide arrays: expression index computation and outlier detection. Proceedings of the National Academy of Science, USA, 98(1), 31–36 (2001)
11. Liu, X., Milo, M., Lawrance, N.D., Rattray, M.: A tractable probabilistic model for affymetrix probe-level analysis across multiple chips. Bioinformatics 18, 3637–3644 (2005)
12. Milo, M., Fazeli, F., Niranjan, M., Lawrence, N.D.: A probabilistic model for the extraction of expression levels from oligonucleotide arrays. Biochemical Society Transactions 31, 1510–1512 (2003)
13. Öz, M.T., Yılmaz, R., Eyidoğan, F., de Graaff, L., Y¨ucel, M., Öktem, H.A.: Microarray analysis of late response to boron toxicity in barley (hordeum vulgare l.) leaves. Turkish Journal of Agriculture and Forestry 33, 191–202 (2009)
14. Purutçuoğlu, V.: Robust gene expression index. Mathematical Problems in Engineering, 1–12 (2011), doi: 10.1155/2011/182758
15. Purutçuoğlu, V., Wit, E.: Fgx: a frequentist gene expression index for affymetrix arrays. Biostatistics 8(2), 433–437 (2007)
16. Schena, M.: Microarray Analysis. John Wiley and Sons, Hoboken (2003)
17. Stekel, D.: Microarray Bioinformatics. Cambridge University (2003)
18. Tiku, M.L., Akkaya, A.D.: Robust Estimation and Hypothesis Testing. New Age International Ltd. (2004)
19. Wit, E., McClure, J.: Statistics for Microarray Design, Analysis, and Inference, 1st edn. John Wiley and Sons Ltd. (2004)
20. Wu, Z., Irizarry, R.A., Gentleman, R., Murillo, F.M., Spencer, F.: A model-based background adjustment for oligonucleotide expression arrays. Journal of the American Statistical Association 99, 909–917 (2004)
21. Yang, Y.H., Dudoit, S., Luu, P., Lin, D.M., Peng, V., Ngai, J., Speed, T.P.: Normalization for cDNA microarray data: a robust composite method addressing single and multiple slide systematic variation. Nucleic Acids Research 30(4), e15 (2002)

Robust Regression Metamodelling of Complex Systems: The Case of Solid Rocket Motor Performance Metamodelling

Elçin Kartal-Koç, İnci Batmaz, and Gerhard-Wilhelm Weber

Abstract. Computer-based simulation models have been widely used to monitor the performance of complex systems. In spite of recent improvements in computers, use of simulation models may still be computationally impractical. On the other hand, metamodels can be used instead as an alternative approach, response surface metamodels (RSMs), particularly, polynomial regression metamodels (PRMs) being the most commonly preferred ones. However, existence of extreme observations in data may cause statistically invalid inferences if some remedial actions are not taken. One of the effective remedies is to use robust regression metamodels (RRMs), which decrease sensitivity of the results to such data. In this study, performances of several RRMs are compared with those of RSMs with respect to several criteria including accuracy, stability, efficiency and robustness using a validation method for the ballistic performance functions of a solid rocket engine. Furthermore, the effect of sampling strategies on the performance of these metamodels is assessed by collecting data using some traditional and computational design of experiment (DOEs). Results indicate that the performances of RRMs are competing with that of RSM, especially, in computational DOEs. Besides, they are more efficient and do not require an expert support with a capable software.

Elçin Kartal-Koç · İnci Batmaz
Department of Statistics, Middle East Technical University,
06531 Ankara, Turkey
e-mail: kartalelcin@gmail.com, ibatmaz@metu.edu.tr

Gerhard-Wilhelm Weber
Institute of Applied Mathematics, Middle East Technical University,
06531 Ankara, Turkey
e-mail: gweber@metu.edu.tr

A. Byrski et al. (Eds.): Advances in Intelligent Modelling and Simulation, SCI 416, pp. 221–251.
springerlink.com

1 Introduction

Computer-based simulation models have been widely used to monitor the behaviour and performance of real life complex systems which are characterized by many variables interacting with each other. In spite of recent improvements in computer capacities and speed, construction of simulation models may still be computationally expensive and impractical. As an alternative approach, surrogate models, also called metamodels, are proposed to be used instead of expensive computer simulation models. Here, a *metamodel* refers to a mathematical or statistical model of a *simulation model* [25]. The most commonly used metamodelling technique is the multiple linear regression, particularly, polynomial regression metamodels (PRMs) [8, 34]. These metamodels are usually named as response surface metamodels (RSMs) in the simulation literature. Some examples for RSM applications are given in [2, 3, 22, 54]. Other well-investigated metamodels are artificial neural networks (ANNs) [22, 59], kriging metamodels (KMs) [4, 45], radial basis functions (RBFs) [13, 17] and multivariate adaptive regression splines (MARS) [15, 22].

There are many comparative studies involving above metamodels in literature. In Varadarajan et al. [56], ANNs are compared with RSMs for modelling nonlinear thermodynamic behaviour of an engine design. RSMs and ANNs are compared for the design optimisation of solid rocket engine in Kuran [27]. A comparative study including KMs and RSMs is carried out in the study of Simpson et al. [47]. Furthermore, other comprehensive comparisons are done by Yang et al. [61] and Simpson et al. [48]. In the former one, MARS, ANN, RSM and moving least squares (MLS) metamodels are compared through automotive crash analysis. In the latter one, four metamodelling techniques including MARS, RSM, RBF and KM are tested on fourteen test problems. In Staum [51], stochastic kriging, generalized least squares (GLS) regression and KMs are explained and compared for metamodelling stochastic simulations.

In many engineering applications, metamodels have been used for various purposes including design evaluation, optimisation and reliability analysis of systems that involve expensive simulation models. An extensive study of metamodelling techniques is carried by Jin et al. [20] for multiple test problems. In another study, metamodelling techniques are used for optimisation under uncertainty by Jin et al. [19]. Besides, an application of metamodels in multidisciplinary design optimisation is carried out by Sobieszczanki-Sobenski and Haftka [49]. In addition, Barton [1] surveys both local and global metamodel-based optimisation methods. In Ramu et al. [40], selection of design experiments, metamodel selection, sensitivity analysis and optimisation are addressed through a case study on a roof slap . Furthermore, different metamodelling techniques are used in the reliability-based design optimisation studies by Mourelatos et al. [30] and Kuran [27].

Another important application area of metamodelling, optimisation and reliability studies is the *aerospace systems* in which there are many input variables interacting with each other. In mechanical and aerospace systems, a comprehensive review of metamodelling applications is provided by Simpson et al. [47, 48]. Papila et al. [38] present a method for optimising a supersonic turbine for a rocket propulsion system. In another study of Papila [39], ANN and RSMs are established for supersonic turbine design optimisation. Won et al. [60] implement RSM to determine the external geometric parameters in such a way that the drag coefficient of a standoff missile is minimized. Vaidyanathan et al. [55] study the rocket engine component optimisation by using ANNs. On the other hand, Kuran [27] demonstrates an application of surrogate models on a multi-objective reliability based design optimisation problem. In their study, objective and constraint functions are established by using the RSM (i.e. PRM) and ANN metamodels. For the same problem, different versions of metamodelling studies are carried out by Kuran et al. [28] and Kartal [21].

According to Kartal [21], statistically valid PRMs for ballistic performance functions of a solid rocket engine can be developed if a careful residual analysis is conducted. Generally, the least squares (LS) method, which is based on the minimisation of sum of the squares of error terms, is used to estimate the parameters of a PRM [35]. When the error terms are normally distributed, the LS estimates of the parameters have good statistical properties [34]. Unfortunately, in many real life applications, errors are not distributed normally, especially, due to the presence of extreme observations (outliers) in data. Note that experimental studies indicate that almost 10% of observations can be observed as outliers [32]. So, for a successful PRM based RSMs building, a skillful handling of outliers with appropriate techniques is necessary [52, 53, 62]. Moreover, since the heterogenous variance of simulation outputs can violate the white-noise assumption, remedies are discussed in the contex of simulation, principally, GLS and variance-stabilising transformations by Kleijnen [26]. But, the use of diagnostic and treatment procedures of outliers often requires abilities beyond those of the average analyst.

In statistics, the robust regression metamodel (RRM) is an important tool for analyzing data contaminated with outliers. With the help of this approach, outliers can be detected and resistant results can be obtained without spending an extra effort such as applying transformation on the response or deletion of outliers. Therefore, in this study, the performance of PRM based RSMs are compared with those of RRMs such as M-estimation, LTS-estimation, S-estimation and MM-estimation [11] for modelling the ballistic performance functions of a solid rocket engine.

In many applications, in addition to the performance, time required for the construction of metamodels is an important concern. This problem can be overcome by providing as much information as possible with the least amount of time by using design of experiments (DOEs) [7, 26, 33]. Among the most popular ones there is central composite design (CCD), face-center cube design (FCD) and Box-Behnken design (BBD). They are called *classical designs*, and are firstly proposed for conducting physical experiments [34]. Note also that they have good properties such as orthogonality, rotability and optimality. On the other hand, Latin hypercube sampling design (LHS) [31], uniform sampling design (US) [14] and orthogonal arrays (OA) [37] are the *modern type designs* used in computer simulation experiments. There are many comparative studies on the use of DOEs for metamodelling purpose. To illustrate, the performances of standard and small-size second-order DOEs are examined thoroughly in the development of PRMs by Batmaz and Tunalı [2] and Batmaz and Tunalı [3], respectively. Another study is conducted by Simpson et al. [48] to compare five different DOEs including LHS, Hammersley Sequence (HS), OA, US and randomly generated data points.

In this study, the sample points required for the construction of metamodels are obtained by using five designs; three of them are classical designs, namely CCD, BBD and FCD, and two of them are modern designs, namely LHS and US. Additionally, the effect of sampling strategies on the performance of the PRM and RRM are examined and compared. In these comparisons, the performances are evaluated with respect to several criteria including accuracy, robustness, stability, efficiency and ease of use with the help of a validation method. To apply this method, 100 new data points are generated by using US. Furthermore, repeated analysis of variance (RANOVA) is employed to detect if there are statistically significant differences between the designs and metamodels developed with respect to various accuracy measures, described in Section 4, and their stabilities.

This paper is organized as follows. Sampling designs, metamodelling techniques, and the validation method and comparison criteria used in the study are presented in Sections 2, 3 and 4, respectively. In Section 5, the application environment, a solid rocket motor, is described. Results and discussion are given in Section 6. In the last section, conclusions and future studies are stated.

2 Sampling Techniques

DOEs play an important role in the construction of a metamodel by providing as much information as possible from a limited number of observations. In literature, DOEs are typically classified in two groups: *classical* and *modern* design. While classical designs refer to physical (laboratory) experiments, modern ones refer to experiments carried out on the computers. The main difference between these two groups is based on the assumption that the random errors occur in laboratory experiments but not in computer experiments [16]. Hence, the sampling strategies of designs are determined accordingly. To exemplify, in the classical DOEs, the sample points are located near or on the boundaries of the design (factor) space in such a way to minimize the variance. In the modern designs, however, samples are placed the interior of the factor space by minimising the bias, the difference between the true and estimated response function. Due to this reason, modern designs are sometimes refer to as *space filling designs* [45]. More information on both groups of designs is provided below.

2.1 Central Composite and Face-Center Cube Design

One of the commonly used classical DOE techniques is the CCD [5]. It is a two-level factorial or fractional factorial design (2^k or 2^{k-p}) augmented with $2k$ axial points located at α-distance from the center of the design. Here, k is the number of variables of interest and p represents the fraction. In the CCDs, the number of sample points is determined by the formula $2^k + 2k + n_0$, which grows by k. The 2^k points represent the *corner* points of a hypercube (for $k > 3$) while $2k$ axial points lie outside of the hypercube, whose α-distances from the center are determined by k. Conventionally, CCDs are analyzed by second-order PRM based RSMs. In these metamodels, main effects and two-factor interactions are estimated by the factorial points. For experiments having random errors, the number of replicates at the center point (n_0), also called center runs, becomes important [34]; they are used to investigate the existence of a curvature in the metamodel. If it exists, the axial points are used for an efficient estimation of quadratic terms. CCDs are efficient designs when dealing with a spherical region of interest. However, when the region of interest is cuboidal, the FCD, a special type of CCD where α=1, becomes more effective second-order design [2]. In this design, axial points take place at the centers rather than outside of the faces of a hypercube as in the case of CCD.

2.2 *Box-Behnken Design*

As another spherical design, BBD is an important alternative to CCD [6]. It belongs to a family of three-level designs in which a two-level factorial design is combined with an incomplete block design. It does not contain any points at the *corner* of the hypercube, defined by the higher and lower factor level combinations. This is why they become advantageous when the corner points are expensive or impossible to test [32].

2.3 *Latin Hypercube Design*

LHS is one of the most popular modern DOEs proposed by McKay et al. [31]. It is constructed by dividing the range of each input variables into n intervals, and then, randomly selecting a point from each interval. Next, the first point selected this way from the first factor is coupled with the ones selected randomly from the other factors, and the process is repeated $(n-1)$ times until a matrix with (nxk) dimension is obtained. Here, n and k are the number of sample points and factors, respectively.

2.4 *Uniform Design*

US design has been used as a *space filling* technique since firstly proposed by Fang et al. [14]. It is characterized by providing a uniform distribution for design points over the entire design space, which is achieved by minimising the discrepancy between design points. Note that US is very similar to LHS, if the domain of the experiment is finite. Hence, design points are not selected randomly but from the center of the cells [48].

3 Metamodelling Techniques

In general, the form of an approximation function can be written as below

$$\hat{y}(X) = f(X, y(X)), \tag{1}$$

where X is the $n \times (k+1)$ input matrix in which columns and rows represent columns of ones and input variables, and cases, respectively; $y(X)$ is a measured response, and f is a user-specified function. The form of the functions for f is typically polynomials, splines, kriging and neural networks. In the following sections, the main characteristics of the metamodels used in this study, namely PRM and RRM, are described.

3.1 Polynomial Regression Metamodels

In many RSM applications, PRMs have been used extensively [5, 22, 48, 57]. Because they are approximating the behaviour of the response in terms of the input variables of interest effectively with the following formula

$$y(X) = f(X) + \varepsilon. \tag{2}$$

Here, X and $y(X)$ are as defined in (1); $f(X)$ is a known polynomial function of the input matrix, X, and ε is the random error term, which is assumed to be normally distributed with zero mean and constant variance, σ^2. Specifically, a second-order PRM can be expressed as follows

$$y(x) = \beta_0 + \sum_{i=1}^{k} \beta_i x_i + \sum_{i=1}^{k} \beta_{ii} x_i^2 + \sum_{i} \sum_{j>i} \beta_{ij} x_i x_j + \varepsilon. \tag{3}$$

To estimate the parameters in (3), generally, the LS estimation method is used. The objective function of the method is given below

$$\arg\min_{\beta} = Q(\beta) = \arg\min_{\beta} = \sum_{i=1}^{n} \varepsilon_i^2 \tag{4}$$

where, $\varepsilon_i = (y_i - x_i^T \beta)$ and denotes the ith error term. Next, significance of the metamodel and its parameters are determined with the analysis of variance (ANOVA).

In order to obtain statistically valid results from the LS fit of PRM, it is very important to validate the following assumptions [32]:

- Observations are random samples obtained from the population.
- The error terms, ε_i $(i = 1, ..., n)$, are normally distributed random variables.
- The expected value of each error term, ε_i $(i = 1, ..., n)$, is zero; that is, $E(\varepsilon_i) = 0$.
- The variance of error terms is constant at all levels of X, denoted as $\mathrm{Var}(\varepsilon_i) = \sigma^2$ $(i = 1, ..., n)$
- The error terms, ε_i $(i = 1, ..., n)$, are statistically independent.

Although randomisation in simulation can ensure the first assumption, the rest should be checked. Whenever the assumptions are not validated, remedial measures such as transformation on the response should be taken. Tunalı and Batmaz [53] emphasize the importance of diagnostic checks such as normality, variance homogeneity and lack of fit tests in the development of PRMs. Unfortunately, most of the PRMs studies do not provide any information regarding the validity checks of the above assumptions in their studies (e.g. [20, 48]).

One of the important reasons for having non normal error distribution is the existence of outliers in data. As stated before, observing up to 10% of data as outliers in real life data may be unavoidable. On the other hand, in simulation data, extreme settings of input variables in the computer experiments may result in outliers. Therefore, it is important to distinguish whether the extreme points are outliers or just important extreme cases. An observation, y_i $(i = 1, ..., n)$, is defined as an *outlier* if

$$\left|\frac{e_i}{\sigma}\right| = |r_i| \geq c\sigma \tag{5}$$

where $e_i = (y_i - \hat{y}_i)$. Here, e_i and r_i represent the ordinary and standardized residual associated with the ith response, respectively; c is an integer multiple of sigma. In another words, an observation may be consider as an outlier when the absolute value of the standardized residual associated with it is greater than or equal to some multiple of the standard error, usually taken as three.

Outliers in PRMs are treated in different ways [44]. One of them is to transform the response before metamodelling [41]. Nevertheless, this approach may cause difficulties in the interpretation of the results. Another one is to remove outlying observation from the analysis. This approach, on the other hand, may result in serious loss of information. Deleting outliers may also cause completely ignoring them at all. Therefore, in statistics, alternative metamodelling techniques, called as robust methods, have been developed to overcome disadvantages of the above mentioned approaches and dampen the effects of these observations on building metamodels. More detailed information on the robust estimation techniques is given in the following section.

3.2 Robust Regression Metamodels

PRM based RSMs are commonly preferred metamodelling techniques in simulation studies. However, as stated above, presence of extreme observations in data may cause statistically invalid inferences if some remedial actions are not taken. One of the effective remedies is to use RRMs, which decrease sensitivity of the results to such data [44]. In literature, many robust methods have been proposed for this purpose. Among them Huber M-estimation, high breakdown value estimation and combination of these two methods are the most commonly used ones. In this study, four such methods, namely, M-estimation, LTS-estimation, S-estimation and MM-estimation are considered, and are briefly described in the following sections.

3.2.1 M-Estimate

M-estimation method is proposed by Huber [18] for the first time, and since then, it has been extensively used to model data with outliers. In order to obtain M-estimate, the following function of errors is minimized [11]:

$$\arg\min_{\beta} Q(\beta) = \sum_{i=1}^{n} \rho(\frac{\varepsilon_i}{\sigma}), \tag{6}$$

where $\varepsilon_i = y_i - x_i^T \beta$. If σ is known, after taking the derivatives of $Q(\beta)$ with respect to β, $\hat{\beta}$ is calculated as a solution of the system of k equations given below

$$\sum_{i=1}^{n} \psi(r_i) x_{ij} = 0, \quad j = 1, ..., k, \tag{7}$$

where $\psi = \rho'$. If ρ is convex, then, $\hat{\beta}$is the unique solution. The system of equations in (7) is solved by iteratively re-weighted least square (IRWLS) method.

The weight function w(x) can be represented with the following formula,

$$w(x) = \frac{\psi(x)}{x}. \tag{8}$$

When σ is unknown, the following robust estimate of σ is calculated iteratively:

$$\hat{\sigma}^{m+1} = \text{median}_{i=1}^{n} \left| y_i - x_i^T \hat{\beta}^{(m)} \right| / (c_0) \tag{9}$$

where $c_0 = \Phi^{-1}(0.75) = 0.6745$.Note that this value of c_0 makes the $\hat{\sigma}$ an approximately unbiased estimator of σ if n is large and error distribution is normal.

3.2.2 LTS-Estimate

The least trimmed squares (LTS) estimate is proposed by Rousseeuw [42]. This estimate can be obtained by minimising the following expression:

$$\arg\min_{\beta} Q(\beta) = \sum_{i=1}^{n} \varepsilon_{(i)}^2, \tag{10}$$

where $\varepsilon_{(1)}^2 \leq \varepsilon_{(2)}^2 \leq ... \leq \varepsilon_{(n)}^2$ are the ordered squared errors. Note that the breakdown value for the LTS-estimator is $\frac{n-h}{n}$, where $h = [(3n + k + 1)/4]$, n is the number of observations and k is the number of predictor variables.

3.2.3 S-Estimate

The S-estimate is developed by Rousseeuw and Yohai [42], and its objective function is defined as below

$$\arg\min_{\beta} Q(\beta) = \arg\min_{\beta} S(\beta), \tag{11}$$

where, the dispersion $S(\beta)$, is calculated as a solution of the following equation

$$\frac{1}{n-k}\sum_{i=1}^{n}\chi\left(\frac{\varepsilon_i}{S}\right)=\theta, \tag{12}$$

where $\theta=\int\chi(s)d\Phi(s)$ such that $\hat{\beta}$ and $S(\hat{\beta})$ are asymptotically consistent estimate of θ and σ, respectively, for the Gaussian regression model. And, usually, Tukey and Yohai functions are used for $\chi(.)$ [11]. For S-estimate, the breakdown value is $\frac{\beta}{\sup_s\chi(s)}$.

3.2.4 MM-Estimate

MM estimation is proposed by Yohai [63] as an improvement of high breakdown estimators with respect to efficiency. MM estimators are calculated in three stages. First, an initial high breakdown estimate, $\hat{\beta}'$, is calculated by using LTS estimate. Then, M-estimate of scale, $\hat{\sigma}'$, with 50% breakdown is computed by using Tukey or Yohai functions [11]. Finally, the MM-estimator, $\hat{\beta}$, is defined as any solution of

$$Q(\beta)=\sum_{i=1}^{n}\rho\left(\frac{\varepsilon_i}{\hat{\sigma}'}\right), \tag{13}$$

which satisfies the condition that $Q(\hat{\beta})\leq Q(\hat{\beta}')$.

4 Validation Method and Comparison Criteria Used

In this section, the validation method and the criteria used for evaluating the performance of metamodels developed as well as DOEs used for collecting data are briefly described.

4.1 Validation Method

The performances of the metamodels developed are usually assessed for the training data. Nevertheless, these performances may be biased since they are obtained from the same data used for developing the metamodel itself. So, the performance evaluations of metamodels under unseen data become very important. In this study, 100 additional new data is generated for each response variable by using US. Thus, there are five test data sets. And, comparison criteria are applied on both training and test data sets. It is naturally expected to have worse results on the test data than on the training data. But, results obtained from the test data are more realistic than training data. Of course, any method performing similar on both data sets is more desirable. Stability criterion, described in Section 4.2.2, tries to determine the change in the performance of a metamodel with respect to a measure when training and test data sets are used.

4.2 Comparison Criteria

Performances are evaluated and compared with respect to several criteria including accuracy, stability, robustness and efficiency. They are briefly described below.

4.2.1 Accuracy

Accuracy criterion measures the prediction capability of the model for the new response values. The equations for these metrics are given in the following equations.

i) Mean Absolute Error (MAE)

$$MAE = \frac{1}{n}\sum_{i=1}^{n} |y_i - \hat{y}_i|, \tag{14}$$

where y_i and $\hat{y}_i$ are the ith $(i = 1,\ldots,.n)$ actual and predicted value of the response for observation of a real simulation model, respectively. It gives the average magnitude of the error. Hence, smaller MAE values determine the better metamodels.

ii) Root Mean Square Error (RMSE)

Root mean square error (RMSE) is the square root of the mean square error (MSE), and defined as

$$RMSE = \sqrt{\frac{1}{n}\sum_{i=1}^{n} (y_i - \hat{y}_i)^2}. \tag{15}$$

which represents the departure of the metamodel from the real simulation model. It gives more weight on inaccurate estimates. Therefore, as the RMSE values become smaller, the better fit is obtained for the metamodel.

iii) Multiple Coefficient of Determination (R^2)

$$R^2 = 1 - \left(\frac{\sum_{i=1}^{n} (y_i - \hat{y}_i)^2}{\sum_{i=1}^{n} (y_i - \bar{y})^2}\right). \tag{16}$$

R^2 indicates how much variation in response is explained by the metamodel built. The higher the R^2, the better the metamodel fits to the data.

iv) Percentage of Residuals within Three Standard Deviation of Mean (PWI)

Proportion of residuals within some user-specified range can be defined with the sum of indicator variables over all observations. The indicator variables take value of one if the absolute value of the difference between the actual and predicted response is within some user-specified threshold. In this study, PWI measures the proportion of residuals within three standard deviation of the mean.

4.2.2 Stability

A metamodel is stable when it performs just as well on both seen (training) and unseen (test) data sets. The stability can be measured as

$$\min\left\{\frac{M_{train}}{M_{test}}, \frac{M_{test}}{M_{train}}\right\}, \tag{17}$$

where M_{train} and M_{test} are the performance measure values for the test and the training, respectively. Closer values of this measure to one indicate higher stability [36].

4.2.3 Robustness

Robustness can be defined as the capability of achieving similar performance on different problems [20] with different scales, nonlinearity properties and noisy behaviour. Robustness is measured as the spread of an accuracy measure obtain from different data sets. In this study, metamodels are established to approximate the behaviour of six performance functions for which sample points are selected by using five experimental designs. Therefore, robustness of metamodels can be examined with respect to different response variables and sampling strategies.

4.2.4 Efficiency

Efficiency can be interpreted according to both computational ease and independence from expert. A metamodel is *efficient* if it requires less effort to be constructed. Moreover, *independence from expert* is the degree to which the metamodel is built and tested without the help of experts.

5 The Case: Solid Propellant Rocket Motor Performance Metamodelling

5.1 Application Environment

Performance of a solid rocket motor as a complex system is governed by various variables such as propellant properties, propellant grain geometry and nozzle geometric parameters. The reliability of a solid rocket motor using RSMs is investigated by Bozkaya et al. [9]. In their study, the number of variables governing the ballistic reliability of a solid rocket motor is determined. In the same environment, Kuran et al. [28] predict the settings of the design (input) variables for the optimum ballistic performance of a solid rocket motor with star grain by Taguchi design. In another study, Kuran [27] discusses the reliability based design optimisation of ballistic performance for a solid rocket motor. In that study, ballistic performance functions are modeled by utilising PRM and ANNs. In Kartal et al. [24], however, multicriteria optimisation is studied for the performance of solid rocket motors. In addition to these surrogate models, Kartal [21], Kartal et al. [22, 23] and Dönmez et al. [12] model and compare the ballistic performance functions with RRM and MARS. As a conclusion of these comparison studies, PRM and RRMs are proposed for approximating the ballistic performance functions of a solid rocket motor. Ballistic performance calculations are carried out by a code with which one-dimensional steady conservation equations of mass, momentum and energy are solved to find the pressure, density and velocity distributions along the rocket motor case. Details of ballistic calculations can be found elsewhere [9].

There are 21 variables used for determining the ballistic performance functions (see Table 1), which are propellant properties, environmental conditions, propellant grain geometry parameters and nozzle geometric parameters. In a screening study, Bozkaya [10] reduces the number of variables, which govern the performance functions, to 10 (see Table 1). These include the geometric properties of the propellant, which are effective on the ballistic performance of a solid propellant motor with a wagon wheel, and the ballistic properties of the propellant such as the burning rate constant, enthalpy and density of the propellant. Detailed information about the range and the distributions of the variables are given in Kuran [27].

Table 1 Input and output variables.

Input: geometric properties of the propellant	**Output: performance functions**
Radius of Grain (R)	Total impulse (TI)
Throat Radius	Maximum acceleration (MA)
Grain Geometry Parameter (L1)	Launch acceleration (LA)
Grain Geometry Parameter (L2)	Maximum chamber pressure (MCP)
Grain Geometry Parameter (R1)	Sliver fraction (SF)
Grain Geometry Parameter (R2)	Maximum temperature (MT)
Grain Geometry Parameter (XI)	
Input: ballistic properties of the propellant	
Burning rate constant	
Enthalpy of the propellant	
Density of the propellant	

5.2 *Model Development*

In this study, six ballistic performance functions of a solid rocket motor are tried to be approximated with the geometric and ballistic properties of propellant (see Table 1) by using PRM and RRMs. While developing PRMs, all assumptions are validated. If not, remedial actions are taken as described in Section 3.1. When assumptions are satisfied, the significance of metamodels developed is analyzed with RANOVA.

In this process, as expected, many outlying simulated values are detected by using the equation (5) with $c = 3$. To exemplify, the normal probability plot (NPP) of standardized residuals obtained from the PRM fit to the data collected by CCD for the response TI is presented in Fig. 1. The percentage of outliers detected from the PRM fit to the data obtained from five DOEs for each of the simulated ballistic performance variables are listed in Table 2. Note that, as expected, the percentage

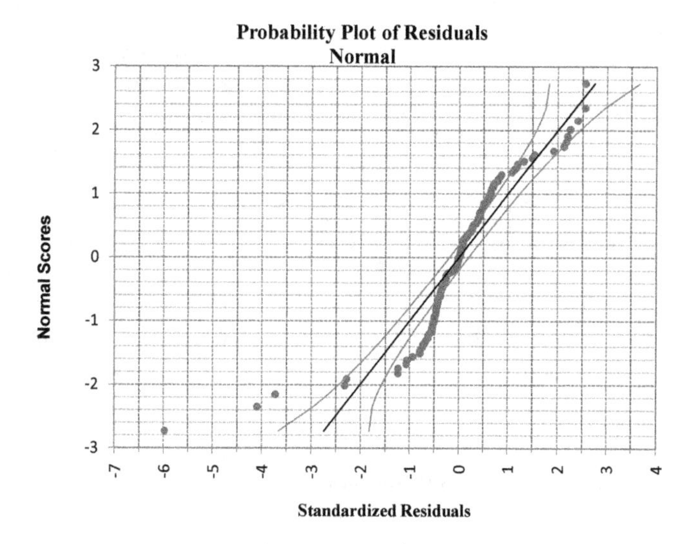

Fig. 1 NPP of residuals from the PRM fit to the data collected by BBD for the response TI.

Table 2 Percentage of outliers in data sets.

	TI(%)	MA(%)	SF(%)	LA(%)	MCP(%)	MT(%)
BBD	11.8	0.0	0.0	5.0	5.0	0.0
CCD	5.5	1.4	8.9	2.7	2.7	4.1
FCD	2.7	0.0	3.4	0.7	0.0	1.3
US	1.3	1.3	0.0	1.3	1.3	0.0
LHS	0.6	0.6	0.0	3.1	0.0	0.6

of outliers is as high as 11.8%. The simulated data are also used to build RRMs including four different estimation techniques, namely M-estimation, MM-estimation, S-estimation and LTS-estimation.

Note that the statistical software MINITAB™[29], SPSS®16.0 [50] and ROBUSTREG procedures of SAS®9.1 [46] are used in the metamodel development and analysis.

Here, the simulation data are collected by using both classical (i.e. CCD, FCD, BBD) and modern designs (i.e. LHS and US) for each performance function listed as outputs in Table 1. Thus, there are 30 simulated data sets. The number of computer experiments is determined by the order of the polynomial constructed and by the type of DOE. In this study, since quadratic RSMs are formed, the number of design points should be at least $[(k+1)(k+2)/2]$, where k is the number of input variables. In CCD and FCD, one-quarter fractional factorial design is utilized, and 149 sampling points are generated from each. Besides 161 data points are generated by BBD. In addition, to be comparative with classical designs, 160 design points are generated by using LHS and US. Representative design space in terms of two

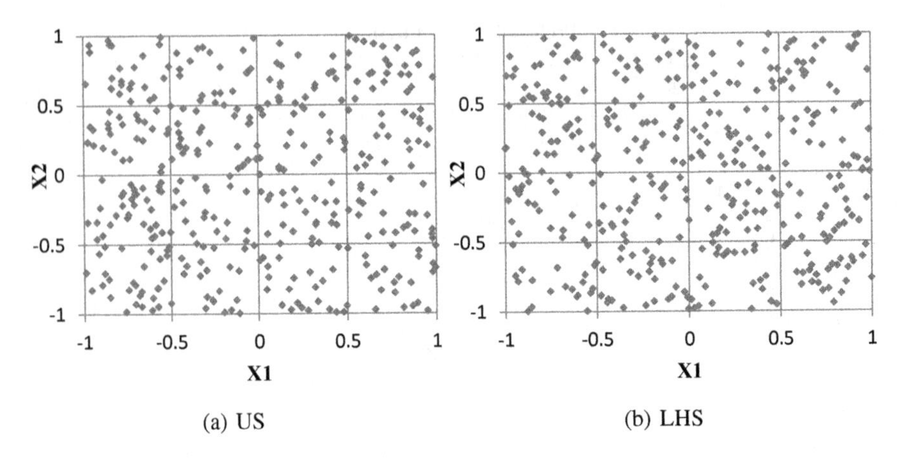

Fig. 2 Design space of the computer experiments.

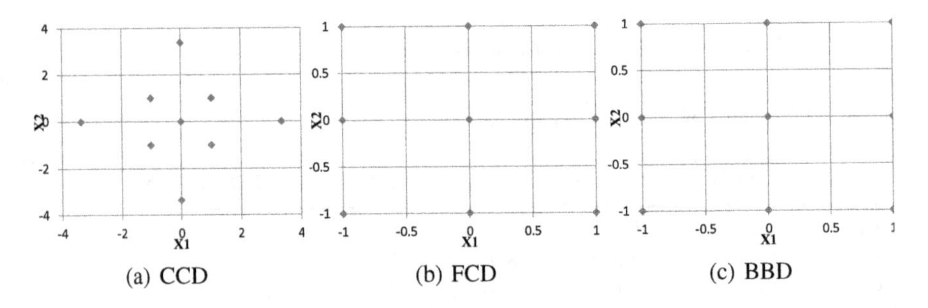

Fig. 3 Design space of the classical designs.

input variables (e.g. x_1 and x_2) for computer experiments and classical designs are visualized in Fig. 2 and Fig. 3, respectively. As it is seen in the figures, most of the design space remains unexplored in classical designs, while whole sampling space is used by the computer experiments.

In order to test the accuracy of surrogate models, the validation technique is applied as described in Section 4.1. The accuracy of metamodels is determined by calculating the MAE, R^2, RMSE and PWI measures (see Section 4.2.1 for their definitions) for each metamodel developed by using the training and test data sets. In addition, stabilities of these measures are calculated to compare the performance of the models on both training and testing data set. As well as the accuracy criterion, the performances of metamodels are also compared with respect to robustness, efficiency and other features including simplicity, ease of use of the metamodel and independency from the expert. The comparison results are presented and discussed in the following section.

6 Results and Discussion

6.1 Overall Performance

The mean and standard deviation values of all accuracy measures for the training and test data sets are presented for each response in Table 3 and Table 4 and for all responses in Table 5, as well as those of the stability of the measures. Note that zero values (0.000) displayed for standard deviations in these tables indicate the values less than 10^{-4}. In these tables in any data set, greater mean for R^2 and PWI and small mean for MAE and RMSE indicate better performance of the associated method while smaller standard deviation indicates more robust method. In the mean time, stability measure close to one indicates better stability. Depending on the results presented in Table 3 and Table 4, the following conclusions may be drawn:

- PRMs perform better than the RRM with respect to RMSE, R^2 and PWI measures in training and with respect to R^2 and PWI in test data sets; robust metamodels seem to perform equally well among themselves.
- PRMs are more stable with respect to R^2 and PWI. Also, the stability of the robust methods with respect to other two measures varies among themselves.
- In training data, the systematic pattern observed in R^2; all metamodels are robust with respect to this measure in half of the data. Besides, PRM is more robust regarding R^2 and PWI.
- Similar behaviour is observed in test data with respect to R^2 and PWI. All metamodels are robust according to R^2 and PWI measures in half of the data.
- The stability of PRMs with respect to R^2 and PWI is more robust. Among robust methods, stabilities of M-estimate is more robust.

Overall performances of methods with respect to four measures considered (see Table 5) can be summarized as follows:

- In train data set, LTS performs better and is more robust with respect to R^2 and PWI. Similarly, M-estimate and PRM perform better and are more robust with respect to MAE and RMSE. And, PRM is statistically better than the robust methods S, MM and LTS regarding RMSE (p-values<0.05).
- In test data, PRM overperforms in terms of all measures, and is more robust with respect to all except RMSE. Besides, all robust methods perform as well as PRM with respect to MAE and PWI. They are also as robust as PRM with respect to PWI.

Furthermore, S- and M-estimate are best performing among robust methods regarding R^2, being S-estimate statistically better than M-estimate (p-value=0.049).

Table 3 Performances of metamodels with respect to different responses.

			TRAIN					TEST					STABILITY				
			S	MM	M	LTS	PRM	S	MM	M	LTS	PRM	S	MM	M	LTS	PRM
y_1	MAE	Mean	*0.009	*0.009	*0.009	*0.009	*0.009	*0.437	*0.437	*0.437	0.441	*0.437	0.022	*0.023	0.022	0.022	0.022
		Std	**0.015	**0.015	0.016	**0.015	**0.015	**0.366	**0.366	**0.366	0.370	0.367	0.020	0.021	0.021	0.020	**0.019
	RMSE	Mean	0.049	*0.047	*0.047	*0.047	*0.047	*0.517	*0.517	*0.517	0.520	*0.517	*0.117	0.115	0.116	0.114	0.115
		Std	**0.089	0.090	0.090	0.090	0.090	**0.433	**0.433	**0.433	0.436	0.434	**0.107	0.109	0.110	0.108	0.109
	R^2	Mean	0.991	*0.992	0.991	0.991	*0.992	*1.000	*1.000	*1.000	0.991	*1.000	0.992	0.992	0.992	*1.000	0.992
		Std	**0.019	**0.019	**0.019	**0.019	**0.019	**0.000	**0.000	**0.000	0.019	**0.000	0.019	0.019	0.019	**0.000	0.019
	PWI	Mean	0.974	0.982	*0.983	*0.983	*0.983	*0.994	*0.994	*0.994	*0.994	*0.994	0.980	0.988	*0.989	*0.989	*0.989
		Std	0.025	0.010	**0.009	**0.009	**0.009	**0.013	*0.013	**0.013	**0.013	**0.013	0.013	0.010	**0.007	**0.007	**0.007
y_2	MAE	Mean	*0.087	0.090	*0.087	0.088	0.090	0.432	0.426	0.429	0.437	*0.421	0.350	0.371	0.331	0.340	*0.383
		Std	**0.069	0.072	**0.069	0.070	0.072	0.371	0.367	0.368	0.374	**0.363	0.285	0.313	**0.240	0.274	0.327
	RMSE	Mean	0.344	0.356	0.358	0.363	*0.304	0.534	0.526	0.533	0.545	*0.523	0.574	0.586	*0.609	0.599	0.518
		Std	0.377	0.376	**0.371	0.376	0.387	0.458	0.453	0.457	0.466	**0.451	0.301	0.286	**0.275	0.287	0.341
	R^2	Mean	0.768	0.760	0.763	0.755	*0.788	0.824	*0.891	0.781	0.856	0.888	0.737	*0.770	0.727	0.723	0.750
		Std	0.411	0.409	**0.399	0.410	0.424	**0.199	0.120	0.229	0.186	**0.119	0.406	0.411	**0.383	0.415	0.409
	PWI	Mean	0.979	0.978	0.981	0.975	*0.991	0.996	0.996	*0.998	0.996	0.996	0.983	0.982	0.983	0.979	*0.992
		Std	0.017	0.019	0.016	0.022	**0.003	0.009	0.009	**0.004	0.009	0.009	0.012	0.017	0.012	0.020	**0.003
y_3	MAE	Mean	*0.016	0.017	0.017	0.019	0.017	0.402	*0.400	0.401	0.401	*0.400	0.030	0.031	0.031	*0.036	0.031
		Std	**0.022	**0.022	0.023	0.025	0.023	0.234	**0.232	**0.232	0.233	**0.232	0.033	*0.032	0.035	0.038	0.035
	RMSE	Mean	*0.088	0.090	*0.088	0.091	*0.088	0.503	*0.501	0.502	0.502	0.502	0.121	0.123	0.121	*0.127	0.120
		Std	0.130	0.128	0.130	**0.127	0.130	0.288	**0.287	0.288	0.288	**0.287	0.167	0.164	0.167	**0.162	0.168
	R^2	Mean	*0.980	*0.980	*0.980	0.979	*0.980	*1.000	*1.000	*1.000	0.999	*1.000	*0.980	*0.980	*0.980	*0.980	*0.980
		Std	**0.037	**0.037	**0.037	0.038	**0.037	**0.000	**0.000	**0.000	0.002	**0.000	**0.037	**0.037	**0.037	**0.037	**0.037
	PWI	Mean	0.971	0.973	0.972	0.970	*0.991	*0.998	0.996	0.996	0.996	0.996	0.973	0.977	0.976	0.974	*0.989
		Std	0.012	0.013	0.013	0.021	**0.010	**0.004	0.005	0.005	0.005	0.005	0.014	0.016	0.015	0.021	**0.009

* indicates better performance with respect to mean; ** indicates better performance with respect to standard deviation.

Table 4 Performances of metamodels with respect to different responses (cont.).

			TRAIN					TEST					STABILITY				
			S	MM	M	LTS	PRM	S	MM	M	LTS	PRM	S	MM	M	LTS	PRM
y_4	MAE	Mean	0.018	0.016	*0.013	0.014	0.014	*0.386	0.387	0.392	0.393	0.394	*0.038	0.034	0.029	0.031	0.030
		Std	0.016	0.015	**0.011	0.012	**0.011	*0.242	0.243	0.248	0.248	0.250	0.021	0.019	**0.011	0.012	**0.011
	RMSE	Mean	0.051	0.044	0.027	0.028	*0.022	*0.478	0.479	0.484	0.486	0.487	*0.085	0.073	0.048	0.052	0.041
		Std	0.060	0.055	0.021	0.022	**0.015	**0.297	0.299	0.305	0.305	0.306	0.077	0.070	0.015	0.016	**0.009
	R^2	Mean	0.995	0.996	*0.999	*0.999	*0.999	0.996	0.998	0.998	0.997	*0.999	0.996	0.998	*0.999	0.998	*0.999
		Std	0.009	0.008	**0.001	**0.001	**0.001	0.004	0.003	**0.002	0.005	**0.002	0.005	0.005	**0.001	0.004	**0.001
	PWI	Mean	0.974	0.974	0.966	0.976	*0.985	*1.000	*1.000	*1.000	*1.000	*1.000	0.974	0.974	0.966	0.976	*0.985
		Std	0.013	0.013	0.013	0.011	**0.009	**0.000	**0.000	**0.000	**0.000	**0.000	0.013	0.013	0.013	0.011	**0.009
y_5	MAE	Mean	0.068	0.094	0.068	*0.067	0.068	0.431	0.439	0.432	*0.430	0.436	0.222	*0.255	0.215	0.214	0.201
		Std	0.028	0.084	0.028	**0.026	0.034	0.282	0.290	0.279	**0.275	0.280	0.132	0.132	0.121	0.124	**0.100
	RMSE	Mean	0.288	0.403	0.288	0.289	*0.247	0.580	0.581	0.577	*0.574	0.576	*0.509	0.435	0.508	0.506	0.3711
		Std	0.352	0.609	**0.340	0.353	0.381	0.354	0.352	0.350	**0.344	0.369	0.365	**0.263	0.353	0.351	0.357
	R^2	Mean	0.820	0.841	0.827	*0.892	0.823	*0.944	0.805	0.931	0.821	0.940	0.818	*0.922	0.830	0.884	0.814
		Std	0.367	0.319	0.346	**0.205	0.381	**0.040	0.341	0.063	0.305	0.067	0.353	**0.114	0.325	0.201	0.360
	PWI	Mean	0.969	0.964	0.972	0.972	*0.990	0.990	*0.992	*0.992	*0.992	*0.992	0.978	0.972	0.980	0.980	*0.993
		Std	0.014	0.011	0.015	0.016	**0.009	**0.000	0.004	0.004	0.004	0.004	0.012	0.008	0.012	0.013	**0.005
y_6	MAE	Mean	*0.000	0.001	0.001	0.001	*0.000	*0.447	*0.447	*0.447	*0.447	*0.447	*0.036	0.018	0.031	0.029	0.008
		Std	**0.000	0.001	0.001	0.001	0.001	**0.408	**0.408	**0.408	**0.408	**0.408	0.054	0.024	0.053	0.038	**0.009
	RMSE	Mean	*0.002	*0.002	*0.002	*0.002	*0.002	0.540	*0.539	*0.539	0.540	*0.539	*0.037	0.021	0.033	0.031	0.010
		Std	**0.003	**0.003	**0.003	**0.003	**0.003	**0.492	**0.492	**0.492	**0.492	**0.492	0.052	0.025	0.051	0.036	**0.010
	R^2	Mean	*1.000	*1.000	*1.000	*1.000	*1.000	*1.000	*1.000	*1.000	*1.000	*1.000	*1.000	*1.000	*1.000	*1.000	*1.000
		Std	**0.000	**0.000	**0.000	**0.000	**0.000	**0.000	**0.000	**0.000	**0.000	**0.000	**0.000	**0.000	**0.000	**0.000	*0.000
	PWI	Mean	0.949	0.971	0.955	*0.981	0.822	*0.990	*0.990	*0.990	*0.990	*0.990	0.959	*0.981	0.959	0.971	0.824
		Std	0.050	0.026	0.052	**0.022	0.376	**0.022	**0.022	**0.022	**0.022	**0.022	0.055	**0.006	0.054	0.022	0.373

* indicates better performance with respect to mean; ** indicates better performance with respect to standard deviation.

Table 5 Overall performances of metamodels (in means and standard deviations).

		TRAIN					TEST					STABILITY				
		S	MM	M	LTS	PRM	S	MM	M	LTS	PRM	S	MM	M	LTS	PRM
MAE	Mean	0.033	0.038	*0.032	0.033	0.033	*0.423	*0.423	*0.423	0.425	*0.423	*0.116	0.122	0.110	0.112	0.112
	Std	**0.045	0.058	**0.045	**0.045	0.047	0.296	0.296	*0.295	0.297	*0.295	0.175	0.191	**0.160	0.169	0.19
RMSE	Mean	0.137	0.157	0.135	0.137	*0.118	0.525	*0.524	0.525	0.528	*0.524	*0.240	0.226	0.239	0.238	0.196
	Std	**0.241	0.318	**0.241	0.244	**0.241	0.361	*0.360	0.361	0.362	0.362	0.292	**0.269	0.297	0.295	0.273
R^2	Mean	0.926	0.928	0.927	*0.936	0.930	0.961	0.949	0.952	0.944	*0.971	0.920	*0.944	0.921	0.931	0.922
	Std	0.227	0.215	0.219	**0.194	0.231	0.100	0.155	0.120	0.153	**0.067	0.227	**0.180	0.216	0.200	0.228
PWI	Mean	0.970	0.974	0.971	*0.976	0.960	*0.995	*0.995	*0.995	*0.995	*0.995	0.975	*0.979	0.976	0.978	0.962
	Std	0.025	**0.016	0.024	0.017	.153	**0.011	**0.011	**0.011	**0.011	**0.011	0.024	**0.012	0.025	0.016	0.152

* indicates better performance with respect to mean; ** indicates better performance with respect to standard deviation.

Table 6 Performances of metamodel groups with respect to DOEs groups.

		CLASSICAL DESIGNS						MODERN DESIGNS					
		TRAIN		TEST		STABILITY		TRAIN		TEST		STABILITY	
		Robust	PRM	Robust	PRM	Robust	PRM	Robust	PRM	Robust	PRM	Robust	PRM
MAE	Mean	0.049	*0.048	0.653	*0.652	*0.075	0.074	0.011	*0.010	*0.079	*0.079	*0.175	0.170
	Std	0.055	**0.054	0.084	**0.083	**0.079	0.081	0.017	**0.015	**0.060	0.063	**0.243	0.281
RMSE	Mean	0.213	*0.182	*0.806	*0.806	*0.231	0.206	0.034	*0.022	0.105	*0.101	*0.243	0.180
	Std	0.314	**0.296	*0.096	0.098	**0.290	0.303	0.051	**0.028	0.083	**0.081	0.279	**0.233
R^2	Mean	0.884	*0.885	0.922	*0.954	*0.884	0.872	0.997	*0.999	0.995	*0.997	*0.997	*0.997
	Std	**0.264	0.292	0.165	**0.083	**0.254	0.286	0.006	**0.002	0.011	**0.006	0.006	**0.005
PWI	Mean	0.973	*0.988	*0.993	*0.993	0.978	*0.987	0.973	*0.990	0.997	*0.998	0.976	*0.990
	Std	0.016	**0.009	**0.013	0.014	0.015	**0.012	0.027	**0.006	**0.005	**0.005	0.026	**0.006

* indicates better performance with respect to mean; ** indicates better performance with respect to standard deviation.

- MM-estimate is more stable with respect to MAE, R^2 and PWI measures. In addition, S-estimate is more stable with respect to RMSE, especially, compared to M-estimate (p-value=0.047). All robust methods are more stable then PRM in terms of RMSE. Although not the best, S-estimate is more stable compared to M-estimate (p-value=0.047).
- The stability of MM-estimate with respect to RMSE, R^2 and PWI and that of M-estimate with respect to MAE are more robust compared to those of the other methods.

In addition to above comparisons, performance of methods for each design is evaluated with respect to all measures by classifying them into two groups: RRM and PRM, and classical (BBD, CCD, FCD) and modern (LHS, US) designs (see Table 6). Results may be interpreted as follows:

- In both data sets, PRM performs better and more robust in terms of all measures in all designs. Particularly, in training data set, PRM is better in modern designs (p-value=0.015) with respect to RMSE. Besides, there are statistically significant difference between the design performances with respect to MAE (p-value=0.025) in training data, whereas with respect to MAE and RMSE (p-value$<$0.01) in test data, regardless of metamodels.
- However, robust methods are more stable and their stabilities are more robust, with an exception that PRM is more stable according to RMSE only (p-value=0.020).

6.2 *Performance with Respect to Designs*

Mean and standard deviation of all measures for each method for different designs are given in Table 7 and Table 8. The effect of design type on the performance of methods can be evaluated as below:

- In training data, S-estimate and MM-estimate do not perform well at all in almost all designs. PRM is good with respect to all measures in modern designs, and also good in BBD and FCD with respect to most of the measures; it is also more robust with respect to the same measures. In addition, statistical analysis indicates that CCD is worse than all other designs (p-values$<$0.05) regardless of metamodels with respect to RMSE and R^2. And PRM is performing better than S-, M-, and LTS-estimate with respect to RMSE (p-value$<$0.05) regardless of the type of DOE. Also, BBD is worse than CCD and LHS with respect to PWI regardless of the metamodels considered (p-values$<$0.05). In training data, PRM is more robust than others, particularly, in modern DOEs.
- MM-estimate is more stable but S-estimate is not at all in classical designs. Also, S-estimate and PRM are more stable than the others in US. Moreover, PRM is more stable compared to the others with respect to RMSE (p-value$<$0.01). Stabilities of all measures for PRM are more robust, especially, in modern designs.

Table 7 Performances of metamodels with respect to DOEs.

			TRAIN					TEST					STABILITY				
			S	MM	M	LTS	PRM	S	MM	M	LTS	PRM	S	MM	M	LTS	PRM
BBD	MAE	Mean	0.042	*0.040	*0.040	*0.040	0.041	0.631	0.627	*0.623	0.635	0.627	0.066	0.065	0.065	0.062	*0.067
		Std	0.062	0.062	0.062	**0.061	0.064	0.126	0.127	**0.123	0.131	**0.123	0.091	0.093	0.098	**0.086	0.099
	RMSE	Mean	0.090	0.080	0.086	0.089	*0.060	0.770	*0.765	*0.765	0.776	0.768	*0.116	0.105	0.113	0.111	0.079
		Std	0.133	0.123	0.136	0.136	**0.085	0.131	0.133	**0.130	0.138	**0.130	0.162	0.154	0.172	0.157	**0.107
	R^2	Mean	0.977	0.981	0.978	0.977	*0.990	0.976	*0.980	0.966	0.976	0.979	0.992	*0.991	0.986	0.986	0.988
		Std	0.048	0.041	0.046	0.050	**0.019	0.041	**0.032	0.072	0.038	0.038	**0.011	0.014	0.029	0.021	0.020
	PWI	Mean	0.954	0.958	0.964	0.969	*0.984	*0.982	*0.982	0.983	*0.982	*0.982	0.972	0.976	0.976	0.971	*0.984
		Std	0.014	0.017	**0.011	0.021	0.013	**0.019	**0.019	0.020	**0.019	**0.019	0.017	**0.016	0.019	0.023	0.019
CCD	MAE	Mean	0.058	0.079	*0.056	0.058	0.058	0.676	0.683	0.683	0.683	*0.684	0.085	*0.111	0.081	0.084	0.083
		Std	**0.044	0.087	0.045	**0.044	0.048	0.073	0.076	0.066	0.064	**0.062	**0.059	0.114	0.060	0.061	0.065
	RMSE	Mean	0.427	0.521	0.405	0.411	*0.410	*0.838	0.841	0.843	0.843	0.849	*0.473	0.406	0.450	0.444	0.441
		Std	*0.415	0.585	0.423	0.431	0.437	0.097	0.096	0.084	**0.083	0.089	0.416	**0.344	0.441	0.431	0.438
	R^2	Mean	0.675	0.693	0.689	*0.737	0.674	0.915	0.830	0.896	0.839	*0.936	0.684	*0.771	0.704	0.744	0.680
		Std	0.448	0.426	0.435	**0.390	0.458	0.154	0.322	0.183	0.284	**0.098	0.437	**0.371	0.412	0.393	0.449
	PWI	Mean	0.989	0.986	0.982	0.986	*0.990	0.998	*1.000	*1.000	*1.000	0.998	0.989	0.986	0.982	0.986	*0.990
		Std	**0.006	**0.006	0.015	0.011	**0.006	0.004	**0.000	**0.000	**0.000	0.004	**0.006	**0.006	0.015	0.011	**0.006
FCD	MAE	Mean	*0.044	0.047	0.045	*0.044	0.046	0.649	0.647	0.650	0.648	*0.645	0.068	*0.073	0.069	0.069	0.072
		Std	**0.054	0.060	0.055	0.055	0.057	0.050	**0.049	0.050	**0.049	**0.049	0.082	0.092	**0.081	0.085	0.089
	RMSE	Mean	0.100	0.115	0.118	0.115	*0.077	0.808	0.803	0.809	0.808	*0.801	0.124	0.144	*0.145	0.142	0.098
		Std	0.104	0.133	0.132	0.130	**0.076	0.063	**0.059	0.061	0.061	0.060	0.126	0.165	0.157	0.158	**0.098
	R^2	Mean	0.982	0.973	0.972	0.973	*0.990	0.922	0.944	0.907	0.915	*0.946	0.931	*0.963	0.922	0.929	0.949
		Std	0.027	0.048	0.046	0.047	**0.014	0.157	**0.104	0.184	0.180	0.105	0.139	**0.062	0.153	0.149	0.092
	PWI	Mean	0.970	0.968	0.973	0.975	*0.989	*0.998	*0.998	*0.998	*0.998	*0.998	0.971	0.969	0.975	0.977	*0.987
		Std	**0.007	0.014	0.011	0.015	0.008	**0.004	**0.004	**0.004	**0.004	**0.004	0.007	0.013	0.009	0.015	**0.006

* indicates better performance with respect to mean; ** indicates better performance with respect to standard deviation.

Table 8 Performances of metamodels with respect to DOEs (cont.).

			TRAIN					TEST					STABILITY				
			S	MM	M	LTS	PRM	S	MM	M	LTS	PRM	S	MM	M	LTS	PRM
US	MAE	Mean	*0.010	0.011	*0.010	0.011	*0.010	*0.070	*0.070	0.071	0.072	*0.070	*0.233	*0.233	0.211	0.215	0.230
		Std	**0.016	0.017	**0.016	0.017	**0.016	**0.061	0.062	**0.061	0.063	0.062	0.329	0.362	**0.288	0.321	0.378
	RMSE	Mean	0.027	0.028	0.026	0.029	*0.024	0.093	0.093	0.093	0.095	*0.092	0.193	0.181	*0.198	0.192	0.154
		Std	0.400	0.400	0.400	**0.399	0.401	**0.372	**0.372	**0.372	**0.372	**0.372	0.375	0.384	**0.373	0.382	0.381
	R^2	Mean	0.998	0.998	0.998	0.998	*0.999	0.996	0.996	*0.997	0.996	*0.997	*0.997	0.996	*0.997	0.996	*0.997
		Std	0.003	0.003	0.003	0.003	**0.002	0.009	0.010	0.008	0.010	**0.007	0.006	0.007	0.006	0.007	**0.005
	PWI	Mean	0.958	0.979	0.960	0.978	*0.990	*0.998	0.997	0.997	0.997	0.997	0.960	0.982	0.964	0.981	*0.990
		Std	0.048	0.009	0.049	0.015	**0.005	**0.004	0.005	0.005	0.005	0.005	0.048	0.006	0.050	0.010	**0.004
LHS	MAE	Mean	0.012	0.012	0.012	0.011	*0.010	*0.086	*0.086	0.087	0.087	0.088	*0.130	0.129	0.124	0.129	0.110
		Std	0.021	0.021	0.021	0.020	**0.016	**0.066	**0.066	**0.066	**0.066	0.068	0.165	0.166	0.164	0.156	**0.149
	RMSE	Mean	0.042	0.041	0.041	0.041	*0.021	0.117	0.11	0.117	0.117	*0.111	0.296	0.291	0.291	*0.301	0.207
		Std	0.070	0.069	0.068	0.067	**0.027	0.095	0.095	0.095	0.094	**0.087	0.379	0.377	0.379	0.380	**0.306
	R^2	Mean	0.996	0.996	0.996	0.996	*0.999	0.994	0.994	0.994	0.994	*0.998	0.997	0.997	0.997	0.997	*0.998
		Std	0.009	0.009	0.009	0.009	**0.002	0.015	0.015	0.014	0.014	**0.005	0.005	0.005	0.005	0.005	**0.004
	PWI	Mean	0.977	0.978	0.978	0.972	*0.991	0.997	0.997	0.997	0.997	*0.998	0.980	0.981	0.981	0.975	*0.991
		Std	0.015	0.015	0.015	0.020	**0.008	0.005	0.005	0.005	0.005	**0.004	0.012	0.013	0.013	0.019	**0.008

* indicates better performance with respect to mean; ** indicates better performance with respect to standard deviation.

6.3 Interactions among Methods and Designs

The investigation on the interaction of methods and designs result in the following findings (see Fig. 4, 5):

- In training data;
  - All methods show similar performance in different designs with respect to MAE, RMSE and R^2.
  - All methods perform worse with respect to MAE, RMSE and R^2 in CCD. In the same design, MM-estimates are particularly worse in performance compared to others according to MAE and RMSE.
  - Designs other than CCD perform almost equally well with respect to RMSE and R^2.
  - All methods are performing the best in modern designs with respect to all measures, except PWI.
  - When PWI measure is considered, there are some interactions between designs and methods; PRM performs the best in terms of PWI among the others.
- In test data;
  - All methods perform the same with respect to all measures in all designs.
  - While all methods perform the best in modern designs, they perform worse in CCD with respect to MAE, RMSE and R^2.
  - Methods perform better in CCD and worse in BBD in terns of PWI; they perform almost equally well in the other designs with respect to the same measure.
- All methods have similar stabilities in terms of MAE, RMSE and R^2.
- While methods are the most stable in UD and CCD with respect to MAE and RMSE, respectively; they are the most stable in BBD, LHS and UD with respect to R^2.
- The stability of the methods differs with respect to designs when PWI is considered; PRM is the most stable one in all designs.

In addition to above comparisons, interacting effects are also investigated between grouped designs and metamodels. Results may indicate the followings:

- In training data no interaction exists between the design and method groups. That is, the performance of the metamodel estimation methods is the same regardless of the design type used.
- In both training and test data, all metamodelling methods perform better in modern designs with respect to all accuracy measures.
- In addition, both method groups are more stable in modern designs in terms of MAE and R^2.
- However, there is an interaction between the method and design groups according to RMSE and PWI. While robust methods are more stable in all designs with respect to RMSE, PRM is more stable in all designs with respect to PWI.

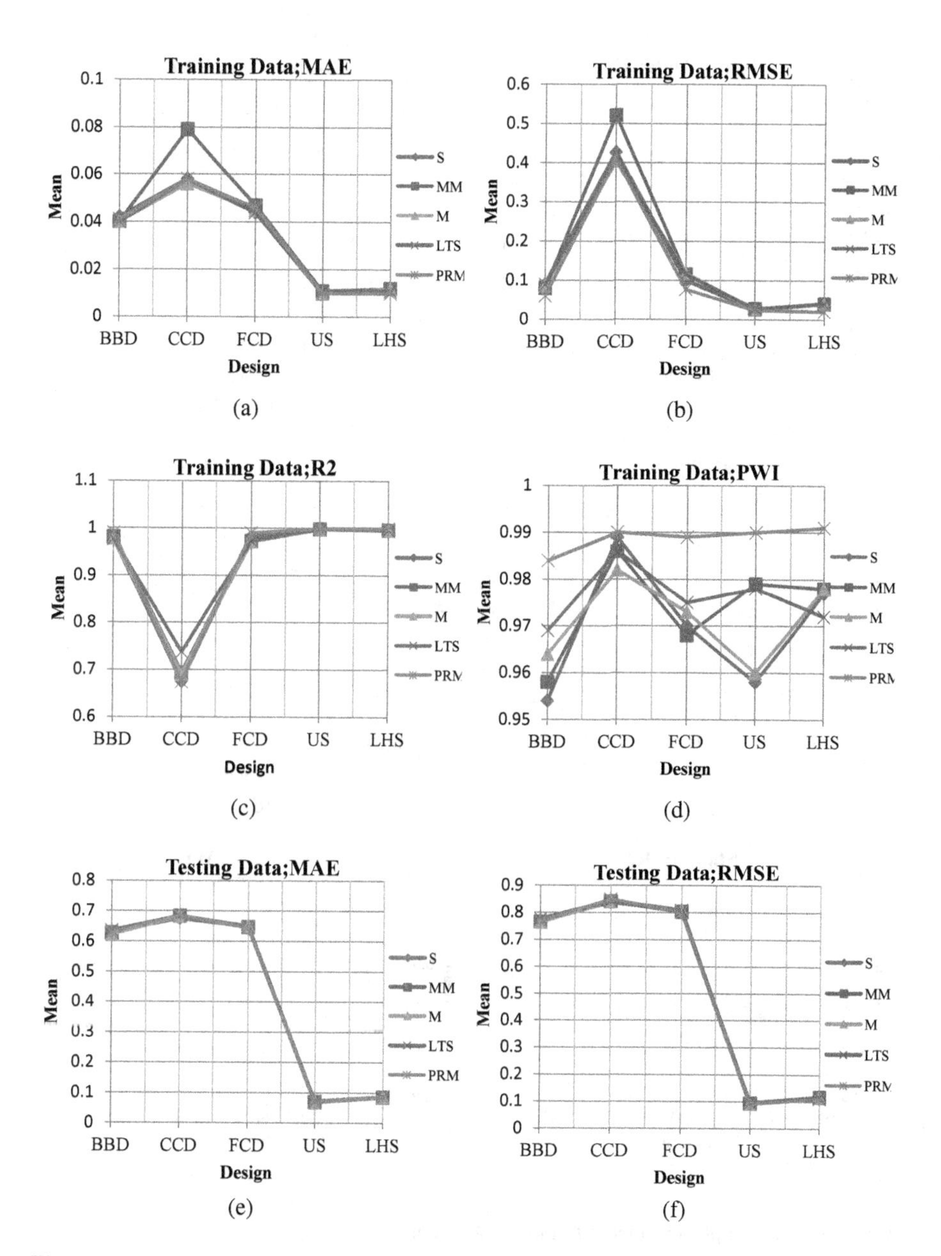

Fig. 4 Interaction plots of methods and designs.

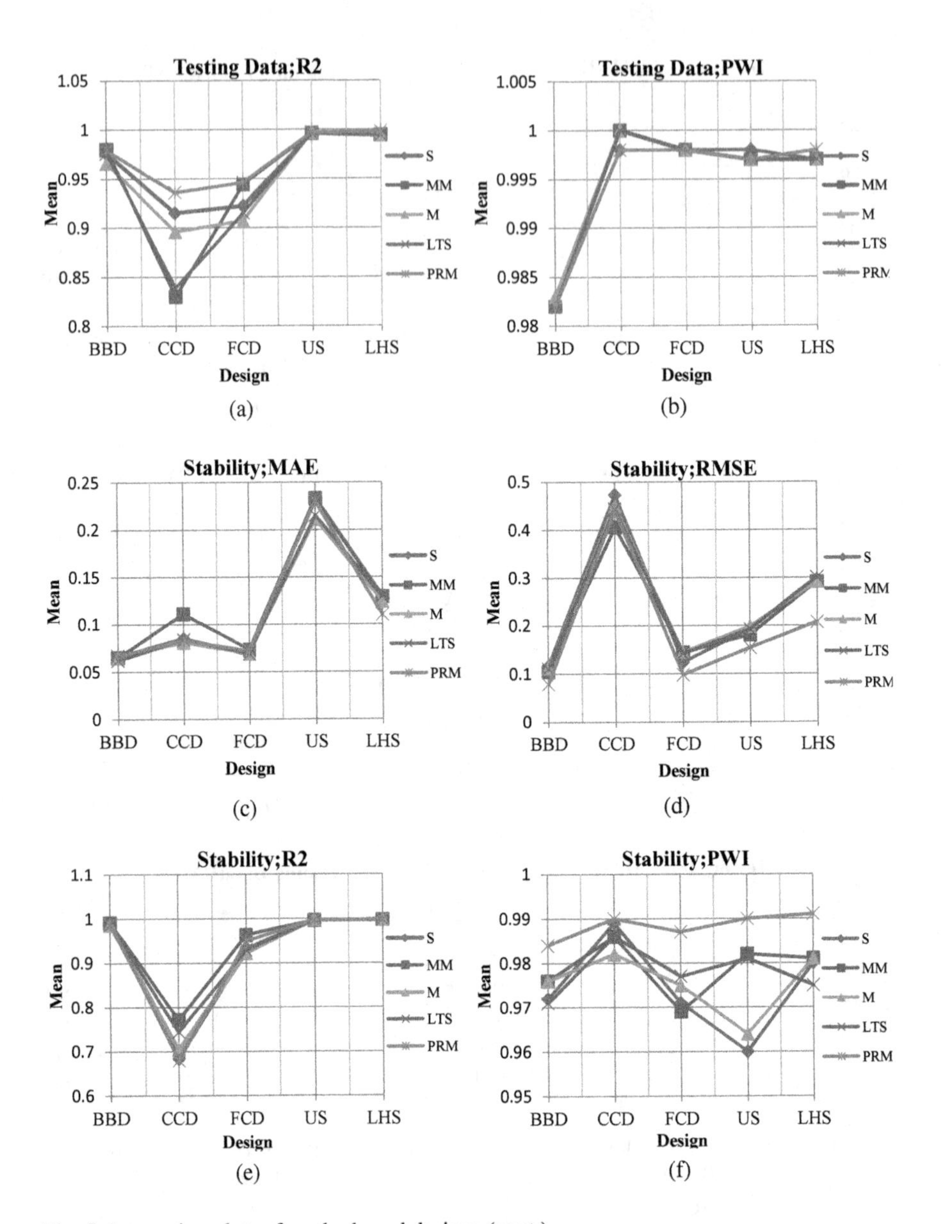

Fig. 5 Interaction plots of methods and designs (cont.)

6.4 Performance with Respect to Efficiency

As mentioned before the efficiency of metamodels can be measured by the time needed for model construction. The problem scale, sample size and computer platforms are some of the variables that govern the time. In our case, the number of input variables used for all metamodels are the same ($k = 10$). In addition, the sizes of samples are particularly determined to be very close to each others (n ranges from 149 to 161). To have a more realistic application environment, metamodels are constructed on the same computer platform (Intel®Core^{TM}2 Duo CPU T7250@2.00 GHz 2.00 GB RAM) by using the SAS®9.1 [46].

The time required for the construction of PRMs depends on model estimation involving parameter estimation and ANOVA, and the residual analysis. If the model is correct, and the assumptions are satisfied, the model estimation procedure takes less than one second for each metamodels. However, the residual analysis conducted for assumption checks constitute most of the time of the metamodel development. If the assumptions are not satisfied, remedial measures should be taken. For example, after outlier detection, an appropriate transformation should be investigated. Hence, finding an appropriate transformation or searching for a new model is another time consuming procedure.

However, RRM technique is capable of dealing with outliers automatically and produces more stable (resistant) estimates for parameters. The time required for parameter estimation, and conducting robust ANOVA is less than five seconds for each metamodels in the presence of outliers. Therefore, in the overall, it can be said that RRM development is less time consuming, that is more efficient, than PRM construction.

6.5 With Respect to Other Features

The metamodelling techniques can also be evaluated with respect to the other features such as simplicity, ease of use, transparency and need for expertise. As far as simplicity is concerned, the PRM estimation is very straightforward with a statistical software support. However, it requires a carefull analysis of residuals, which can not be implemented automatically. This process is exposed to the user interfere in case of assumptions' violation. In this sense, PRM modelling can be considered to be more complicated procedure than that of RRM. On the other hand, M-and MM-estimation methods require some weight functions and scale estimation methods to be determined by the user. Besides, S-estimation, functions used in the calculation of breakdown value should be specified. Nevertheless, with the use of software such as ROBUSTREG procedure in SAS [46], implementation of robust estimation methods becomes very easy by directly using default options provided. This way, RRMs can be developed automatically. Furthermore, both techniques provide transparency in terms of the functional relationship and input factor contributions; so that these models can easily be interpreted with the exception that a power transformation is applied on the response. In that case, RRM may provide us more easy interpretation of the fitted model.

7 Conclusion and Feature Studies

The comparative study presented in this paper has provided insightful observations into performance of RSMs utilized with both LS and RRMs under different modelling criteria. The main goal of the study is basically to emphasize the importance of the LS regression assumptions in the construction of RSMs and to introduce a statistically popular approach to be used in this area. In the study, it is stressed that the reliability of a simulation metamodel depends on the validity of the LS estimation assumptions. If the simulation metamodel is constructed without applying any diagnostic checks for assumptions, the prediction based on this model may not be valid. However, in many studies, including simulation metamodels, these issues are not discussed in detail. On this account, by introducing a new method to this area, it is aimed to make more accurate and efficient predictions in cases where the assumptions are violated.

In the Section 6, performances of metamodels and designs used for approximating the relationships between the performance functions and the propellant properties are rigorously compared. These comparisons indicate that all methods performs almost the same in each design; while modern designs provide similar best performance, US being the best, classical ones provide similar worse performance, CCD being the worst. According to R^2 measure, MM- and LTS-estimates perform the same in all designs while others perform similar, especially, in CCD and FCD. Besides, all methods perform better in modern designs while perform worse in CCD. When PWI measure is considered, on the other hand, all methods perform the same in all designs, CCD being the best. Furthermore, all methods are more stable in terms of MAE and RMSE in US and CCD, respectively. Also, all methods are more stable in modern designs and BBD with respect to R^2; they are not stable at all in CCD. Moreover, the stability of all methods varies with respect to PWI. Note that PRM is the most stable one compared to others in all designs according to this measure.

On the other hand, the comparison of methods and design groups shows that all methods perform better in modern designs according to all accuracy measures considered, and they are more stable in modern designs with respect to MAE and R^2. However, robust methods and PRM are more stable in all designs with respect to RMSE and PWI, respectively.

To conclude, all methods considered are competing with each other with respect to accuracy and stability, particularly, in modern designs. In both training and test data sets, PRM performs better and more robust in terms of all measures in all designs. However, robust methods are more stable and their stabilities are more robust.

At this point, it should be emphasized that the detected outliers are removed from the dataset to satisfy the PRM assumptions. In general, this may lead to a significant loss of information, although it is not too much problem in this case. In the study, it is also stated that the experimental design used for gathering data plays an important role when building metamodels. Therefore, a similar comparative study, which takes into account the accuracy and robustness, is conducted for design of experiments. It is also concluded that the modern designs, namely, LHS and US designs may lead to better performance for all methods than the classical designs. The main reason of this result may be the fact that samples are taken from the interior of a design space in computer experiments. In classical designs, however, the sample points are located at some certain selected points. So, although generally classical designs are used in combination with PRMs, computer experiments may also be proposed because of their better performance.

In terms of efficiency for model construction, PRM can be very time consuming, especially for data contaminated with outliers. Besides, analysis of such data may require an expert knowledge and support. However, robust regression modelling can be considered as a good alternative for PRM based RSMs because of its more efficient estimation performance in the presence of outliers. Since the outliers are automatically handled within the robust regression methodology, there is no need for an additional time used for outlier analysis. This compactness makes the robust regression more applicable for response surface methodology.

Considering simplicity, if PRM is valid, its implementation for prediction is very simple and and time saving. There are no parameters required to be specified by a user. But, some weight functions or scale estimation methods should be specified by the user for developing RRMs unless any skillfull software is utilized. Furthermore, with their final form of equations, both LS and robust regression models are transparent models in terms of the function relationship and factor contribution.

This study can be further extended to several directions. First, similar comparisons can be conducted under the same designs using some computational models such as MARS and an improvement of it, CMARS [58]. Also, multiresponse optimisation can be applied by using both RRMs and PRMs built in this study.

Acknowledgements. We would like to acknowledge Bayındır Kuran for leading us to study on this particular case of solid rocket motors and for his valuable contributions on this study. We also thank M. Sinan Hasanoğlu for providing us the simulated data and Fatma Yerlikaya-Özkurt for preparing our manuscript in LaTeX.

Abbreviations

ANN Artificial neural networks
ANOVA Analysis of variance
BBD Box-Behnken design
CCD Central composite design
DOE Design of experiments
FCD Face-center cube design
GLS Generalized least squares
HS Hammersley sequence
IRWLS Iteratively reweighted least squares
KM Kriging metamodels
LA Launch acceleration

LHD Latin hypercube design
LS Least squares
LTS Least trimmed square
MA Maximum acceleration
MAE Mean absolute error
MARS Multivariate adaptive regression splines
MCP Maximum chamber pressure
MLS Moving least squares
MSE Mean square error
MT Maximum temperature
NPP Normal probability plot
OA Orthogonal arrays
PRM Polynomial regression metamodels
PWI Percentage of residuals within three standard deviation of mean
R^2 Multiple coefficient of determination
RBF Radial basis functions
RMSE Root mean square error
RRM Robust regression metmodels
RSM Response surface metamodels
TI Total impulse
UD Uniform design.

References

1. Barton, R.R.: Simulation optimization using metamodels. In: Rossetti, M.D., Hill, R.R., Johansson, B., Dunkin, A., Ingalls, R.G. (eds.) Proceeding of the 2009 Winter Simulation Conference (2009)
2. Batmaz, İ., Tunali, S.: Second order experimental designs for simulation metamodeling. Transactions of the Society for Computer Simulation 78, 699–715 (2002)
3. Batmaz, İ., Tunali, S.: Small response surface designs for metamodel estimation. European Journal of Operational Research 145, 455–470 (2003)
4. Booker, A.J., Dennis Jr., J.E., Frank, P.D., Serafini, D.B., Torczon, V., Trosset, M.W.: A rigorous framework for optimization of expensive functions by surrogates. Structural Optimization 17(1), 1–13 (1999)
5. Box, G.E.P., Wilson, K.B.: On the experimental attainment of optimum conditions. Journal of the Royal Statistical Society 13, 1–45 (1951)
6. Box, G.E.P., Behnken, D.W.: Some new three-level designs for the study of quantitative variables. Technometrics 2, 455–475 (1960)
7. Box, G.E.P., Hunter, W.G., Hunter, J.S.: Statistics for Experimenters: An Introduction to Design, Data Analysis, and Model Building. John Wiley and Sons, New York (1978)
8. Box, G.E.P., Draper, N.R.: Emprical Model-building and Response Surfaces. Wiley, New York (1987)
9. Bozkaya, K., Sumer, B., Kuran, B., Ak, M.A.: Reliability analysis of solid rocket motor based on response surface method and monte carlo simulation. In: 41st AIAA/ASME/SAE/ASEE Joint Propulsion Conference and Exhibit, Tucson, Arizona (2005)

10. Bozkaya, K., Kuran, B., Hasanoglu, M.S., Yıldırım, C., Ak, M.A.: Effects of production variations on the reliability of a solid rocket motor. In: 42nd AIAA/ASME/SAE/ASEE Joint Propulsion Conference & Exhibit, Sacramento, California (2006)
11. Chen, C.: Robust Regression and outlier detection with the ROBUSTREG procedure. In: Proceedings of the Twenty-Seventh Annual SAS Users Group International Conference. SAS Institute Inc., Cary (2002)
12. Dönmez, A., Kartal, E., Batmaz, İ., Kuran, B.: Comparison of linear and robust regression methods for metamodling complex systems. In: 23rd European Conference on Operational Research, Bonn, Germany (2009)
13. Dyn, N., Levin, D., Rippa, S.: Numerical procedures for surface fitting of scattered data by radial functions, SIAM J. Sci. Statist. Comput. 7, 639–659 (1986)
14. Fang, K.T.: Experimental design by uniform distribution. Acta Mathematice Applicatae Sinica 3, 363–372 (1980)
15. Friedman, J.: Multivariate adaptive regression splines. Annals of Statistics 19, 1–141 (1991)
16. Guinta, A.A., Wojtkkiewicz, S.F., Eldred, M.S.: Overview of modern design of experiments methods for computational simulations. In: 41st Aerospace Sciences Meeting and Exhibit, Reno, Nevada (2003)
17. Hardy, R.L.: Multiquadratic equations of topography and other irregular surfaces. Journal of Geophysical Research 76, 1905–1915 (1971)
18. Huber, P.J.: Robust regression: Asymptotics, conjectures and Monte Carlo. Ann. Stat. 1, 799–821 (1973)
19. Jin, R., Du, X., Chen, W.: The use of metamodeling techniques for optimization under uncertainity. Structural and Multidisciplinary Optimization 25(2), 99–116 (2003)
20. Jin, R., Chen, W., Simpson, T.W.: Comparative studies of metamodeling techniques under multiple modeling criteria. Structural and Multidisciplinary Optimization 23(1), 1–13 (2001)
21. Kartal, E.: Metamodeling complex systems using linear and nonlinear regression methods. MS thesis (2007) (unpublished)
22. Kartal, E., Batmaz, İ.: Comparison of linear and nonlinear metamodeling techniques for complex systems. In: COMPSTAT 2008: International Conference on Computational Statistics, Porto, Portugal (2008)
23. Kartal, E., Batmaz, İ., Kuran, B.: Robust metamodeling of complex systems. In: Workshop on Recent Developments in Financial Mathematics and Stochastic Calculus, METU, Ankara (2008)
24. Kartal-Koç, E., Batmaz, İ., Köksal, G.: Multicriteria optimisation for the performance of solid rocket motor. In: 23rd European Conference on Operational Research, Bonn, Germany (2009)
25. Kleijnen, J.P.C.: Statistical Tools for Simulation Proctitioners. Marcel Dekker, NY (1987)
26. Kleijnen, J.P.C.: Design and Analysis of Simulation Experiments. Springer, NY (2008)
27. Kuran, B.: Reliability based design optimization of a solid rocket motor using surrogate models. In: 43rd AIAA/ASME/SAE/ASEE Joint Propulsion Conference and Exhibit., Cincinnati, Ohio (2007)
28. Kuran, B., Hasanoglu, M.S., Bozkaya, K.: Robust design optimization for multiple responses using response surface methodology and taguchi approach: Solid rocket motor application. In: 9th AIAA Non-Deterministic Approaches Conference, Honolulu, Hawaii (2007)
29. MINITAB: Statistical and Process Management Software for Six Sigma and Quality Improvement, version 14 (2008), http://www.minitab.com

30. Mourelatos, Z.P., Zhou, J.: A design optimization method using evidence theory. Journal of Mechanical Design 128(4), 901–908 (2006)
31. McKay, M.D., Bechman, R.J., Conover, W.J.: A Comparison of three methods for selecting values of input variables in the analysis of output from a computer code. Technometrics 21(2), 239–245 (1979)
32. Montgomery, D.C., Peck, E.A., Vining, G.G.: Introduction to Linear Regression Analysis. Wiley & Sons, NY (2001)
33. Montgomery, D.C.: Design and Analysis of Experiments. John Wiley & Sons, NY (2001)
34. Myers, R.H., Montgomery, D.C.: Response Surface Methodology: Process and Product Optimization Using Designed Experiments. Wiley & Sons, NY (2002)
35. Neter, J., Wasserman, W., Kutner, M.H.: Applied Linear Statistical Models. Irwin, Boston (1996)
36. Osei-Bryson, K.M.: Evaluation of decision trees: A multi-criteria approach. Computers Operations Research 31, 1933–1945 (2004)
37. Owen, A.B.: Orthogonal arrays for computer experiments, integration and visualization. Statistica Sinica 2, 439–452 (1992)
38. Papila, N., Shyy, W., Griffin, L., Huber, F., Tran, K.: Preliminary design optimization for a supersonic turbine for rocket propulsion. In: 36th AIAA/ASME/SAE/ASEE, Joint Propulsion Conference and Exhibit, Huntsville, Alabama (2000)
39. Papila, N., Shyy, W., Griffin, L.W., Dorney, D.J.: Shape optimization of supersonic turbines using response surface and neural network methods. Journal of Propulsion Power 18, 509–518 (2001)
40. Ramu, M., Raja, V.P., Thyla, P.R., Gunaseelan, M.: Design optimization of complex structures using metamodels. Jordan Journal of Mechanical and Industrial Engineering 4(5), 653–664 (2010)
41. Rawling, O.J.: Applied Regression Analysis: A Search Tool. Wadsworth, Belmont (1988)
42. Rousseeuw, P.J.: Least median of squares regression. Journal of the American Statistical Association 79, 871–880 (1984)
43. Rousseeuw, P.J., Yohai, V.: Robust Regression by means of S estimators, robust and nonlinear time series analysis. Lecture Notes in Statistics, vol. 26, pp. 256–274. Springer, New York (1984)
44. Rousseeuw, P.J., Leroy, A.M.: Robust Regression and Outlier Detection. Wiley- Interscience, NY (1987)
45. Sacks, J., Welch, W.J., Mitchell, T.J., Wynn, H.P.: Design and analysis of computer experiments. Statistical Science 4(4), 409–435 (1989)
46. SAS: Business Analytics and Business Intelligence Software. Version 9.1 (2008), `http://www.sas.com`
47. Simpson, T.W., Mauery, T.M., Korte, J.J., Mistree, F.: Comparison of response surface and kriging models for multidisciplinary design optimization. In: 7th AIAA/USAF/-NASA/ISSMO Symposium on Multidisciplinary Analysis & Optimization, St. Louis, MO, vol. 1, pp. 381–391 (1998)
48. Simpson, T.W., Perlinski, J., Koch, P.N., Allen, J.K.: Metamodels for computer-based engineering design, survey and recommendations. Engineering with Computers 17(2), 129–150 (2001)
49. Sobieszczanski-Sobieski, J., Haftka, R.T.: Multidisciplinary aerospace design optimization: Survey of recent developments. Structural Optimization 14, 1–23 (1997)
50. SPSS: Statistical Package for the Social Sciences. Version 15.0. (2010), `http://www-01.ibm.com/software/analytics/spss/`

51. Staum, J.: Better simulation metamodeling: The why, what and how of stochastic kriging. In: Rossetti, M.D., Hill, R.R., Johansson, B., Dunkin, A., Ingalls, R.G. (eds.) Proceeding of the 2009 Winter Simulation Conference, Austin, TX (2009)
52. Taylan, P., Yerlikaya-Özkurt, F., Weber, G.W.: An approach to mean shift outlier model (MSOM) by Tikhonov regularization and conic programming. Preprint at IAM, METU, Submitted to Intelligent Data Analysis guest (eds.) (2011)
53. Tunalı, S., Batmaz, İ.: Dealing with the least squares regression assumptions in simulation metamodeling. International Journal of Computers and Industrial Engineering 38, 307–320 (2000)
54. Tunalı, S., Batmaz, İ.: A metamodeling methodology involving both quantitative and qualitative variables. European Journal of Operational Research 150, 437–450 (2003)
55. Vaidyanathan, R., Papila, N., Shyy, W., Tucker, K.P., Griffin, L.W., Haftka, R.T., Fitz-Coy, N.: Neural network and response surface methodology for rocket engine component optimization. In: 8th AIAA/USAF/NASA/ISSMO Symposium on Multidisciplinary Analysis and Optimization, Long Beach, CA (2000)
56. Varadarajan, S., Chen, W., Pelka, C.: The robust concept exploration method with enhanced model approximation capabilities. Engineering Optimization 5, 787–809 (2000)
57. Venter, G., Haftka, R.T., Starnes Jr., J.H.: Construction of response surfaces for design optimization applications. In: 6th AIAA/USAF/NASA/ISSMO Symposium on Multidisciplinary Analysis and Optimization, Bellevue, WA, vol. 1, pp. 548–564 (1996)
58. Weber, G.W., Batmaz, İ., Köksal, G., Taylan, P., Yerlikaya-Özkurt, F.: A New Contribution to Nonparametric Regression with Multivariate Adaptive Regression Splines Supported by Continuous Optimization. Inverse Problems in Science and Engineering (2011), doi:10.1080/17415977.2011.624770
59. Werbos, P.J.: Generalization of backpropagation with application to a recurrent gas market model. Neural Networks 1(4), 339–356 (1988)
60. Won, H., Levine, S., Pfaender, H., Mavris, D.N.: Using response surface metamodels to optimize the aerodynamic performance of a high speed stanoff missile within a multidisciplinary environment. In: Aircraft Technology, Integration and Operations (ATIO), Los Angeles, CA (2002)
61. Yang, R.J., Gu, L., Liaw, L., Gearhart, C., Tho, C.H., Liu, X., Wang, B.P.: Approximations for safety optimization of large systems. In: Renaud, J.E. (ed.) ASME 2000 Design Engineering Technical Conferences-Design Automation Conference, ASME, Paper No. DETC-2000/DAC-14245, Baltimore, MD (2000)
62. Yerlikaya-Özkurt, F., Taylan, P., Weber, G.W.: Mean shift outlier model with MARS and continuous optimization. In: IFORS 2011, Melbourne, Australia (2011)
63. Yohai, V.J.: High breakdown point and high efficiency robust estimates for regression. Annals of Statistics 15, 642–656 (1987)

Application of Analytic Programming for Evolutionary Synthesis of Control Law—Introduction of Two Approaches

Roman Šenkeřík, Zuzana Oplatková, Ivan Zelinka, and Roman Jašek

Abstract. This research deals with an evolutionary synthesis of control law for Logistic equation, which is a discrete chaotic system. The novelty of the research is that an Analytic Programming (AP), which is a tool for symbolic regression, is used for the synthesis of feedback controller for chaotic system. This work introduces and compares two approaches representing blackbox type cost function, as well as not-blackbox type cost function. These two approaches are used for the purpose of stabilisation of the higher periodic orbits, which stand for oscillations between several values of chaotic system. The work consists of the descriptions of analytic programming as well as chaotic system and used cost functions. For experimentation, Self-Organising Migrating Algorithm (SOMA) and Differential Evolution (DE) were used.

1 Introduction

There is a growing interest about the interconnection between evolutionary techniques and control of chaotic systems. The first steps were done in [23, 17, 18], where the control law was based on the Pyragas method, which is Extended delay feedback control (ETDAS) [13]. These papers were concerned with tuning several parameters inside the control technique for chaotic system. Compared to this, presented research also shows a possibility for generating the whole control law (not only to optimize several parameters) for the purpose of stabilisation of a chaotic

Roman Šenkeřík · Zuzana Oplatková · Roman Jašek
Tomas Bata University in Zlín, Faculty of Applied Informatics,
Nam. T.G. Masaryka 5555, 760 01 Zlín, Czech Republic
e-mail: {senkerik,oplatkova,jasek}@fai.utb.cz

Ivan Zelinka
Technical University of Ostrava, Faculty of Electrical Engineering and Computer Science,
17. listopadu 15, 708 33 Ostrava-Poruba, Czech Republic
e-mail: ivan.zelinka@vsb.cz

A. Byrski et al. (Eds.): Advances in Intelligent Modelling and Simulation, SCI 416, pp. 253–268.
springerlink.com

systems. The synthesis of control law is inspired by the Pyragas' delayed feedback control TDAS and ETDAS [6, 12].

These two methods are very advantageous for evolutionary computation, due to the amount of accessible control parameters, which can be easily tuned by means of evolutionary algorithms (EA).

Instead of EA utilisation, analytic programming (AP) is used in this research. AP is a superstructure of EAs and is used for synthesis of analytic solution according to the required behaviour [21]. Control law from the proposed system can be viewed as a symbolic structure, which can be synthesized according to the requirements for the stabilisation of the chaotic system. The advantage is that it is not necessary to have some "preliminary" control law and to estimate its parameters only. This system will generate the whole structure of the law even with suitable parameter values.

This work is focused on the expansion of AP application for synthesis of a whole control law instead of parameters tuning for existing and commonly used method control law to stabilize desired Unstable Periodic Orbits (UPO) of chaotic systems.

This work is a cumulation of previous work [17, 14] focused either on tuning of parameters for an existing control method [17] or synthesis of whole control laws [14] focused on stabilisation of simple $p-1$ orbit—stable state. In general, this research is concerned to stabilize $p-2$ UPO—higher periodic orbits (oscillations between two values) and represents the expansion of preliminary studies [15, 16]. Furthermore it introduces two possible approaches. Firstly, the simple cost function utilising the position of desired UPO is described. Afterwards, blackbox approach cost function, thus without knowledge about exact UPO position in the chaotic attractor, is introduced. This means that AP will synthesize suitable control law based only on the demanded type of chaotic system behaviour and not based on the position of UPO.

Firstly, AP is explained, and then a problem design is proposed. The next sections are focused on the description of used cost function and evolutionary algorithms. Results and conclusion follow afterwards.

2 Analytic Programming

Basic principles of the AP were developed in 2001 [22]. Until that time only Genetic Programming (GP) and Grammatical Evolution (GE) had existed. GP uses Genetic Algorithms (GA) while AP can be used with any EA, independently on individual representation. To avoid any confusion, based on the nomenclature according to the used algorithm, the name - Analytic Programming was chosen, since AP represents synthesis of analytical solution by means of EA. Various applications of AP are described in [21].

The core of AP is based on a special set of mathematical objects and operations. The set of mathematical objects is a set of functions, operators and so-called terminals (as well as in GP), which are usually constants or independent variables. This set of variables is usually mixed together and consists of functions with different number of arguments. Because of a variability of the content of this set, it is termed the "general functional set"—GFS. The structure of GFS is created by subsets of functions according to the number of their arguments. For example GFSall is a set of all functions, operators and terminals, GFS3arg is a subset containing functions with only three arguments, GFS0arg represents only terminals, etc. The subset structure presence in GFS is vitally important for AP. It is used to avoid synthesis of pathological programs, i.e. programs containing functions without arguments, etc. The content of GFS is dependent only on the user. Various functions and terminals can be mixed together [22].

The second part of the AP core is a sequence of mathematical operations, which are used for the program synthesis. These operations are used to transform an individual of a population into a suitable program. Mathematically stated, it is a mapping from an individual domain into a program domain. This mapping consists of two main parts. The first part is called Discrete Set Handling (DSH) (Fig. 1) [22, 7, 9] and the second one stands for security procedures which do not allow synthesising pathological programs. The method of DSH allows handling arbitrary objects including nonnumeric objects like linguistic terms {hot, cold, dark,...}, logic terms (True, False) or other user defined functions. In the AP, DSH is used to map an individual into GFS and together with security procedures creates the above-mentioned mapping, which transforms arbitrary individual into a program.

Fig. 1 depicts the following procedure. Numbers in the individual of any evolutionary algorithm are used as indices into discrete set of parameters. These parameters like real numbers, linguistic terms and others are applied in the cost function to find out the quality of the solution. The indices in the individual serve in the evolutionary process but not for obtaining final quality of the solution.

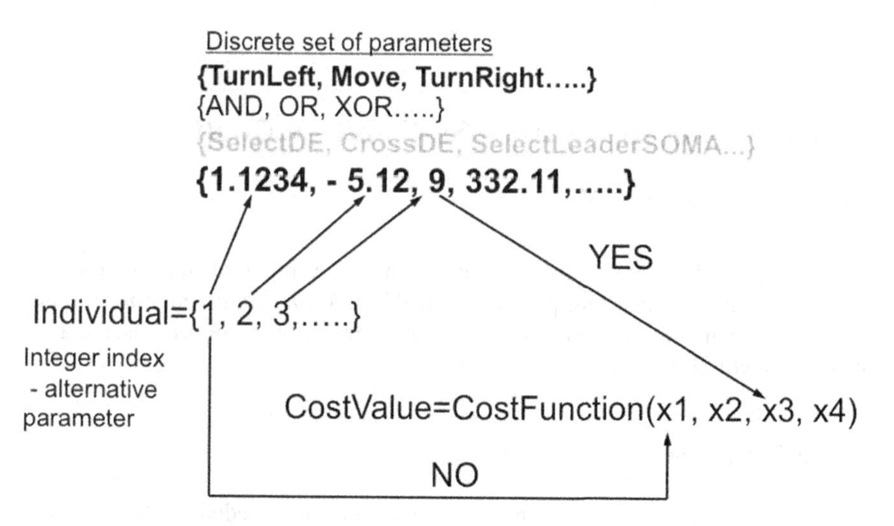

Fig. 1 Discrete set handling

AP needs some EA [9] that consists of a population of individuals for its run. Individuals in the population consist of integer parameters, i.e. an individual is an integer index pointing into GFS. The creation of the program can be schematically observed in Fig. 2. The individual contains numbers which are indices into GFS. According to this process, index equal to one in Fig. 2 means the first operator in GFS, which is an operator plus. As operator plus needs two arguments, following two indices in the individual are used—sixth Sin plus seventh Cos. Sin and Cos functions need one argument therefore other indices are applied in the similar manner as described above. The security option ensures that items closer to the end in the individual are replaced from GFS with lower number of arguments. The detailed description is represented in [22, 7, 9].

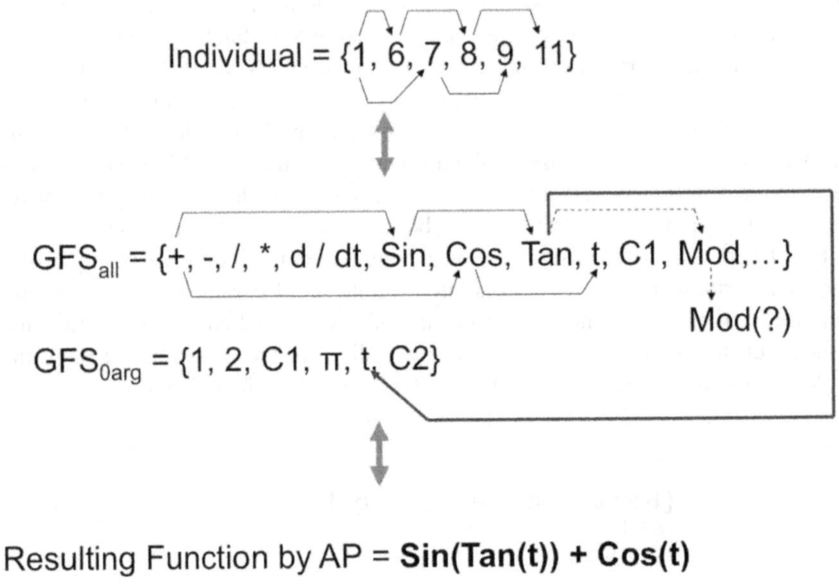

Individual = {1, 6, 7, 8, 9, 11}

GFS_{all} = {+, -, /, *, d / dt, Sin, Cos, Tan, t, C1, Mod,...}

Mod(?)

GFS_{0arg} = {1, 2, C1, π, t, C2}

Resulting Function by AP = **Sin(Tan(t)) + Cos(t)**

Fig. 2 The main principle of AP

AP exists in 3 versions—basic without constant estimation, AP_{nf}—estimation by means of nonlinear fitting package in *Wolfram Mathematica* environment and AP_{meta}—constant estimation by means of another evolutionary algorithms; meta implies meta-evolution.

3 Problem Design

The brief description of used chaotic system, original feedback chaos control method ETDAS, and two used cost functions is given here. The ETDAS control technique was used in this research as an inspiration for synthesising a new feedback control law by means of evolutionary techniques.

3.1 Selected Chaotic System

The chosen example of chaotic system was the one-dimensional Logistic equation as in form (1):

$$x_{n+1} = rx_n(1 - x_n) \tag{1}$$

The Logistic equation (Logistic map) is a one-dimensional discrete-time example of how a complex chaotic behaviour can arise from very simple non-linear dynamical equation [5]. This chaotic system was introduced and popularised by the biologist Robert May [8]. It was originally introduced as a demographic model of a typical predator—prey relationship. The chaotic behaviour can be observed by varying the parameter r. When $r = 3.57$, this is the beginning of chaos, at the end of the period-doubling behaviour. When $r > 3.57$, the system exhibits chaotic behaviour.

The example of this behaviour can be clearly seen from the bifurcation diagram in Fig. 3.

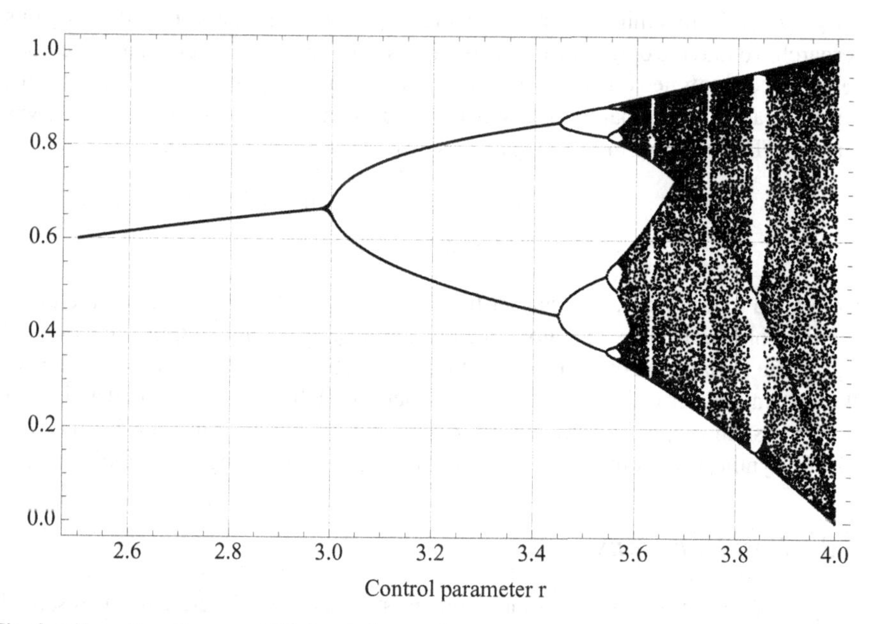

Fig. 3 Bifurcation diagram of the Logistic equation

3.2 ETDAS Control Method

This work is focused on explanation of application of AP for synthesis of a whole control law instead of demanding tuning of EDTAS method control law to stabilize desired UPO. In this research, desired UPO is only $p-2$ (higher periodic orbit—oscillation between two values). ETDAS method was obviously an inspiration for preparation of sets of basic functions and operators for AP. The original control method—ETDAS has form (2):

$$F(t) = K[(1-R)S(t-\tau_d) - x(t)]$$
$$S(t) = x(t) + RS(t-\tau_d) \quad (2)$$

where: K and R are adjustable constants, F is the perturbation; S is given by a delay equation utilising previous states of the system and is a time delay. The original control method—ETDAS in the discrete form suitable for one-dimensional logistic equation has the form (3):

$$x_{n+1} = rx_n(1-x_n) + F_n$$
$$F_n = K[(1-R)S_{n-m} - x_n]$$
$$S_n = x_n + RS_{n-m} \quad (3)$$

Where: m is the period of m-periodic orbit to be stabilized. The perturbation F_n in equations (3) may have arbitrarily large value, which can cause diverging of the system outside the interval $\{0, 1.0\}$. Therefore, F_n should have a value between $-F_{max}, F_{max}$. In this initial study a suitable F_{max} value was taken from the previous research. To find the optimal value also for this parameter is in future plans. Previous research concentrated on synthesis of control law only for $p-1$ orbit (a fixed point). An inspiration for preparation of sets of basic functions and operators for AP was simpler TDAS control method (4) and its discrete form given in (5).

$$F(t) = K[x(t-\tau) - x(t)] \quad (4)$$
$$F_n = K(x_{n-m} - x_n) \quad (5)$$

Compared to this work, the data set for AP presented in the previous research [14] required only constants, operators like plus, minus, power and output values x_n and x_{n-1}. Due to the recursive attributes of delay equation S utilising previous states of the system in discrete ETDAS (3), the data set for AP had to be expanded and cover longer system output history, thus to imitate inspiring control method for the successful synthesis of control law securing the stabilisation of higher periodic orbits.

3.3 Cost Functions

Proposal for the cost function comes from the simplest Cost Function (CF) presented in [17]. The core of CF could be used only for the stabilisation of $p-1$ orbit. The idea was to minimize the area created by the difference between the required state and the real system output on the whole simulation interval τ_i.

But another universal cost function had to be used for stabilising of higher periodic orbit and having the possibility of adding penalisation rules. It was synthesized from the simple CF and other terms were added. In this case, it is not possible to use the simple rule of minimising the area created by the difference between the required and actual state on the whole simulation interval τ_i, due to many serious reasons, for example: degrading of the possible best solution by phase shift of periodic orbit.

This CF is in general based on searching for desired stabilized periodic orbit and thereafter calculation of the difference between desired and found actual periodic orbit on the short time interval τ_s (approx. 20–50 iterations) from the point, where the first min. value of difference between desired and actual system output is found. Such a design of CF should secure the successful stabilisation of either $p-1$ orbit (stable state) or higher periodic orbit anywise phase shifted. The CF_{Basic} has the form (6):

$$CF_{Basic} = pen_1 + \sum_{t=\tau_1}^{\tau_2} |TS_t - AS_t| \tag{6}$$

where TS: target state, AS: actual state, τ_1: the first minimal value of difference between TS and AS, τ_2: the end of optimisation interval $(\tau_1 + \tau_s)$, $pen_1 = 0$ if $\tau_i - \tau_2 \geq \tau_s$ or $pen_1 = 10 \cdot (\tau_i - \tau_2 < \tau_s)$ (i.e. late stabilisation).

Other cost functions (CF2) had to be used for the stabilising of the chaotic system in "blackbox mode", i.e. without exact numerical value of target state. In this case, it is not possible to use the simple rule of minimising the area created by the difference between the required and actual state on the whole simulation interval τ or its arbitrary part.

Our approach is based on searching for periodic orbits in chaotic attractor and stabilising the system on these periodic orbits by means of applying the optimal feedback perturbation F_n. It means that this new CF did not take any numerical target state into consideration, but the selected target behaviour of system. Therefore, the new CF is based on the searching for optimal feedback perturbation F_n securing the stabilisation on any type of selected UPO ($p-1$ orbit—stable state, $p-2$ orbit—oscillating between two values etc.). The slight disadvantage of this approach is that for each UPO (i.e. different behaviour) a different CF is needed.

The proposal of CF2 used in the case of $p-2$ orbit is based on the following simple rule. The iteration $y(n)$ and $y(n+2)$ must have the same value. But this rule is also valid for the case of $p-1$ orbit, where in discrete systems, the iteration $y(n)$ and $y(n+1)$ of output value must be the same. Thus another condition had to be added. It says that in the case of $p-2$ orbit there must be some difference between the n and $n+1$ output iteration. Considering the fact of minimising the CF the value this condition had to be rewritten into this suitable form (7):

$$\frac{1}{|y(n+1) - y(n)| + c} \tag{7}$$

where c: small constant $1 \cdot 10^{-16}$ which was added to prevent the evolutionary optimisation from crashing, since upon finding the suboptimal solution stabilized at $p-1$ orbit it returns the division by zero, $y(n)$ is output value in discrete time step in chaotic system. The CF_2 has the form (8).

$$CF_2 = p1 + \sum_{t=0}^{\tau} |y(n+2) - y(n)| + \frac{1}{|y(n+1) - y(n)| + c} \tag{8}$$

where $p1$: penalisation.

In the proposed CF there had to be included penalisation, which should avoid the finding of solutions, where the stabilisation on saturation boundary values $\{0,1\}$ or oscillation between them (i.e. artificial $p-2$ orbit) occurs. This penalisation was calculated as the sum of the number of iterations, where the system output reaches the saturation boundary value.

4 Used Evolutionary Algorithms

Analytic programming needs any population based evolutionary algorithm for its run. This research used two evolutionary algorithms: Self-Organising Migrating Algorithm (SOMA) [19] and Differential Evolution (DE) [11].

SOMA is a stochastic optimisation algorithm that is modeled on the social behaviour of cooperating individuals. Numerous applications either of canonical [2, 1, 3] or special version [4] of SOMA have proven that these heuristics are suitable for solving a difficult class of problems.

DE is a population-based optimisation method that works on real-number-coded individuals. Both algorithms were chosen because it has been proven that they have the ability to converge towards the global optimum.

4.1 Self Organising Migration Algorithm—SOMA

SOMA works with groups of individuals (population) whose behaviour can be described as a competitive-cooperative strategy. The construction of a new population of individuals is not based on evolutionary principles (two parents produce offspring) but on the behaviour of social group, e.g., a herd of animals looking for food. This algorithm can be classified as an algorithm of a social environment. To the same group of algorithms, sometimes called swarm intelligence, Particle Swarm Optimisation (PSO) algorithm can also be put in. In the case of SOMA, there is no velocity vector as in PSO, only the position of individuals in the search space is changed during one generation, here called 'migration loop'.

The rules are as follows: In every migration loop the best individual is chosen, i.e. individual with the minimum cost value, which is called the Leader. An active individual from the population moves in the direction towards the Leader in the search space. The movement consists of jumps determined by the Step parameter, until the individual reaches the final position given by the PathLength parameter. This procedure is depicted in Fig. 4(a). For each step, the cost function for the actual position is evaluated and the best value is saved. At the end of the crossover, the position of the individual with minimum cost value is chosen. If the cost value of the new position is better than the cost value of an individual from the old population, the new one appears in new population. Otherwise the old one remains there. The main principle is depicted in Figs. 4(b) and 4(c) and the crossover is described by equation (9):

$$x_{i,j}^{ML+1} = x_{i,j,START}^{ML} + (x_{L,j}^{ML} - x_{i,j,START}^{ML}) \cdot t \cdot PRTVector_j \tag{9}$$

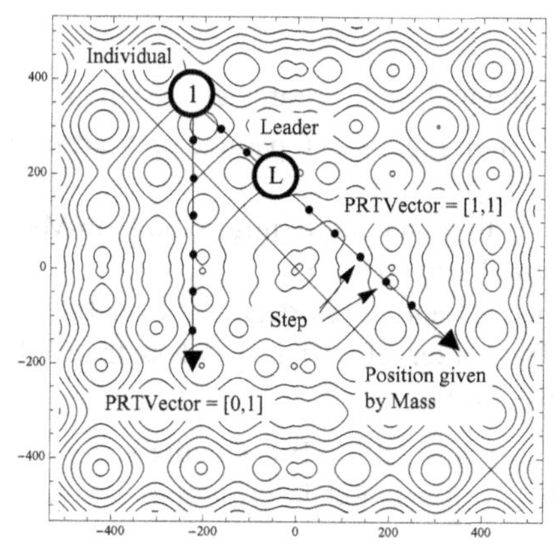

(a) Crossover in SOMA, PathLength is replaced here by Mass

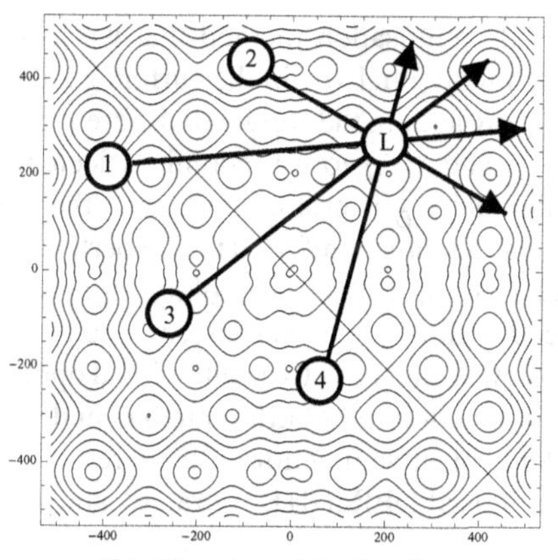

(b) Choosing of the Leader

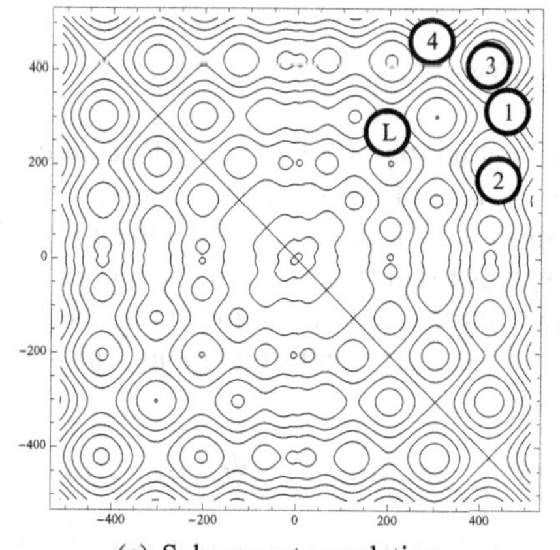

(c) Subsequent population

Fig. 4 Principles of SOMA

where:

$x_{i,j}^{ML+1}$ value of i-th individual's j-th parameter, in step t in migration loop $ML+1$,

$x_{i,j,START}^{ML}$ value of i-th individual's j-th parameter, Start position in actual migration loop,

$x_{L,j}^{ML}$ value of Leader's j-th parameter in migration loop ML,

t step, $t \in < 0, byStepto, PathLength >$,

$PRTVector$ vector of ones and zeros dependent on PRT parameter of SOMA. If random number from interval $< 0,1 >$ is less than PRT, then 1 is saved to PRTVector, otherwise it is 0.

The detailed principle is described in [19]. For the source codes in Mathematica, Matlab and C++ together with detailed description please refer to [20].

4.2 Differential Evolution

DE is a population-based optimisation method that works on real-number-coded individuals [11]. For each individual $x_{i,G}$ in the current generation G, DE generates a new trial individual $x'_{i,G}$ by adding the weighted difference between two randomly selected individuals $x_{r1,G}$ and $x_{r2,G}$ to a randomly selected third individual $x_{r3,G}$ (10). The resulting individual $x'_{i,G}$ is crossed-over with the original individual $x_{i,G}$. The fitness of the resulting individual, referred to as a perturbed vector $u_{i,G+1}$, is then compared with the fitness of $x_{i,G}$. If the fitness of $u_{i,G+1}$ is greater than the fitness of $x_{i,G}$, then $x_{i,G}$ is replaced with $u_{i,G+1}$; otherwise, $x_{i,G}$ remains in the population as $x_{i,G+1}$. DE is quite robust, fast, and effective, with global optimisation ability. It does not require the objective function to be differentiable, and it works well even with noisy and time-dependent objective functions. Please refer to (10), [11] and [10] for the detailed description of used DERand1Bin strategy and all other DE strategies.

$$u_{i,G+1} = x_{r1,G} + F \bullet (x_{r2,G} - x_{r3,G}) \tag{10}$$

5 Simulation Results

As described in the last paragraph of Section 2, AP requires an EA for its run. In this work AP_{meta} version was used. SOMA ATO strategy [19] was used within main AP process, which means to find a suitable form of the control law. DE was used in meta-evolution process, thus to find optimal values of constants in the evolutionary synthesized control law. Settings of EA parameters for both processes were based on performed numerous experiments and simulations with AP_{meta} (see Tables 1, 2).

5.1 Optimisation with Standard—Not Blackbox—Approach

Following Table 3 contains four best examples of synthesized control laws. Since this is an initial study, obtained simulations results depicted in Fig. 5 may evoke the impression, that the chaotic system was not fully stabilized on desired $p-2$ UPO.

Table 1 Parameter set up for SOMA used as the main algorithm in AP_{meta}.

Parameter	Value
PathLength	3
Step	0.11
PRT	0.1
PopSize	50
Migrations	4
Max. CF Evaluations (CFE)	5345

Table 2 Parameter set up for DE used as the second algorithm in AP_{meta}.

Parameter	Value
PopSize	40
F	0.8
CR	0.8
Generations	150
Max. CF Evaluations (CFE)	6000

Table 3 Simulation results for control of Logistic equation with Standard approach.

No.	Control law with coefficients	CF Value	Fig.
1	$F_n = x_{n-6}^{95.2285}$	0.0011	5(a)
2	$F_n = x_{n-6} - \frac{x_{n-6}}{-0.264459 + x_{n-1} + x_n}$	0.0084	5(b)
3	$F_n = -0.0169(-x_{n-6} - 23.4428)x_{n-5}(x_{n-2} - x_n)$	0.0206	5(c)
4	$F_n = x_{n-6} x_n^{\frac{62.739 x_{n-5} x_{n-3}}{x_{n-8} x_{n-6} x_{n-1}} 54.3298 x_{n-5} x_{n-4}^{33.7148}}$	0.1546	5(d)

Due to the properties of unmodified cost function taken over from previous research, the system was stabilized only on short time interval and than escaped to either chaotic behaviour or artificial controlled $p-4$ orbit. More about this phenomenon is written in conclusion section. Table 3 covers final output from AP—simplified synthesized control law with coefficients estimated by means of second algorithm DE, corresponding CF value, and identification of figure with simulation results.

5.2 *Optimisation with Blackbox Approach*

The Table 4 contains examples of synthesized control laws. Obtained simulation results were classified into 3 groups, based on level of approaching to real $p-2$ UPO. More about this phenomenon is written in conclusion section. Table 4 covers identification number of UPO approaching level group, final output from AP—simplified synthesized control law with coefficients estimated by means of second algorithm

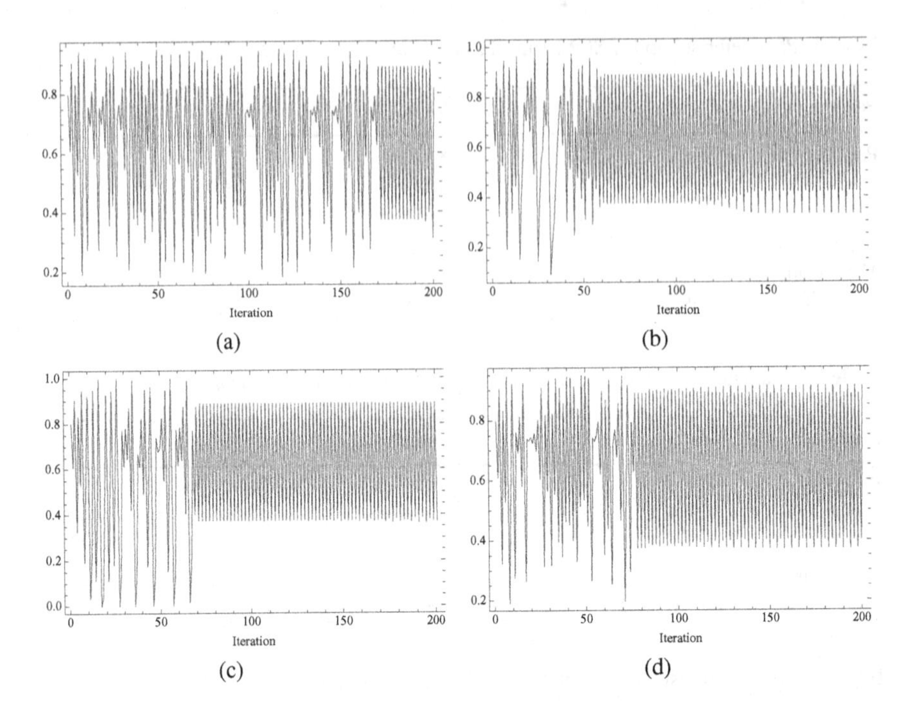

Fig. 5 Simulation results for the best new synthesized control laws—standard approach

Table 4 Simulation results for control of Logistic equation with Blackbox approach.

No.	Control law with coefficients	CF Value	Orbit Values	Fig.
1	$F_n = 0.39286x_{n-4}x_{n-2}^2$	195.881	0.98–0.44	6(a)
2	$F_n = -\frac{x_{n-6}-50.0535}{x_{n-7}+47.367} - x_{n-3}$	198.685	0.98–0.44	6(b)
3	$F_n = x_{n-1}^{19.464}$	149.061	0.94–0.21	6(c)
4	$F_n = x_n^{\frac{12.5103x_{n-2}}{0.09052-x_{n-2}+x_{n-1}}}$	188.251	0.91–0.36	6(d)

DE, corresponding CF value, orbit values between which system oscillates, and identification of figure with simulation results. Simulation output representing successful stabilisation of chaotic system is depicted in Fig. 6.

5.3 Comparison of Both Approaches

Following Table 5 contains brief statistical comparison of both approaches for 50 runs. Presented data confirms the robustness of evolutionary computation with AP, since there is no high deviation of CF values present in the results. Noticeable difference of CF values between the both approaches is caused by simple fact. The blackbox type CF evaluates the difference between target state and actual system

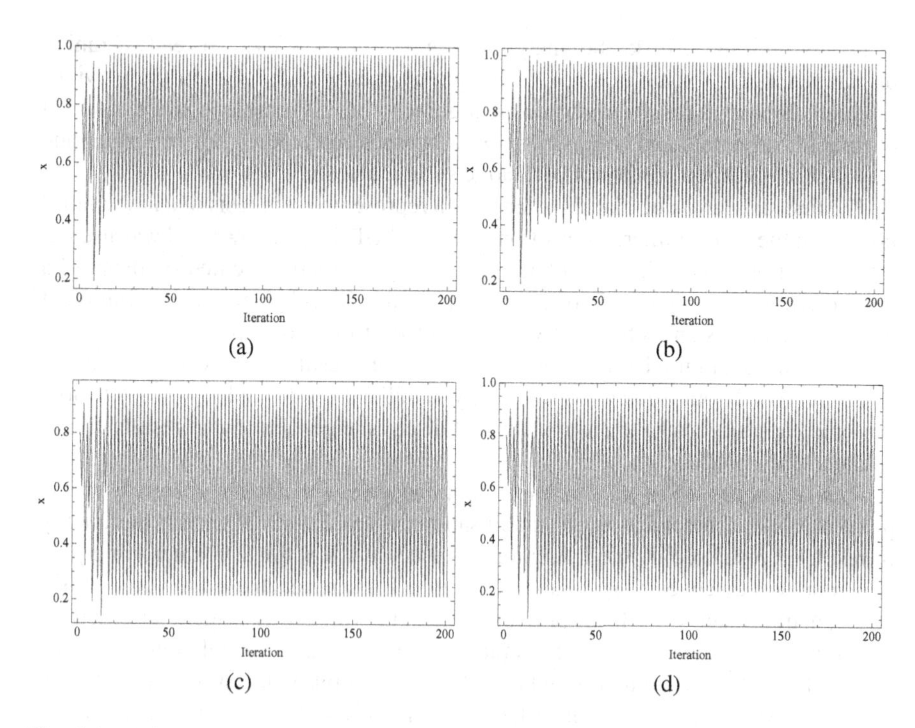

Fig. 6 imulation results for the best new synthesized control laws—blackbox approach

Table 5 Simple statistical comparison for both approaches.

	Standard approach	Blackbox approach
Min. CF Value	0.0011	149.004
Max. CF Value	0.2479	347.57
Avg. CF Value	0.0186	203.633
CF Value Median	0.0062	200.448
Max CFE	32070000	32070000

output on the whole simulation interval, which means that initial chaotic transient is included in the final CF value, whereas the standard type CF evaluates this difference only from the point of stabilisation on desired UPO.

6 Conclusion

This research introduces two possible approaches for the evolutionary synthesis of control law by means of Analytic Programming (AP), thus optimisation of stabilisation of Logistic equation, which was selected as an example of discrete chaotic system.

In this presented work, the analytic programming was used instead of tuning of parameters for existing control technique by means of EA's as in the previous research. Presented results reinforce the argument that AP is able to solve this kind of difficult problems and to produce a new synthesized control law in a symbolic way securing desired behaviour of chaotic system and stabilisation.

The first standard (not-blackbox) approach represented synthesis of a new control law, when the exact numerical position of desired UPO was known and was used as a cost function input, and the second blackbox approach represented synthesis of a new control law without the knowledge of desired UPO, when only the definition of type of chaotic system behaviour was used as a cost function input.

Since this is an initial study, obtained simulations results may evoke the impression, that the chaotic system was not fully stabilized on desired $p-2$ UPO. Both presented approaches have advantages and disadvantages. Presented results can be summarized as follows.

Standard approach is easy to implement and gives satisfactory results from the point of view of precision of the stabilisation on real $p-2$ UPO. But the quality of results is restricted by the limitations of used cost function. An interesting phenomenon was discovered in simulation results. This approach used the unmodified cost function taken over from previous research, which was in general based on searching for desired stabilized periodic orbit and thereafter calculation of the difference between desired and found actual periodic orbit on the short time interval. Therefore AP synthesized control laws, which followed this simple fact and secured the stabilisation on part of this short time interval, and than the system freely escaped to either chaotic behaviour or artificial controlled p-4 orbit. It is very interesting, that these control laws are able to stabilize the chaotic system on optional artificial periodic orbits. Most of common control method was developed for stabilisation only on real UPO with low energy costs, thus, when the system enters the UPO, there is no perturbation. On the other hand, the above described phenomenon lends weight to the argument, that AP is a powerful symbolic regression tool, which is able to strictly and precisely follow the rules given by cost function and synthesize any symbolic formula, in the case of this research—the feedback controller for chaotic system.

The blackbox approach brings the advantage of avoidance of mathematical analysis of chaotic systems, but also in this case an interesting phenomenon was discovered in simulation results. Since there was no information about exact position of $p-2$ orbit in the chaotic attractor transferred into evolutionary process and cost function was designed to operate in blackbox mode, thus on the basis of selection of desired system behaviour, AP synthesized control laws, which can be classified based on level of approaching to real $p-2$ UPO. It is very interesting, that these control laws are able to fully stabilize the chaotic system on optional artificial periodic orbits and keep the chaotic system on this artificial UPO.

Presented approaches were tested only on the simpler discrete chaotic systems. The complexity of the method, when used for more complex chaotic behaviour (e.g., Lorenz system), will significantly increase, since the very time demanding process of solving of set of differential equations will have to be included in every cost function evaluation.

The question of energy costs and more precise stabilisation will be included into future research together with the development of better cost functions, different AP data set, and performing of numerous simulations to obtain more results and produce better statistics, thus to confirm the robustness of this approach. To obtain a one solution for evolutionary approach, more than 32 millions CF evaluations were required.

Acknowledgements. This work was supported by the grant NO. MSM 7088352101 of the Ministry of Education of the Czech Republic and by grants of Grant Agency of Czech Republic GACR 102/09/1680 and by European Regional Development Fund under the project CEBIA-Tech No. CZ.1.05/2.1.00/03.0089.

References

1. Coelho, L.D.: Self-organizing migrating strategies applied to reliability-redundancy optimization of systems. IEEE Transactions on Reliability 58(3), 501–510 (2009)
2. Coelho, L.D.: Self-organizing migration algorithm applied to machining allocation of clutch assembly. Mathematics and Computers in Simulation 80(2), 427–435 (2009)
3. Coelho, L.D., Mariani, V.C.: An efficient cultural self-organizing migrating strategy for economic dispatch optimization with valve-point effect. Energy Conversion and Management 51(12), 2580–2587 (2010)
4. Davendra, D., Zelinka, I., Senkerik, R.: Chaos driven evolutionary algorithms for the task of pid control. Computers & Mathematics with Applications 60(4), 1088–1104 (2010)
5. Hilborn, R.C.: Chaos and Nonlinear Dynamics: An Introduction for Scientists and Engineers. Oxford University Press (2000)
6. Just, W.: Principles of time delayed feedback control. In: Schuster, H.G. (ed.) Handbook of Chaos Control. Wiley-Vch (1999)
7. Lampinen, J., Zelinka, I.: New ideas in optimization. In: Mechanical Engineering Design Optimization by Differential Evolution. McGraw-Hill (1999)
8. May, R.M.: Stability and Complexity in Model Ecosystems. Princeton University Press (2001)
9. Oplatkova, Z., Zelinka, I.: Investigation on evolutionary synthesis of movement commands. Modelling and Simulation in Engineering (2009)
10. Price, K., Storn, R.M.: Differential evolution homepage, `http://www.icsi.berkeley.edu/~storn/code.html` (accessed September 30, 2011)
11. Price, K., Storn, R.M., Lampinen, J.A.: Differential evolution: A practical approach to global optimization. Natural Computing Series. Springer (1995)
12. Pyragas, K.: Continuous control of chaos by self-controlling feedback. Physics Letters A 170, 421–428 (1992)
13. Pyragas, K.: Control of chaos via extended delay feedback. Physics Letters A 2006 (1995)
14. Senkerik, R., Oplatkova, Z., Zelinka, I., Davendra, D.: Synthesis of feedback controller for three selected chaotic systems by means of evolutionary techniques: Analytic programming. Mathematical and Computer Modelling (2010), doi:10.1016/j.mcm.2011.05.030

15. Senkerik, R., Oplatkova, Z., Zelinka, I., Davendra, D., Jasek, R.: Evolutionary synthesis of control law for higher periodic orbits of chaotic logistic equation. In: 25th European Conference on Modelling and Simulation. European Council for Modelling and Simulation, pp. 452–458 (2011)
16. Senkerik, R., Oplatkova, Z., Zelinka, I., Davendra, D., Jasek, R.: Synthesis of feedback control law for stabilization of chaotic system oscillations by means of analytic programming — preliminary study. In: 5th Global Conference on Power Control and Optimization (2011)
17. Senkerik, R., Zelinka, I., Davendra, D., Oplatkova, Z.: Evolutionary Design of Chaos Control in 1D. In: Zelinka, I., Celikovsky, S., Richter, H., Chen, G. (eds.) Evolutionary Algorithms and Chaotic Systems. SCI, vol. 267, pp. 165–190. Springer, Heidelberg (2010)
18. Senkerik, R., Zelinka, I., Davendra, D., Oplatkova, Z.: Utilization of soma and differential evolution for robust stabilization of chaotic logistic equation. Computers & Mathematics with Applications 60(4), 1026–1037 (2010)
19. Zelinka, I.: Soma — self organizing migrating algorithm. In: Babu, B., Onwubolu, G. (eds.) New Optimization Techniques in Engineering. Springer (2004)
20. Zelinka, I.: Soma homepage, `http://www.fai.utb.cz/people/zelinka/soma/` (accessed September 30, 2011)
21. Zelinka, I., Davendra, D., Senkerik, R., Jasek, R., Oplatkova, Z.: Analytical programming — a novel approach for evolutionary synthesis of symbolic structures. In: Kita, E. (ed.) Evolutionary Algorithms. InTech (2011)
22. Zelinka, I., Oplatkova, Z., Nolle, L.: Boolean symmetry function synthesis by means of arbitrary evolutionary algorithms-comparative study. International Journal of Simulation Systems, Science and Technology 6(9), 44–56 (2005)
23. Zelinka, I., Senkerik, R., Navratil, E.: Investigation on evolutionary optimization of chaos control. Chaos, Solutions & Fractals 40(1), 111–129 (2009)

Complex Automata as a Novel Conceptual Framework for Modeling Biomedical Phenomena

Witold Dzwinel

Abstract. We show that the complex automata (CxA) paradigm can serve as a robust general framework which can be applied for developing advanced models of biological systems. CxA integrates particle method (PM) and cellular automata (CA) computational techniques. Instead of developing complicated multi-scale models which consist of many submodels representing various scales coupled by a scales-bridging mechanism, we propose here a uniform, single scale, coarse grained computational framework for which information about finer scales is inscribed in CA rules and particle interactions. We demonstrate that our approach can be especially attractive for modeling biological systems, e.g., intrinsically complex phenomena of growth such as cancer proliferation fueled by the process of angiogenesis and *Fusarium Graminearum* wheat infection. We show that these systems can be discretized and represented by an ensemble of moving particles, which states are defined by a finite set of attributes. The particles may represent spherical cells and other non-spherical fragments of more sophisticated structures, such as, transportation system (vasculature, capillaries), pathogen individuals, neural network fragments etc. The particles interact with their closest neighbors via semi-harmonic central forces mimicking mechanical resistance of the cell walls. The particle motion is governed by both the Newtonian laws and cellular automata rules employing the attributes (states) of neighboring cells. CA rules may reflect e.g., cell life-cycle influenced by accompanying biological processes while the laws of particle dynamics and the character of collision operators simulate the mechanical properties of the system. The ability of mimicking mechanical interactions of tumor with the rest of tissue and penetration properties of *Fusarium graminearum*, confirms that our model can reproduce realistic 3-D dynamics of these complex biological systems.

Witold Dzwinel
AGH University of Science and Technology, Al. Mickiewicza 30, 30-059 Kraków, Poland
e-mail: dzwinel@agh.edu.pl

A. Byrski et al. (Eds.): Advances in Intelligent Modelling and Simulation, SCI 416, pp. 269–298.
springerlink.com

1 Introduction

New challenges in systems biology involve searching for new modeling paradigms which allow for simulating multi-scale systems within a unified framework. As shown in Fig. 1, the multi-scale simulations are large-scale simulations involving many spatio-temporal scales and many heterogeneous modeling approaches.

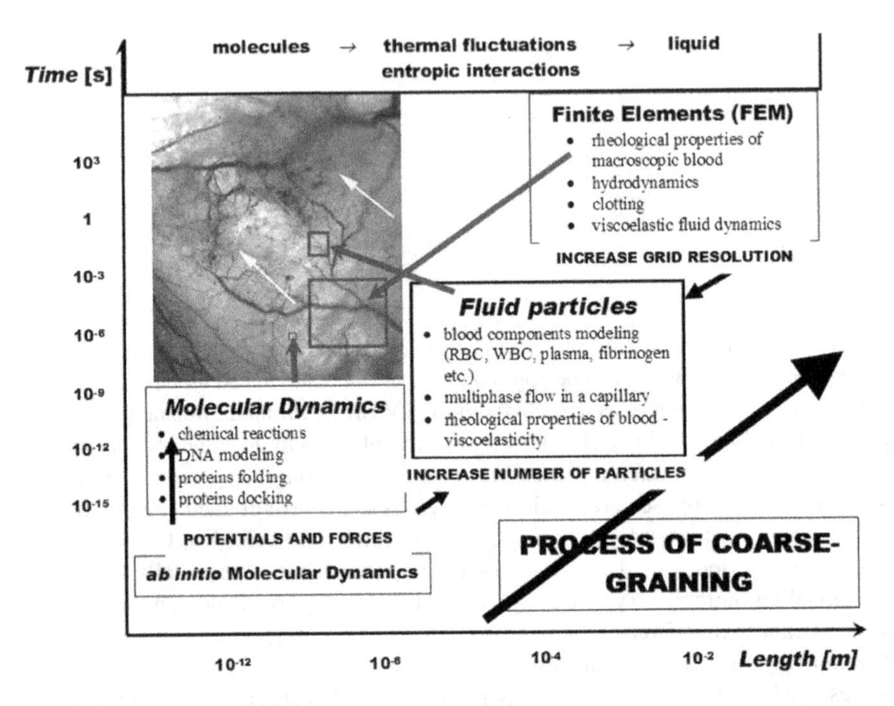

Fig. 1 The scale separation diagram of blood system involving multiple scales and various computational models [14]

This fact poses at least two serious problems. The first one refers to matching and bridging heterogeneous models representing different scales (e.g. discrete and continuum), while the second one, is connected with merging different parallelisation strategies. Moreover, the common strategy of continuum and discrete models bridging by running finer scale models only in the region of interest (ROI) involving more degrees of freedom (see Fig. 2), may fail for biological systems, where coupling between fine and coarser scales can be very tight. This may cause that the finest spatio-temporal scales will still decide about the computational complexity despite the existence of well separated coarse grained modes.

The interesting alternative to the continuous/discrete approaches is the development of a universal computational framework, which could be matched to the following spatio-temporal scales through the process of successive coarse graining. It can be understood as a numerical equivalent of some renormalisation procedures

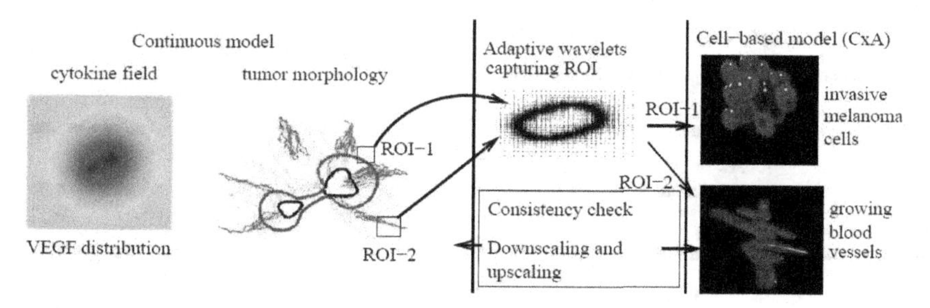

Fig. 2 The continuous model treats macroscopic tumor morphology and cytokine distribution fields. The cell-based model handles individual cell proliferation and motion and is only applied in Regions of Interest (RPI). The connection between two models is made by the adaptive wavelets [40] which are far denser in ROI [41].

used from many years in physics for simplification of formal mathematical models. Thus the successive coarse graining can be defined as the approximation process limiting the number of DoF (degrees of freedom) and the frequency of their motion starting from the smallest to the largest scales of interest.

Signal decomposition and multiresolution are good metaphors of the notions of successive coarse graining and multi-scaling. The principles of signal decomposition are demonstrated in Fig. 3.

Every signal can be decomposed onto approximations (A) and details (D) on successive resolution levels using a set of basis functions with compact support (such as wavelets, RBF etc.). Then, the signal is equal to the sum of approximation on a given resolution level and all the details from the finest levels. By cutting off the least important details, i.e. all of these having the weights below a certain threshold, the signal can be reconstructed using approximation (A3) and only a fraction of the most important details from the finer resolutions. This brings about the question of how this smart method can be applied in developing multi-scale models.

To find the analogies between signal decomposition and developing multi-scale models we should define first a homogeneous computational environment, which allows for defining principal modeling procedures in the scope of the same conceptual framework. We consider here two computational frameworks, namely, cellular automata and particle method. Cellular automata (CA) paradigm is a simplistic model of computation. The seminal Wolfram's book "New Kind of Science" [46] advocates that CA paradigm can be treated as a universal paradigm and a metaphor of reality. Cellular automata model is very compact and elegant indeed. CA is defined as a triple $CA = (a(t), S_A, f_A)$. It consists of a complete number NR of update rules f_A, which govern the deterministic evolution of a lattice of CA states $a(t)$ in a discrete time t. The states are labeled by a finite alphabet $\{S_A\}$. For the simplest 1-D: $CA \rightarrow S_A = \{0,1\}$, and the update rule f_A is a function of the closest neighbors only $f_A : \{SA\}^3 \rightarrow \{SA\} \wedge a_n(t+1) = f_A[a_{n-1}(t), a_n(t), a_{n+1}(t)]$. The number of rules is strictly limited, i.e. $N_R = \sum_{i=[0,7]} 2^i \alpha_i = 256, \alpha_i \in \{0,1\}$. The transition principles are shown in Fig. 4 below. This way, every transition rule has its unique number

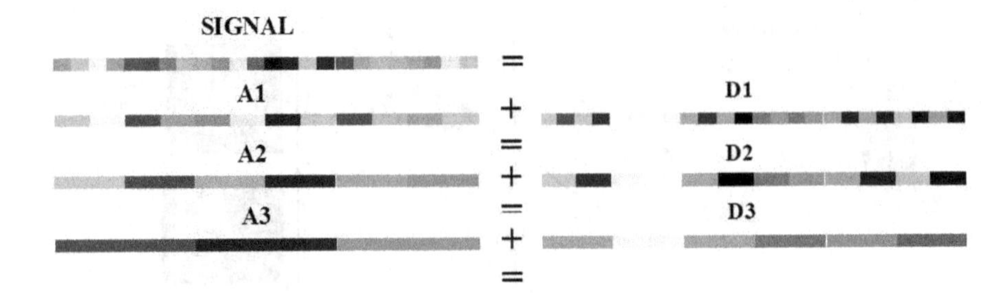

(a) Signal decomposition and details on various resolution levels

(b) Signal decomposition and details after eliminating the least important details

Fig. 3 Signal decomposition onto approximation (A) and details (D) on various resolution levels (1–3). The signal is equal to the sum of approximation on a given level and all the details from the finest levels (e.g. SIGNAL=A3+D3+D2+D1). After eliminating the least important details, the signal can be reconstructed using approximation (A3) and only a fraction of the most important details from the finer scales.

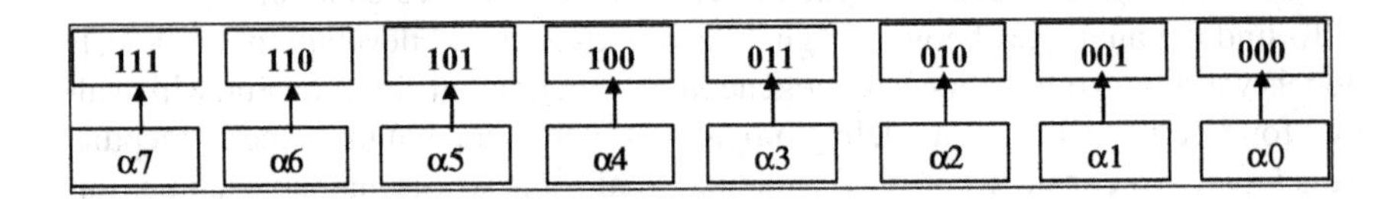

Fig. 4 Transition principles for 1-D CA.

from 0 to 255. The number of possible rules increases with growing alphabet, the number of neighbors and dimensionality of the CA environment.

The scope of CA applications is very broad, ranging from microscopic to macroscopic phenomena (e.g., [46, 7]). So, the natural question is whether the CA paradigm represents truly multi-scale properties allowing for development coarser CA models through approximation of the finer ones. In the first section of this paper we address two important issues:

1. How to construct the coarse graining procedure to retain physically important information from smaller scales?
2. What type of CA can/cannot be coarse grained?

We postulate, according to the results obtained by Israeli and Goldenfeld [24], a general principle of coarse graining which can reflect the signal decomposition procedure from Fig. 3. Then we discuss its usefulness in coarse graining of very different computational framework, namely, Particle Model (PM).

The robustness of CA is still qualitative — metaphoric. Although some CA clones such as lattice gas and lattice Boltzmann gas [7], are able to describe many dynamical properties of physical systems, they simulate mechanical interaction, such as inertia, in a very simplistic and counterintuitive way. This is unlike another broadly used modeling tool — the model of interacting particles or Particle Model (called also Particle Method or Discrete Element Method (DEM) (e.g. [12, 10, 15, 38]) for which mechanical interactions are its intrinsic property. The Particle Model is a discrete, off-grid and very general paradigm of modeling, which has its roots in N-body modeling and well known Molecular Dynamics (MD) method (also the Non-equilibrium Molecular Dynamics NEMD). Its broad scope of application was described in [14]. The principles of the method are shown in Fig. 5.

The system of discrete particles is defined by boundary and initial conditions and by interactions between particles represented by a collision operator Ω_{ij}. The

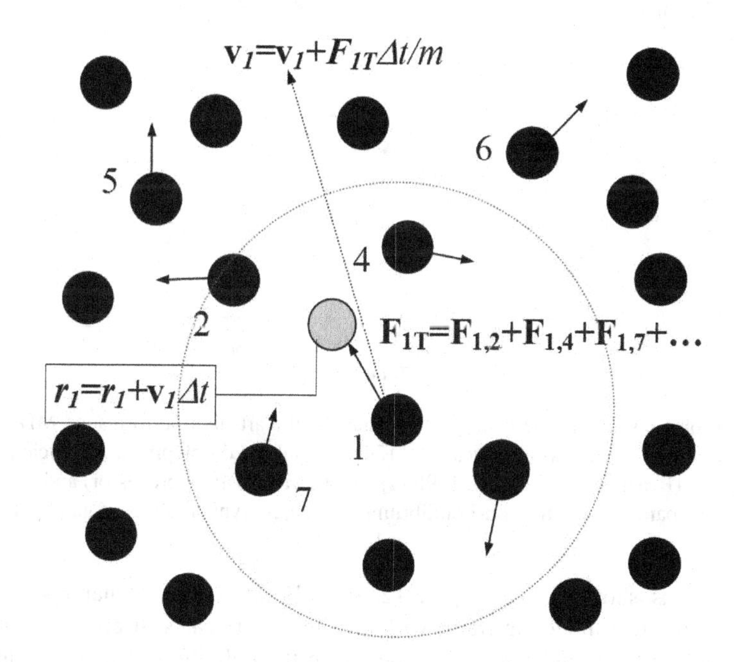

Fig. 5 The diagram presenting principles of Particle Method

particle system evolves according to the Newtonian equations of motion, which can be described in a discrete form as follows:

$$\Delta P_i^n = \sum_j^N \Omega_{ij} \cdot e_{ij}^n \Delta t, \quad \Delta r_i^n = \frac{P_i^n}{m} \Delta t \tag{1}$$

where r_i is the position of particle, P_i is its momentum and t the integration time-step, N is the number of particles in the interaction range.

As shown in Fig. 6, the state-of-the-art supercomputers allow for simulating more than a trillion atoms in a million time-steps using highly efficient MD parallel codes [20]. This particle ensemble corresponds to spatial 3-D scales of a few micrometers and time scales of ten nanoseconds. This may suggest that the particle approach is not appropriate for simulating larger systems due to high computational demands.

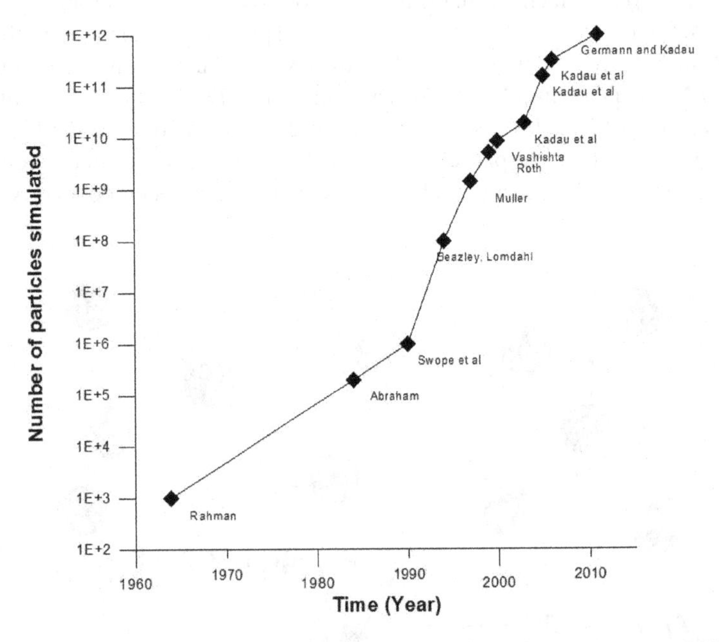

Fig. 6 The plot displaying the history of the state-of-the-art high performance MD simulations. The abrupt speed-up after year 1990 is the result of development of efficient MD parallel codes (Beazley and Lomdahl, 1996 [2] — on TMCM5 1024 processor) and development of a new paradigm of the Non-Equilibrium Molecular Dynamics (see, e.g., [14, 11]).

However, as shown in (e.g., [14, 12, 9, 35]), Molecular Dynamics can be used as an efficient modeling framework also in larger scales. It appears that the spatio-temporal scale under interest depends on the definition of particle (quark, atom, molecule, granule, cluster, chunk of matter, object, individual, the planet, galaxy etc.) and the type of particle-particle interactions (hard/soft: long/short range, central/ non-central, two-body/many-body, dissipative/ conservative, stochastic/ deterministic etc.) [14, 9]. Thus the definition of particle and the respective type of particle-particle interactions can be treated as the results of a successive coarse graining procedure. However, such the procedure cannot be as precisely defined as it is for cellular automata.

The main weakness of the particle model is the difficulty to represent important microscopic degrees of freedom in the form of particle interactions. For example, modeling crowd dynamics using only the metaphor of mechanical collisions of particles [35, 21] (where a particle corresponds to an individual) can neglect some important "microscopic" phenomena such as development of panic, anger, falls and blockage due to local crush etc.. Even if some of these factors can be represented by a sort of stochastic force (such as panic factor in [21]), additional thresholding conditions dependent on the current properties of neighboring individuals should be arbitrary defined. The problem becomes even more acute in modeling biological systems. Assuming that a particle represents a cell, the microscopic processes such as chemical signaling, chemotaxis, haptotaxis, oxygen and proteins diffusion influencing cell behavior and its functions cannot be mimicked by a simple mechanical force. On the other hand, just mechanical interactions between cells can be a crucial factor for some type of growth, such as solid cancer proliferation.

In [33] the authors propose complex automata (CxA) model as a multiscale paradigm consisting of a set of single scale CA representing processes operating on different spatio-temporal scales. The authors define adequate coupling templates between the scales to model and simulate multi-scale phenomena. The CA considered are typically based on the Lattice Boltzmann Models (LBM) [7]. Besides generalized CA to represent single scale models, the CxA approach also includes Agent Based Models (ABM) [33].

Our approach is different. Instead of developing multi-scale model which consists of many submodels representing various scales coupled by a scale-bridging mechanism, we propose a uniform coarse grained model in which information about finer scales is inscribed in CA rules and particle interactions. In the following section we demonstrate that by coupling cellular automata and particle model we can develop a new computational framework which possesses the advantages of the two. By using as examples two modeling targets: proliferation of cancer and invasion of *Fusarium Graminearum* — a pathogen attacking cereal crops — we demonstrate how the concept of complex automata woks in modeling realistic phenomena. In the conclusions we propose some validation issues for CxA model.

2 Coarse Graining

2.1 Cellular Automata

The concept of coarse-graining has been introduced to cellular automata by Israeli and Goldenfarb in [24]. Let us the original CA be defined as $A = (a(t), S_A, f_A)$ and its coarse-graining equivalent $B = (b(t), S_B, f_B)$. The projection (or mapping) function

$P : S_A^N \to S_B$ will be used to map the block of N cells from A into exactly one cell of B. The block of N cells from A, A_N, is called a supercell. Then the condition that has to be satisfied by automata B and projection function P in order to provide coarse-graining of A is as follows:

$$P \cdot \underbrace{f_A \cdot \ldots \cdot f_A}_{N} \cdot a = f_B \cdot P \cdot a \tag{2}$$

Where $P \cdot a$ denotes that the whole lattice a is divided onto blocks of size N, and then we apply projection P for each block separately. The notation $f_A \cdot a$ means, that we apply the local transformation f_A of automata A to every cell in the lattice a. The expression (2) says, that running automata A for N times and mapping the result using P, gives the same final CA configuration as applying P at first, and then running automata B only once. This has to be satisfied for any starting configurations of A.

In [24] Israeli and Goldenfarb show a simple procedure for finding coarse grained configuration of a given automata. This very inefficient procedure has to be described in order to understand some basic properties of coarse graining of cellular automata. Let us define the N-th supercell automata as:

$$A_N = \left(a^N(t), S_A^N, f_{A^N}\right) \tag{3}$$

This new automata operates over blocks of N cells from $a(t)$ lattice. Let us consider 1-dimentional automata with neighborhood of size $k = 3$. The local function is then: $\{SN\}^3 \to S^N$. We can compute easily the value of f_{A^N} for some $x \in \{S^N\}^3$. This could be done by converting x into $3N$-element lattice of automata A, and running automata A exactly N times:

$$y = \underbrace{f_A \cdot \ldots \cdot f_A}_{N} \cdot x \tag{4}$$

Now we may also choose the alphabet of coarse grained automata B to fit into the alphabet of A_N:

$$S_B = S_{A^N} \tag{5}$$

Since there is a nonsense to consider S_B to be larger than $(\#S_A)^N$. That is because for $S_B \equiv S_{A^N}$ mapping function $P(\cdot)$ is injective, and we would not have any benefits using larger alphabet. Utilising all given definitions (see, Eq. (2)), we could rewrite the rule that need to be satisfied by coarse-grained automata B and its local function:

$$f_B\left[P(x_1);P(x_2);P(x_3)\right] = P\left(f_{A^N}\left[x_1;x_2;x_3\right]\right) \tag{6}$$

We need to keep in mind, that $P(\cdot)$ does not has to be injective, and there is a possibility that $(P(y_1);P(y_2);P(y_3)) = (P(x_1);P(x_2);P(x_3))$ for another triple of N-element blocks $(y_1;y_2;y_3) \neq (x_1;x_2;x_3)$. In that case we will get the same result, for both triples: $f_B\left[(P(y_1);P(y_2);P(y_3))\right] = f_B\left[(P(x_1);P(x_2);P(x_3))\right]$:

$$\forall (x,y|P(x_1) = P(x_i)) : P(f_{A^N}\left[x_1;x_2;x_3\right]) = P(f_{A^N}\left[y_1;y_2;y_3\right]) \tag{7}$$

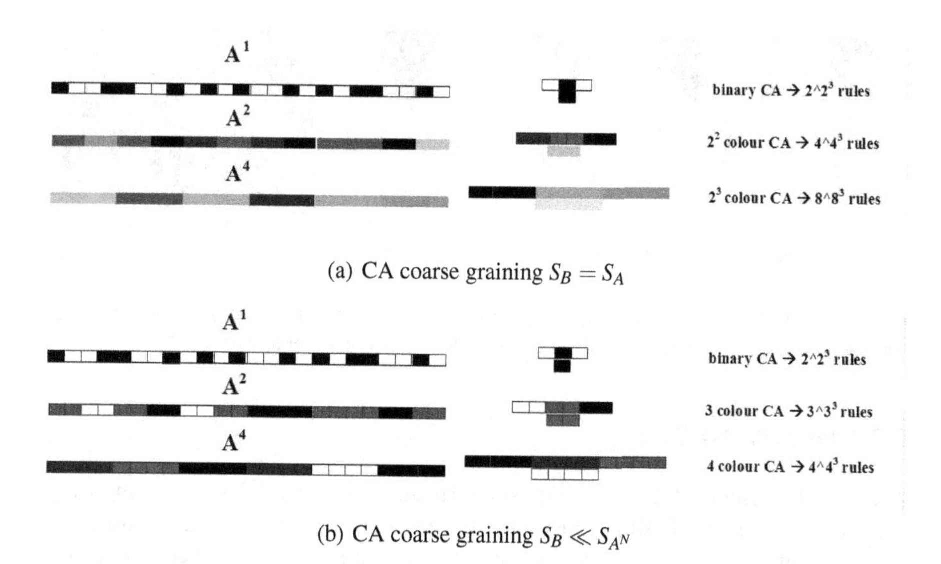

(a) CA coarse graining $S_B = S_A$

(b) CA coarse graining $S_B \ll S_{A^N}$

Fig. 7 The diagram presenting the process of coarse graining of CA. Information about finer scales can be preserved in new rules, created from greater alphabet than the original one.

The process of coarse-graining eliminates degrees of freedom representing local processes without loosing global features of CA evolution. For example, as shown in Fig.6a,b, the rule 128 is the coarse grained version of the rule 146. In [6], the full diagram of coarse graining of all basic 1-D automata was presented. The complex automata (such as rule 110) can be coarse grained only to trivial rules, i.e., 0 and 255. As shown in [18], also the coarse-graining of other rules generating chaotic behavior in 2-D, such as logistic equation, may cause serious artifacts. Instead of chaotic map, the cascade of reverse bifurcation can be observed behind the accumulation point.

The mapping function $P(\cdot)$ is responsible for information loss. As shown in Fig. 7(a), only when $S_B = S_A$ no information is being lost, since $P(\cdot)$ is injective. For such S_B, coarse-graining always exists, and it could be easily derived from Eq. (6) because Eq. (7) will be satisfied only for $x = y$. It turns out, that for $S_B \equiv S_{A^N}$ coarse-graining problem is trivial, so it is reasonable to consider alphabet S_B being much smaller than (see Fig. 7(b)). As shown in Fig. 8(c), increasing the alphabet (i.e. assuming that $S_A \subset S_B \subset S_{A^N}$) the fine-scale information can be partially reconstructed.

The main conclusions can be summarized as follows:

1. Coarse graining of the most of CA to simpler but not trivial automata is possible.
2. Coarse-grained degrees of freedom maybe simple and predictable.
3. Undecidable, chaotic CA cannot be coarse-grained,
4. Finer, physically important DoF, can be incorporated to the coarse-grained model by increasing the alphabet $S_B > S_A$ and changing the rule set.

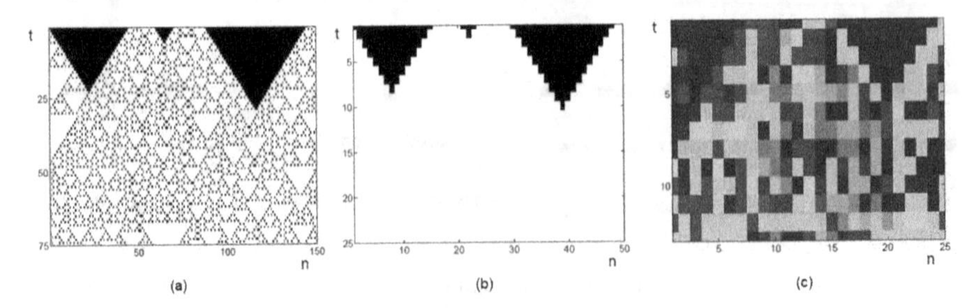

Fig. 8 The effect of coarse graining of rule 146 a) the original rule 146 b) its coarse grained version: rule 128, c) coarse grained version of rule 146 using larger alphabet [24].

2.2 Particle Method

The Particle method (PM) in its simplistic form, defined by particles interacting with each other via two-body short ranged force (see Eq. (1) and Fig. 6), can be used directly for simulation of populations consisting of rather small number of objects. Moreover, the objects must be precisely defined as separate and integrated entities. In such the case, despite the global response of the system is measured in terms of space and time averages, just a local scenario is the focal point of interest. For example, the average speed of crowd flow passing the corridor (see Fig. 9(a)) and its spatial distribution depend on the width of exit door, the existence of crossings, stairs etc. Due to high variability caused by individual behavior, such the simulations should be repeated many times to analyze various possible scenarios. By increasing the number of interacting entities the individual behavior becomes less and less important. The system can be described in terms of stochastic processes and fluctuation-dissipation principles. For example, in the classical molecular dynamics simulations fluctuation of system averages (e.g., kinetic and potential energies) are used to estimate such the important thermodynamic quantities like temperature, pressure, entropy, viscosity, conductivity etc. Moreover, an emergent behavior can be observed such as phase separation shown in Fig. 9(b). By further increase of the number of interacting objects towards to the upper values from the plot in Fig. 6, the number of possible scenarios stops growing so fast with the number of objects. The emergent macroscopic behavior, connected with local system anisotropy, begins to dominate over individual and stochastic terms. As shown in Fig.8c, simulating the flow past a plate using more than a million particles and averaging the velocity of particles on a rectangular grid, we can observe the process of eddies formation characteristic for macroscopic flows. As it was shown in [12, 10, 11], the same computational tool which is used for simulating particle systems in thermodynamic equilibrium, can be generalized easily to enable the simulation of nonequilibrium systems. The simulation of large particles ensembles involving nonequilibrium molecular dynamics (NEMD) provides a consistent microscopic basis for the irreversible macroscopic Second Law of Thermodynamics. The new idea, key to the nonequilibrium development, was the replacement of the external thermodynamic

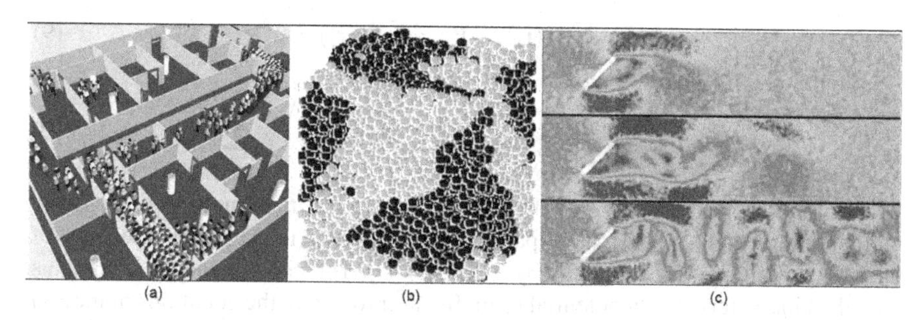

Fig. 9 Three types of scenario simulated using PM approach for increasing number of interacting particles: (a) crowd simulation may depend strongly on individual motion of a single object [16]; (b) phase separation is the emergent effect of stochastic processes (not only) connected with thermal fluctuations; (c) flow of one million of particles past a plate mimics macroscopic behavior of fluids.

environment by internal control variables. The strong anisotropy of simulation conditions, such as those used for simulation presented in Fig. 9(c), can be treated as a typical nonequilibrium molecular dynamics scenario.

Because large-scale NEMD simulation can bridge time scales dictated by the fast modes of motion together with the slower modes, which determine the viscosity, it can capture the effects of varying molecular topology on fluid rheology. These effects may come from chemical reactions or mixing with complicated velocity fields. However, in order to capture spatio-temporal scales of micrometers and microseconds occurring, e.g.,in vascular system (see Fig. 1), we need billions or more of MD particles simulated in hundred of millions of timesteps [31, 32].

Mesoscopic regimes of this capillary system require the fast modes of motion to be coarse grained. At this level, the particles will represent clusters of atoms or molecules, so-called, dissipative particle (DP) [17]. The authors of [17] have shown how to link and pass the averaged properties of molecular ensemble onto DPs by using a systematic coarse-graining procedure. The dissipative particles are represented then by cells defined on the Voronoi lattice with variable masses and volumes (see Fig. 10). The Voronoi cells allow for a very clear statement of the problem of coupling continuum equations and molecular dynamics. This is important when the continuum description breaks down in certain regions such as the contact line between two fluids and a solid, or the singularity of the tip in propagating fracture. Entire representation of all the MD particles (atoms or molecules) can be achieved in a general way by introducing an approximation radial function:

$$f_k(\mathbf{r}-\mathbf{r}_k) = \frac{\theta(\mathbf{r}-\mathbf{r}_k)}{\sum_l \theta(\mathbf{r}-\mathbf{r}_l)} \tag{8}$$

where the positions $\mathbf{r}_k$ and $\mathbf{r}_l$ define the centers of dissipative particles, r is an arbitrary position, and $\theta(\mathbf{r})$ is the Gaussian function. According to [17], the mass, momentum, and internal energy E_k of the k-th dissipative particle are then approximated as:

$$M_k = \sum_i f_k(\mathbf{r}_i), \quad \mathbf{P}_k = \sum_i f_k(\mathbf{r}_i) m\mathbf{v}_i \tag{9}$$

$$\frac{M_k U_k^2}{2} + E_k = \sum_i f_k(\mathbf{r}_i)\left(\frac{m\mathbf{v}_i^2}{2} + \frac{1}{2}\sum_{j \neq i} \phi_{MD}(\mathbf{r}_{ij})\right) \equiv \sum f_k(\mathbf{r}_i)\varepsilon_i \tag{10}$$

where $\mathbf{v}_i$ is the velocity of i-th MD particle having identical masses m, $\mathbf{P}_k$ is the momentum of the k-th dissipative particle, and $\phi_{MD}(\mathbf{r}_{ij})$ is the potential energy of the MD particle pair i, j separated by a distance $\mathbf{r}_{ij}$. The particle energy ε_i contains both the kinetic term and a potential term. In order to derive the equations of motion for dissipative particles the time derivatives of Eqs. (10) must be resolved [17]. Finally, after averaging over the velocities, masses and interactions on the Voronoi lattice, we obtain as follows:

$$\frac{d\mathbf{P}_k}{dt} = M_k\mathbf{g} + \sum_l \langle \dot{M}_{kl} \rangle \frac{\mathbf{U}_k + \mathbf{U}_l}{2} - \sum_l L_{kl}\left(\frac{p_{kl}}{2}\mathbf{e}_{kl} + \frac{\eta}{r_{kl}}[\mathbf{U}_k l + (\mathbf{U}_{kl} \cdot \mathbf{e}_{kl}) e_{kl}]\right) + \sum_l \widetilde{\mathbf{F}}_{kl} \tag{11}$$

where $\mathbf{U}_{kl}$ — relative velocity of dissipative particles k,l; p_{kl} — a pressure term between k and l dissipative particles resulting from conservative MD interactions; L_{kl} — a parameter of the Voronoi lattice; η — the dynamic viscosity of the MD ensemble and the last summation symbolizes the summation over relative fluctuations $\widetilde{\mathbf{F}}_{kl}$ of the coarse-grained representation.

This coarse graining procedure links all the forces between the DP to a hydrodynamic description of the underlying molecular dynamics atoms. The method may be used to deal with situations such as this shown in Fig. 1 in which several different dynamical length scales are simultaneously present. To increase the computational efficiency, the Voronoi cells can be approximated by spheres (see Fig. 10). Using additional simplifications, such as the unification of dissipative particle sizes, we can arrive at a model, which converges to the dissipative particle dynamics (DPD).

In dissipative particle dynamics [23] the two-body interactions between two fluid particles i and j are assumed to be central and short-ranged. The collision operator, $\Omega(r_{ij}, \mathbf{p}_{ij})$, can be defined as a sum of a conservative force F_C, dissipative component F_D and the Brownian force F_B. The Brownian factor represents the coarse

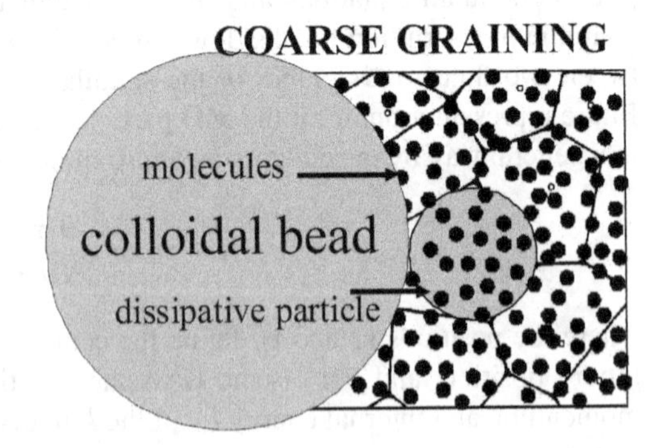

Fig. 10 The coarse-graining procedure. The Voronoi cell contains atoms, which are the closest to the cell center. The centers of all the Voronoi cells correspond to the centers of coarse-grained representation of the system - dissipative particles - which can be approximated then by spheres [17].

grained equivalence of thermal fluctuations. The equations below show the basic formula describing the two-body forces.

$$\mathbf{F}_C = \pi \cdot \omega(r_{ij}) \cdot \mathbf{e}_{ij},\ \mathbf{F}_D = \gamma \cdot m \cdot \omega^2(r_{ij}) \cdot (\mathbf{e}_{ij} \circ \mathbf{v}_{ij})\mathbf{e}_{ij},\ \mathbf{F}_B = \frac{\sigma \cdot \theta_{ij}}{\sqrt{\Delta t}} \cdot \omega(r_{ij}) \cdot \mathbf{e}_{ij} \quad (12)$$

$$\omega(r_{ij}) = \frac{3}{n \cdot \pi r_{cut}^2}\left(1 - \frac{r_{ij}}{r_{cut}}\right),\ \theta_{ij} \in (-1,1)\text{: random number; } n\text{: particle density} \quad (13)$$

$$\Omega(r_{ij}, \mathbf{p}_{ij}) = \mathbf{F}_C + \mathbf{F}_D + \mathbf{F}_B \text{ for } r_{ij} < r_{cut},\ \Omega(r_{ij}, \mathbf{p}_{ij}) = 0 \text{ for } r_{ij} > r_{cut} \quad (14)$$

The value of r_{cut} is the cut-off radius which represents the range of interaction between two interacting DP. DPD model with longer cut-off radius reproduces better dynamical properties of realistic fluids expressed in terms of velocity correlation function [16]. Simultaneously, for a shorter cut-off radius, the efficiency of DPD codes increases as $O(1/r_{cut}^3)$, which allows for more precise computation of thermodynamic properties of the particle system from statistical mechanics point of view. A strong background has been provided to DPD in [15, 16]. Explicit formulas for transport coefficients in terms of the particle interactions were derived. As shown by Marsh et al. [29], for low value of friction (i.e., γ in Eq. (12)), low density case and vanishing conservative interactions the interactions between the dissipative particles produce only small deflections. There exist several other methods, which generalize DPD model, e.g., the fluid particle model (FPM) [15] and the thermodynamically consistent DPD (TC-DPD) [38].

One of the serious drawbacks of DPD is the absence of a drag force between the central particle and the second particle orbiting about the first. To eliminate this deleterious effect, the fluid-particle model introduces a non-central force, which is proportional to the difference between the velocities of the particles. In order to bridge mesoscopic with the macroscopic scales using interacting particles, Serrano and Espanol [38] propose a new thermodynamically consistent dissipative particle model (TC-DPD). It uses the principal advantages of the well known SPH (smoothed particle hydrodynamic) scheme, which is a discrete version of the Navier-Stokes equations [19], and includes thermal fluctuations such as in DPD. This model resolves some problems related to the physical interpretation of the original DPD model. The TC-DPD method at present represents a superset for the classical dissipative particle dynamics, fluid particle model and smoothed particle dynamics models [26]. The scale of interest can be easily matched by controlling the size of fluctuation terms. However, the calculation of fluctuation terms in TC-DPD remains very time consuming.

Much simpler method of scale bridging was described in [19]. Coarse-graining in DPD translates to having a number L of physical molecules be represented by a single DPD particle. A renormalisation or coarse graining is performed that changes the interaction parameters, but does not change the units. The scaling procedures for number of particles N, mass m, space r_c and time τ units, collision operator parameters (see Eqs. (12)–(14): π, λ, σ) in D-dimensional space are as follows:

$$N' = L^{-1} \cdot N, \quad m' = L \cdot m, \quad r'_c = L^{1/D} \cdot r_c, \quad \tau' = L^{1/D} \cdot \tau \tag{15}$$

$$\pi' = L^{1-1/D} \cdot \pi, \quad \gamma' = L^{1-1/D} \cdot \gamma, \quad \sigma' = L^{1-1/2D} \sigma \tag{16}$$

This simple renormalisation scheme allows one to scale a DPD-simulation to any desired length scales. The Authors have proved in [19] that DPD is a scale-free method.

Unlike truly discrete cellular automata, the particle method is a discrete-continuum paradigm in which discrete particles evolve in continuous space and time. However, main coarse-graining principles of CA remain very similar to those applied for particle model. Specifically:

1. Both paradigms are homogeneous, i.e., their principles remain the same in every scale. The coarse-grained CA consists of L finer CA, while the fluid particle is made of L interacting MD particles or finer scale DPD particles.
2. The projection $P(\cdot)$ operator (see Eqs. (6) and (7)) in CA corresponds to the averaging rule (see e.g., Eq. (8)) in PM.
3. The f_B transition function (new set of coarse-grained CA rules) corresponds to a coarse-grained dissipative collision operator Ω and (like in FPM case) a particle motion scheme.
4. Similarly as f_B compared to f_A (set of CA rules on fine-grained level), the collision operator Ω on the coarse-grained particle level can be more complicated than respective interaction scheme on finer levels. For example, additional non-conservative terms in DPD collision operator, such as dissipative and Brownian forces, can represent averaged DoF from conservative atomistic MD scale. Similarly, the ordinary differential equation of motion (such as in molecular dynamics) can be coarse-grained by more demanding stochastic differential equations.

Summarising, the coarse-graining of particle method from microscopic (molecular dynamics) to macroscopic (smoothed particle dynamics) formulation is possible, and promotes PM as a robust and homogeneous multi-scale modeling paradigm.

3 Complex Automata

The complex automata (CxA) principles were formulated by Hoextra and Sloot [22, 39]. This generalized modeling paradigm encompasses CA, lattice Boltzmann gas (LBG) and Agent Based Models (ABM) techniques as building blocks. Decomposition of the simulated system onto N single-scale cellular automata that mutually interact across many spatio-temporal scales is the key idea of CxA. This decomposition can be performed on the basis of Scale Separation Map (SSM) (an example of SSM is shown in Fig. 1) in which each sub-system can be positioned according to its spatial and temporal scales. The processes having well separated spatio-temporal scales can be easily identified as the components of the CxA multi-scale model.

The main problems with CxA model, defined in that way, are its limited feasibility and computability. Even for medium sized systems with two (or three, at most) scales of interest, extraction of a reasonable number of separated regions of interests (ROI), where finer models have to be used, refers to a very restricted number of phenomena such as crack formation [1]. In general, especially in biological systems, the number of ROI involving finer scales and its volume is usually large enough to make the multi-scale simulation extremely demanding (see Fig. 2). Having in mind all the computational and formal problems with coupling heterogeneous models representing different scales, the feasibility of developing a truly useful multi-scale biological model remains a dream of the future.

Development of homogeneous, scale invariant modeling metaphors such as CA and PM, in which the scale under interest is identified by the form of collision operator or set of rules, respectively, represents very competitive option to the multi-scale models involving hierarchy of interacting heterogeneous sub-models operating in different scales. It is mainly due to the efficiency, simplicity and generality. As was shown in Sec. 2.1 and 2.2, the collision operators and rules in coarse grained models may hold various type of information from finer scales. It allows for creation of supermodels made of a few homogeneous imperfect models carrying complementary knowledge about the simulated system as a whole. As was shown in [8], such the supermodel can over perform much more sophisticated multi-scale models and has a nice property of easy data assimilation due to easy coupling of numerical (formal) models with machine learning algorithms[1]. This is because the supermodel approach is methodologically similar to that from machine learning, where ensemble classifier made of simple classifiers become superior over more complex classification algorithms.

As it was shown in the previous sections, cellular automata is advantageous over other modeling approaches in simulating systems where interactions between individuals can be represented by a language instead of mathematical equations. Using more rules, i.e., more complicated language, one can simulate finer scales using coarse-grained CA representation (see Fig. 8 and 7). The same property holds the particle model. The TC-DPD collision operator in macroscale - much more complicated than conservative MD force in atomistic scales (see [38]) - encapsulates in a consistent way averaged degrees of freedom from atomistic scales represented by Wiener stochastic terms.

Summing up, the particle model reconstructs in a natural way mechanical interactions while cellular automata performs better when information exchange between individuals cannot be described only in terms of positions, velocities and forces. Therefore, by coupling the particle model with cellular automata, one can obtain the possibility to reconstruct both mechanical interactions and finer intercellular processes mimicked by CA rules. This way, the uniform coarse-grained complex automata (CxA) model can describe systems involving multiple scales and, simultaneously, avoiding computationally demanding hierarchy of sub-models.

The CxA consists of the following assumptions representing principal simulation steps.

---

[1] SUMO EU FET project webpage,
`http://www.knmi.nl/samenw/sumo/news.html`, 2011.

Assumption 1

The simulated system is made of a set of particles $\Lambda_N = \{O_i : O(\mathbf{r}_i, \mathbf{v}_i, \mathbf{a}_i), i = 1,\ldots,N\}$ where: i: particle index; N: the number of particles; $\mathbf{r}_i, \mathbf{v}_i, \mathbf{a}_i$: particle position, velocity and attributes, respectively. The vector of attributes ai is defined by the particle type, size, and its current state.

Assumption 2

The particle state may depend on time t, concentration of diffusive substances and total pressure exerted on particle i from its closest neighbors.

Assumption 3

The collision operator $\Omega_i(\ldots)$, which is equal to the sum of particle-particle vector interactions $\mathbf{F}_{ij}(|\mathbf{r}_i - \mathbf{r}_j|, \mathbf{v}_i - \mathbf{v}_j, \mathbf{a}_i, \mathbf{a}_j)$ between the central particle i and all the particles j confined in the sphere of radius rcut, defines the total force acting on particle i. The type of particle-particle interaction, $\mathbf{F}_{ij}$, may depend on the current attributes of particles i and j.

Assumption 4

1. The particle dynamics is governed by the Newtonian laws of motion (see Eqs. (1)). The particle positions are shifted just after computing collision operators acting on every particle i. The Eqs. ((1)) are integrated numerically in discrete time-steps Δt.
2. The attributes of particles i are updated according to the state of particles in its neighborhood according to prescribed CA rules.

Assumption 5

The particles attributes may also depend on current solutions of other large-scale models formulated in terms of PDEs (partial differential equations) such as reaction-diffusion or hydrodynamics equations.

The following examples show the advantages of CxA metaphor of modeling.

3.1 Flow with a Thin Layer over a Solid Surface

The complex type of fluid instability is produced in situations with a thin film falling down inclined plane or a vertical wall. It begins with an excess of fluid flowing from an opened gate placed at the top of a dry wall. This allows the viscous fluid of constant volume V to flow down the wall with a straight contact line (see Fig. 11(a)) that moves according to the direction of the gravitational field. Some time after the fluid release (the time depends on the fluid thickness, viscosity, physical properties of the wall surface), a contact line spontaneously develops and produces a series of fingers of fairly constant wavelengths across the slope (see Fig. 11(b)). This flow

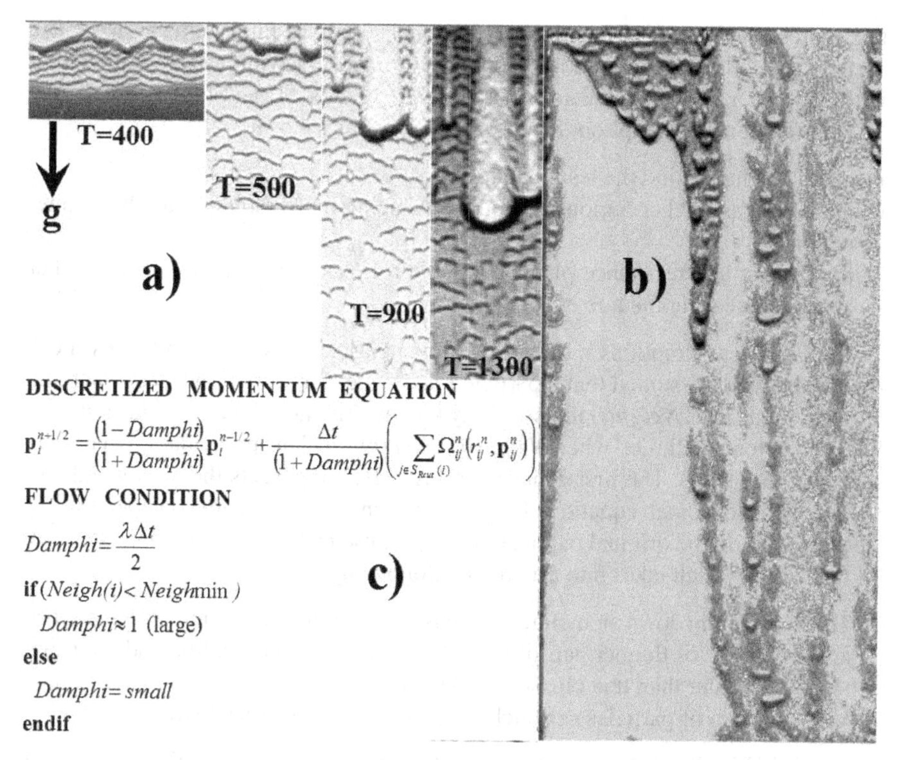

T=400
g
T=500
a)
T=900
T=1300
b)
DISCRETIZED MOMENTUM EQUATION
$\mathbf{p}_i^{n+1/2} = \frac{(1-Damphi)}{(1+Damphi)}\mathbf{p}_i^{n-1/2} + \frac{\Delta t}{(1+Damphi)}\left(\sum_{j \in S_{Rcut}(i)} \Omega_{ij}^n(r_{ij}^n, \mathbf{p}_{ij}^n)\right)$
FLOW CONDITION
$Damphi = \frac{\lambda \Delta t}{2}$
if (Neigh(i)< Neighmin)
Damphi≈1 (large)
else
Damphi= small
endif
c)

Fig. 11 The snapshots from CxA simulations of fluid flow with a thin layer over a solid surface: (a) excessive fluid inflow and smooth wall are assumed; (b) developed wet pattern of high viscosity fluid on dry and coarse wall; (c) the pseudo-code of CxA algorithm.

is caused by the presence of the contact line, which slows down the film drainage. High pressure near the contact line is responsible for ridge production. According to theory and experiments [34], a perturbed capillary ridge has thicker regions of liquid advancing more rapidly than the thinner regions. The larger resistance at the wedge segments of the finger head or during flow initiation results in larger liquid accumulation and, consequently, increases the subsequent rate of spreading. This also results in an increase in the ridge thickness, detected in the gravity-driven fingering experiments. In the case of complete wetting, intermolecular forces, comparable to the main driving force, are powerful enough to exceed viscous dissipation in a wedge and, hence, overcome this accelerating effect. Thus, the contact line not only increases the resistance to the flow, but also provides an appreciable driving force on the fronts of the falling film.

From this scenario, we proposed in [13] a new 2-D numerical particle model of falling sheet evolution, which can be considered as a supplementary one to the EE theory [34]. Let us consider system of the 1.2×10^5 particles which are initially placed at the top of computational box (see the first snapshot from Fig. 11(a)). This region stands for a vertical wall covered by the particle fluid. The white part of the box represents dry wall. There is not additional supply of fluid to the system. The particle move down in gravitational field according to Newtonian laws interacting with each other via DPD collision operator Ω_{ij} (see Eqs. (12)–(14)).

The simulation assumptions can be summarized as follows:

- a ridge forms behind the leading edge,
- the ridge has thicker regions of liquid advancing more rapidly than the thinner regions,
- the contact-line resistance plays a "double role" not only in slowing down but also by increasing the rate of spreading.

To fulfill these assumptions we use the following trick, represented in fact by a CA rule. Namely, we assumed that a particle i undergoes large friction force, when the number of particles ***Neigh(i)*** in its vicinity (i.e., within the sphere of radius rcut, see Eqs. (1)) is too small, i.e., when ***Neigh(i) < Neighmin***. This particular procedure is shown in Fig. 11(c). The first equation in Fig. 11(c) represents the discretized and transformed Newtonian equation of particle motion with DPD collision operator Ω_{ij} (see Eqs. (1)). In the original publication [33] we used a little more complicated CA "if then" rule, which takes into account the following facts:

1. The fluid falling down an inclined plane is transferred literally into growing lobes at the expense of thinner part of the ridge. Therefore, the fluid behind the ridge is moving faster than that closer to its leading edge.
2. Very thin film of particles can stick to the wall and moves very slowly.

As shown in Fig. 11(a),(b) the application of this CA rule together with DPD particle dynamics reveal fine-grain structure of this fluid instability such as (see Fig. 11(a)) i.e.,:

1. 3-D "synchronous" (i.e., the neighboring wave fronts are in phase) or "asynchronous" patterns with transverse modulations appear,
2. Wave-fronts break-ups (for the "synchronous" case), leading to disordered patterns or "herringbone" patterns to appear in patches (for the "asynchronous" case),
3. Spontaneous emergence of avalanches, droplets and rivulets (see Fig. 11(b)).

All of these details are very difficult to simulate within classical fluid dynamical models, due to the critical nature (self-organized criticality) and threshold character of these nonlinear phenomena. Unlike the classical approaches (integrating evolutionary equation [34]), we need not introduce any external and artificial perturbations. All phenomena occur spontaneously due to thermal noise inherent in the nonlinearly interacting particle dynamics. The CxA "if then" rule allows for mimicking local scale phenomena such as surface tension and fluid-wall adhesion. Moreover, purely 2-D simulation can mimic the third dimension — fluid thickness — simulated by large particle density variation.

3.2 Tumor Growth

The complex automata paradigm can be considerably extended introducing more CxA rules and integrating them with continuous models represented by partial differential equations. Below we present only a brief description of our complex automata model simulating tumor progression. More details can be found in [44, 43, 42, 45].

As it is widely known, cancer is one out of major killers in the developed world being responsible for about 20% of deaths in developed countries [25]. Let us skip the complex genetic processes influencing the appearance of the first tumor cells and let us assume that a small cluster of such the cells is ready for proliferation. Typically, further growth of a solid tumor consists of three phases: avascular growth, angiogenesis, vascular growth, metastasis (e.g., [18]).

In avascular phase, the tumor develops due to nutrients diffusion (e.g., O_2) throughout the tissue from neighboring blood capillaries. However, O2 diffusion range is only about 100 μm from the blood vessel thus some of cancer cells are in a chronic shortage of oxygen. Such the hypoxic cells produce and release proteins and other chemical species called tumor angiogenic factors (TAFs) [18]. These signaling compounds diffuse throughout the tissue, and, upon arrival to the blood vessels, they trigger a cascade of events which stimulates the growth of vasculature towards the tumor cluster. In the following phase, vascular one, the tumor having access to unlimited resources of oxygen and other nutrients considerably accelerates its growth. Moreover, through the blood vasculature, the tumor secretes cancerogenic material forming metastases. Thus, whereas in the avascular phase tumors are basically harmless, once they become vascular they are potentially fatal.

As it was shown in many papers (e.g., [5, 27, 3]) computer modeling can allow for answering many principal questions concerning the effects of prescribed chemotherapy or testing new drugs to control the process of tumor growth in all its phases.

There exist many mathematical models of tumor progression in all its phases (e.g. [6, 36, 28]). However, only a few consider mechanical factors of growth. Meanwhile, neglecting all microscopic and mesoscopic biological and biophysical processes, tumor growth is a purely mechanical phenomenon. Due to the effect of tumor directional progression, the surrounded tissue, vasculature and tumor on its own undergo continuous process of remodeling. The tissue and vasculature remodeling due to tumor push on is not only the source of pain (e.g., when tumor push on the nerves in the spinal cord) but it influences the speed of its growth as well. Just tumor remodeling is responsible for its heterogeneity, which influences the drug dosage/rate in chemotherapy. Modeling of mechanical growth involves dissipative interactions between normal, cancerous tissues and vascular network. This kind of tumor dynamics could not be reconstructed by using existing models. As shown in [44], CxA can be used as a robust metaphor which closes this gap.

We assume that a fragment of tissue, is made of a set of particles $\Lambda_N = \{O_i : O(\mathbf{r}_i, \mathbf{v}_i, \mathbf{a}_i), i = 1, \dots, N\}$ where: i: particle index; N: the number of particles, $\mathbf{r}_i, \mathbf{v}_i, \mathbf{a}_i$: particle position, velocity and attributes, respectively. Each particle represents a single cell with a fragment of ECM (extracellular matrix). The vector of attributes $\mathbf{a}_i$ is defined by the particle type tumor cell (TC), normal cell (NC), endothelial cell (EC), cell life-cycle state (see Fig. 12(a)) newly formed, mature, in *hypoxia*, after *hypoxia*, *apoptosis*, *necrosis*, cell size, cell age, *hypoxia* time, concentrations of k=TAF, O_2 (and others) and total pressure exerted on particle i from its closest neighbors. The particle system is confined in the cubical computational box with a constant external pressure. For the sake of simplicity the vessel is constructed of tube-like "particles" EC-tubes—made of two particles connected by a rigid spring (see Fig. 12(b)). We define three types of interactions: particle-particle, particle-tube, and tube-tube. The forces between particles mimic both mechanical repulsion from squashed cells and attraction due to cell adhesiveness and depletion interactions cause by both ECM matrix and the cell. We postulate the heuristics - particle interaction potential $\Omega(d_{ij})$ (Fig. 12(c)) - in the following form:

$$\Omega(d_{ij}) = \begin{cases} a_1 d_{ij}^2 & \text{for } d_{ij} < 0 \\ a_2 d_{ij}^2 & \text{for } 0 < d_{ij} < d_{cut} \\ a_2 d_{cut}^2 & \text{for } d_{ij} \geq d_{cut} \end{cases} \tag{17}$$

where $a_1 > a_2$, $d_{ij} = |\mathbf{r}_{ij} - (r_i + r_j)|$ and $|\mathbf{r}_{ij}|$ is the distance between particles while r_i and r_j are their radiuses.

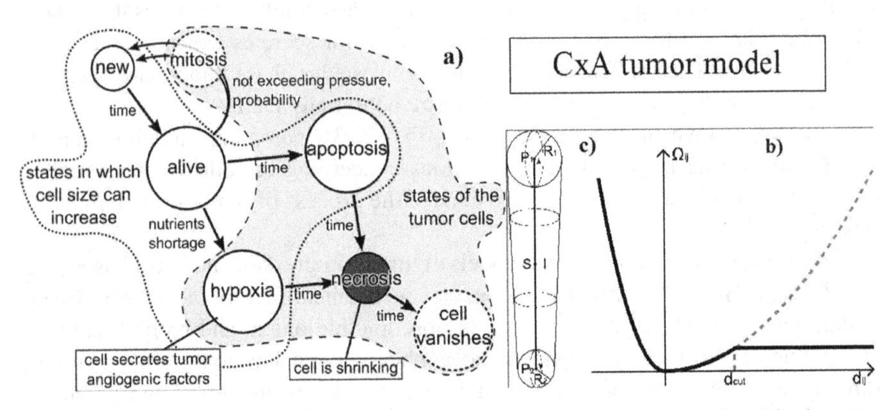

Fig. 12 (a) Tube-like particle made of two spherical "vessel particles". (b) $\Omega(d_{ij})$— the collision operator. (c) The cell life-cycle [45]

We assume that the interactions between spherical particles and EC-tube particles have similar character. However, as shown in [44, 43, 42, 45], additional rules have to be introduced to enable appropriate growth of the vascular network. The particle dynamics is governed by the Newtonian laws (see Eqs. (1)) and DPD collision operator is used for simulating particle-particle interactions.

As shown in Fig. 12(c), both normal and tumor cells change their states from new to apoptotic (or necrotic). After *mitosis*, well oxygenated cells of certain age and size split into two daughter cells with d_{MIN} diameters. The cell diameter increases proportionally to oxygen concentration up to d_{MAX}. Finally, after given time period, the particles die due to programmed cell death (*apoptosis*). For oxygen concentration smaller than a given threshold, the living cell changes its state to *hypoxia* being the source of TAFs. The cells die and become necrotic if they remain in *hypoxia* state too long. We assume that at the beginning, the diameter of necrotic cell decreases twice and, after some time, the cell vanishes. This is contrary to apoptotic cells, which are rapidly digested by their neighbors or by macrophages. Both normal and tumor cells differ considerably in duration of the life cycle phases and, especially, in the period of time they can live in *hypoxia*. The hypoxic cancerous cells can stay alive a few orders of magnitude longer than normal cells.

The life-cycle for EC-tubes is different. They can grow both in length and in diameter. Reduced blood flow, the lack of VEGF (vascular endothelial growth factor), dilation, perfusion and solid stress exerted by the tumor can cause their rapid collapse. Because the EC-tube is a cluster of EC cells, its division onto two adjoined tubes does not represent the process of *mitosis* but is a computational metaphor of vessel growth. Unlike normal and tumor cells, the tubes can appear as tips of newly created capillaries sprouting from existing vessels. The new sprout is formed when the TAFs concentration exceeds a given threshold and its growth is directed to its local gradient.

The distribution of hematocrit is the source of oxygen, while the distribution of tumor cells in *hypoxia* is the source of TAFs. We assume that the cells of any type consume oxygen with the rate depending on both cell type and its current state, while TAFs are absorbed by EC-tubes only. TAFs are washed out from the system due to blood flow.

Because diffusion of oxygen and TAFs through the tissue is many orders of magnitude faster than the process of tumor growth, we assume that both the concentrations and hydrodynamic quantities are in steady state in the time-scale defined by the time-step of numerical integration of equations of motion. To calculate the concentrations of oxygen and TAF we solve the reaction-diffusion equations numerically by using approximation theory. One can estimate a function f at position $\mathbf{r}$ by using smoothing kernels W as follows:

$$f(\mathbf{r}) = \sum_{j=1}^{n} m_j \frac{f_j}{\rho_j} W(\mathbf{r} - \mathbf{r}_j, h) \tag{18}$$

where m_j is the mass, $\mathbf{r}_j$ is the position, ρ_j is the density and f_j is the quantity f for neighbor particle j, respectively. Here, n is the number of neighboring particles within cut of radius $h(|\mathbf{r}\mathbf{r}_j| \leq h)$. When $\mathbf{r} = \mathbf{r}_i$, $f(r)$ is denoted by f_i. The smoothing kernel approximates a local neighborhood r within distance h. Thus, we can estimate the density ρ_i for a particle i at location $\mathbf{r}_i$ by:

$$\rho_i = \sum_{j=1}^{n} m_j W(\mathbf{r}_i - \mathbf{r}_j, h) \tag{19}$$

where j is the index of the neighboring particle. The kernel should be smooth, symmetric and satisfy the following equation:

$$\int_{\Omega} W(\mathbf{r},h)d\mathbf{r} = 1 \tag{20}$$

We used 3D *poly6* kernel proposed by Muller et al.

$$W_{poly6}(\mathbf{r},\mathbf{h}) = \frac{315}{64\pi h^9}\begin{cases}(h^2-|\mathbf{r}^2|)^3 & |\mathbf{r}| \leq h \\ 0 & \text{otherwise}\end{cases} \tag{21}$$

We selected this kernel due to its simplicity. It was shown, that better kernels can be used, however, at the cost of computational efficiency. The Laplacian can be approximated then:

$$\Delta f_i = \sum_{j=1}^{n} \frac{m_j}{\rho_j} f_j \cdot \Delta W(\mathbf{r}_i - \mathbf{r}_j, h) \tag{22}$$

Substituting Laplacian in the reaction-diffusion equation by Eq. (8) we got the following expression for concentrations c_i^K of K={oxygen, TAF} in particle i (χ^K: reaction factor). When K=oxygen, then I=TAF and vice versa.

$$c_i^K = \frac{\left(\frac{\chi_i^I c_i^I}{D} - \sum_{j=1,i\neq j}^{n} \frac{m_j}{\rho_j} c_i^K \cdot \Delta W(\mathbf{r}_i - \mathbf{r}_j, h)\right)}{\left(\frac{m_i}{\rho_i}\Delta W(0,h) - \frac{\chi_i^K}{D}\right)} \tag{23}$$

By solving this equation iteratively in each time-step of Newtonian equation integration, we got approximate concentration of TAF and oxygen in each particle location.

The particles are confined in the cubical computational box of volume V. Because the average kinetic energy in the system is negligible small, from the virial theorem we obtain that:

$$P \approx \frac{1}{3V} \cdot \sum_{i<j}^{N} \mathbf{F}_{ij} \circ \mathbf{r}_{ij} \tag{24}$$

The internal pressure increases due to increasing number of particles (cells). The increase of box volume V compensates the pressure increase above a given threshold.

On the other hand, the blood circulation is slower than diffusion but still faster than *mitosis* cycle. These facts allow for employing fast approximation procedures for both calculation of blood flow rates in capillaries and solving reaction-diffusion equation (see [44]). After initialisation phase, in subsequent time-steps we calculate forces acting on particles, new particle positions, the diffusion of active substances (nutrients, TAFs, pericytes), the intensity of blood flow in the vessels and the states of individual cells triggered by previous three factors and constrained by time clocks of individual cells. All of these modifications of cell states may result in cell *mitosis* or its death. They can also change some cell functions (e.g. those under *hypoxia*) their size and environmental properties (e.g., cancerous cells can secrete acid to eliminate neighboring tissue cells).

Summing up the basic procedures of our CxA particle model consist of: the model initialisation phase, i.e., definition of initial and boundary conditions, and its evolution driven by the following phenomena:

1. Newtonian dynamics of interacting cells, diffusion of oxygen and TAF,
2. cellular life cycle modeled by CxA rules,
3. vessels sprouting and growth,
4. vessels remodeling due to blood flow, vessel maturation and degradation.

As shown in Fig. 13(a), the straight blood vessel which is passing initially throughout cancerous tissue, begin to sprout out. The newly created lumen become functional, i.e., is able to transport blood, when a loop (anastomosis) with another functional vessel creating pressure gradient is formed. Moreover, the capillary has to be covered with adequate quantity of mural cells. Mural cells are vascular support cells that range in phenotype from pericytes to vascular smooth muscle cells [37]. As displayed in Fig. 13(a) and 13(b), the structures of vasculature become very complex and dynamic due to continual vessels maturation and degradation. The newly created vasculature inside the tumor is unstable, vulnerable to rapid changes in blood pressure caused by the lack of previously mentioned factors and mechanical remodeling (see Fig. 13(b)).

Very similar CxA model can be applied for simulation of cereal infection by parasite fungi called *Fusarium Graminearum* (Fg). In the following section we present preliminary results of modeling Fg infection.

3.3 *Fusarium Graminearum Invasion*

Fusarium graminearum is one of the main causal agents of Fusarium head blight (FHB) infection. It attacks cereal crops what results in significant losses. The epidemic, which took place in North America from 1998 to 2000, costs almost $3 billion. Another effect of this plague is the contamination of grain with mycotoxins, which is extremely harmful for animals and humans.

Wheat heads are the most susceptible to infection during anthesis. Other factors favoring infection are high humidity and temperature [4]. Initially, the fungus does not penetrate the epidermis. As shown in [4, 30], at this stage it develops on the external surfaces of florets and glumes and grows towards susceptible sites within the inflorescence. Other roads of colonisation of internal tissue include stomata and underlying parenchyma, partially or fully exposed anthers, openings between the lemma and palea of the spikelet or floret during dehiscence and through the base of the wheat glumes where the apidermis and prenchyma are thinwalled [4, 30]. As every model is a metaphor of some real phenomenon the following assumptions were made:

1. Every *Fusarium* and plant cell is a particle interacting with the other particles.
2. Every cell has a number of attributes that evolve in time.
3. The concentration of nutrients is uniform in a single cell and constant in the specific time step.

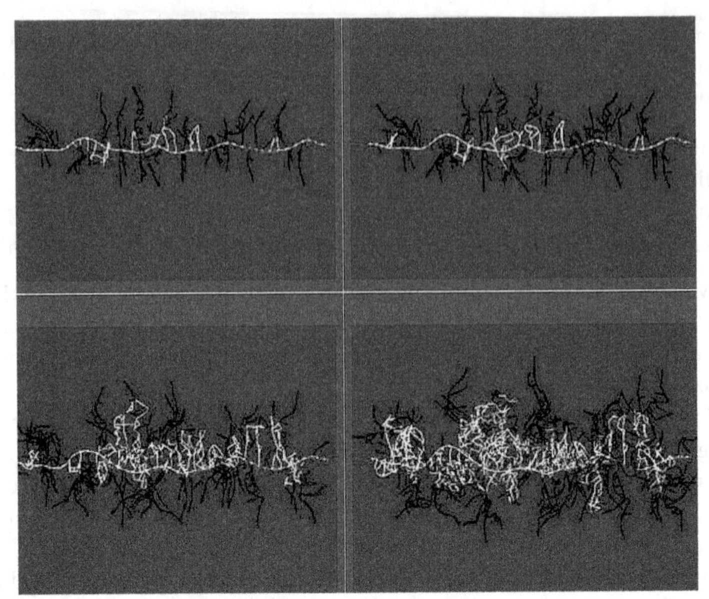

(a) The snapshots from 3-D CxA simulation displaying development of vasculature due to angiogenesis. The white network represents functional blood vessels while the black one shows non-functional vessels. Initially, the single vessel passes throughout the tumor. The tumor cells are invisible in this figure.

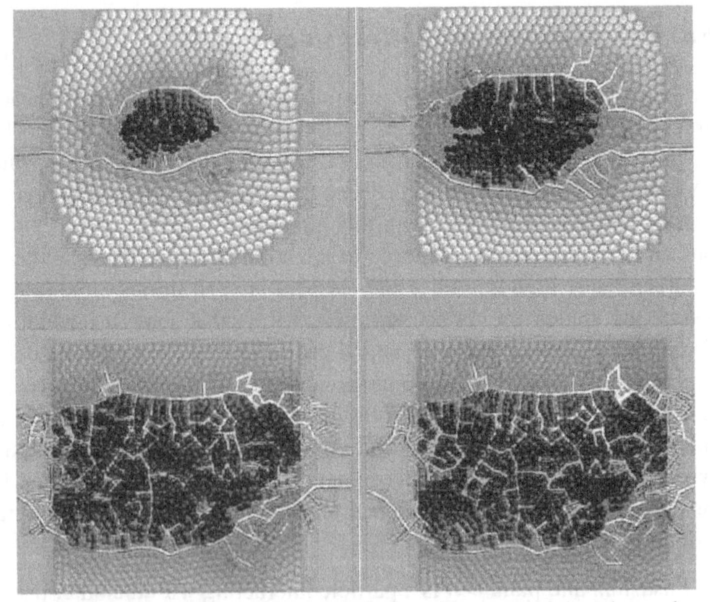

(b) The snapshots from 2-D CxA simulation of tumor growth and vascular network remodeling.

Fig. 13 The snapshots from 3-D and 2-D CxA simulations.

4. Nutrients circulation in the fusarium body, which allows *Fusarium* to proliferate, is the effect of diffusion.
5. Each *Fusarium* and plant cell is in one of three discrete states which models cell-life cycle.

The laboratory experiments from Fig. 14(a),(b) were conducted in vitro in artificial conditions. This means that no nutrients were produced in the course of experiments and the initial amount of food was only consumed by *Fusarium*. They were performed on flat surfaces on the so called Petri dishes. Two types of environment were tested: SNA, which is nutrient-poor and PDA, which is nutrient-rich. Both substances are water solutions. This allows for two important assumptions: the fungus does not encounter much strain from the environment and diffusion does not need to be modeled directly. We may safely assume that the diffusion in water is fast

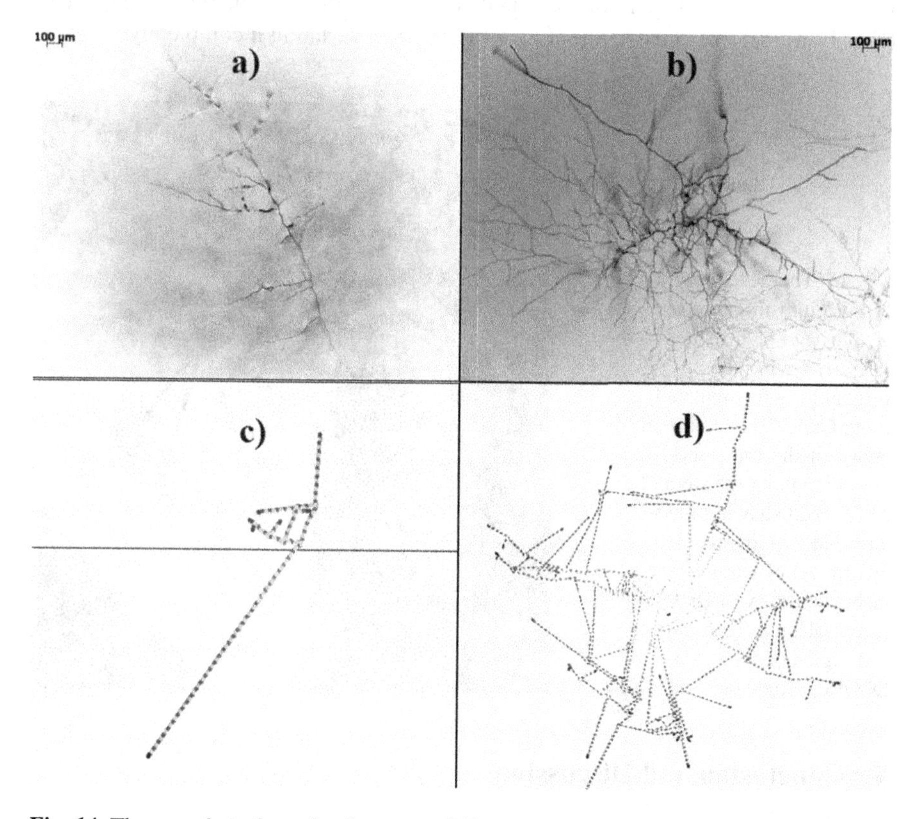

Fig. 14 The snapshots from development of *Fusarium* in SNA (nutrient poor) (a),(c) and PDA (nutrient rich) (b),(d) environment from experiment (a),(b) (courtesy of Dr Shea Miller, Agriculture and Agri-Food Canada, ECORC, Ottawa) and simulation using CxA (c),(d). In (c) Fusarium does not branch at all in the nutrients-poor environment below the horizontal black line. It starts to develop more extensively only after reaching the nutrients rich half space above the black line.

enough to keep uniform nutrients concentration in the whole volume. As a result all fungus cells have identical external nutrients level and there is also no need to model diffusion inside the fungus. In this early modeling stage only model of the hyphae growth and physical behavior has been developed. Due to the absence of plant cells in real experiments interactions with environment were not modeled.

As shown in Fig. 14, the confrontation of simulation results with experimental data is cautiously optimistic. The qualitative character of growth is very similar. However, the structural characters of networks produced by *Fusarium graminearum* and simulation code are clearly different. This can be improved, however, by using higher resolution and playing the parameters responsible for the sprouting phenomenon. Another confrontation of simulation with experiment, displayed in Fig. 15, once again shows good qualitative agreement of the two. Fg spreads mainly through vascular bundles, penetrating also the closest neighborhood. In the case when fungi find the nutrient rich part of the plant, it changes the growth type from linear (Fig. 14(a),(c)) to extensive one (Fig. 14 (b),(d)), devastating it completely.

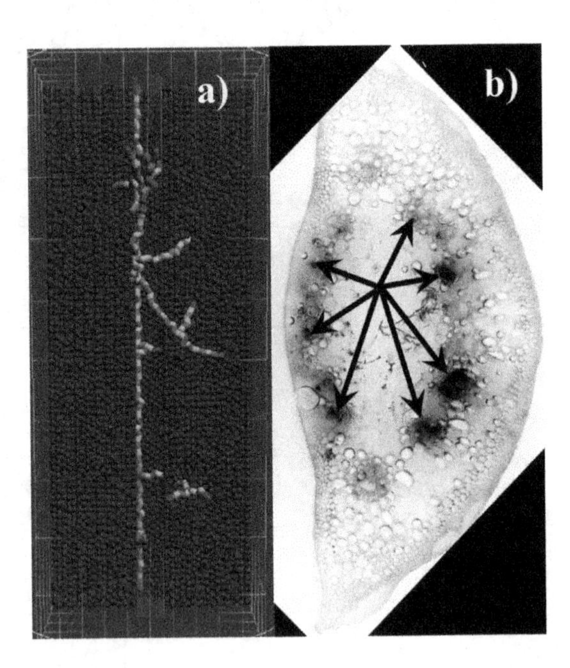

Fig. 15 The development of Fusarium in vascular bundles and rachis - nutrient poor environment: (a) vertical growth simulated using CxA; (b) microscopic image of the cross section throughout the stem of wheat head showing Fusarium (black stains pointed by arrows) penetrating the rachis (courtesy of Dr Margaret Balcerzak, Agriculture and Agri-Food Canada, ECORC, Ottawa).

4 Conclusion and Discussion

The most of biomedical problems, which concentrates on finding remedies against a pathogen expansion, are extremely difficult to simulate. Mainly due to the lack of:

- adequate numerical approaches representing many spatio-temporal scales involved in biological processes, their discrete nature and high nonlinearity;
- the procedures and practices of coupling numerical models with diverse sources of data.

The development of integrated computational models and data coupling procedures is a vital problem in computational sciences, which considerably postpones the successful implementation of computational techniques in biosciences.

We have introduced here a novel modeling concept, called Complex Automata, which integrates the two types of modeling techniques, namely, Particle Method and Cellular Automata. The particle represents here both the component of a dynamical system and plays the role of CA node. We have demonstrated that the multi-scale modeling techniques can be built upon the notion of coarse graining of these two approaches. Instead of developing complicated multi-scale models which consist of many submodels representing various scales coupled by a scales-bridging mechanism, we propose here a uniform, single scale, coarse grained computational framework for which information about finer scales is inscribed in CA rules and particle interactions.

Both the CA rules and the PM collision operator can be matched to the scale under interest by removing or inserting additional degrees of freedom. The number of DoF can be manipulated by changing complexity of particle interactions and by increasing the set of CA rules. The most important advantage of such the approach over other classical multi-scale models is that the spatio-temporal scale considered remains as coarse as possible, while the finer scales are simulated by more complex collision operator and CA rules.

We have shown that the particle method decides about the mechanical properties of the system. Microscopic phenomena involving fluctuations and dissipative behavior can be added in a consistent way exploiting widely known modeling techniques such as DPD or TC-DPD. We demonstrated that some other physical properties, like surface tension, can be simulated by introducing to the interaction model a simple CA rule. Apart from system properties resulting from simple Newtoniam mechanics, other microscopic biological processes can be encapsulated in CA rules. These rules depend on the current configuration of the nearest neighbors and other phenomena, e.g., described by the continuum fields obtained from integrating PDEs.

We have also presented a proof-of-concept of our approach by employing CxA as a metaphor in modeling of two different biological phenomena. The ability of mimicking both mechanical interactions of tumor with the rest of tissue and penetration properties of *Fusarium graminearum*, shows that our model can reproduce realistic 3-D dynamics of these complex biological systems.

We have also presented a proof-of-concept of our approach by employing CxA as a metaphor in modeling of two different biological phenomena. The ability of mimicking both mechanical interactions of tumor with the rest of tissue and penetration properties of Fusarium graminearum, shows that our model can reproduce realistic 3-D dynamics of these complex biological systems.

For identification of weak points of these (i.e., cancer proliferation and Fg infection) and similar pathogens, the contemporary biology employs a bottom-up approach. Manipulation of the level of expression of solitary genes allow for controlling molecular pathways which influence some phenotype behaviors. However, to explain a complex biological phenomenon involving many genes, this process remains very inefficient, expensive and unreliable. The solution space inflates exponentially with the number of genes making impossible to find respective genetic code. We hypothesize that top-down approach exploiting numerical modeling empowered by data assimilation procedures enables more precise selection of these molecular triggers. It can serve as a bidirectional coupling factor between genotype and phenotype.

The general concept of data assimilation with CxA model could be formulated as follows. On the base on observed phenotypic behaviors we can categorize them (e.g. using microscopic images) and associate these categories to genotypes extracted from experimental data (e.g., microarrays) by using machine learning and data analysis tools. On the other hand, running the coarse-grained numerical model (level 1) one can adjust the sets of parameters of the model to the respective phenotype categories. The parameters are associated with some microscopic biochemical processes, which can be scrutinized in molecular level, i.e., much closer to the genotype than emergent phenotypic behavior. To make the prediction more accurate, one can run fine-grained (level 2) model with extended collision operator and larger set of CA rules. The new set of parameters is matched to the data by using previously obtained (level 1) parameters. This procedure can be applied downscale. The concept of data assimilation is being developed now and will constitute the following step in devising reliable data driven CxA models.

We expect that with such the advancement in modeling techniques, new faster and cheaper methods will be developed which allow for attacking biological problems in a more systematic and focused way.

Acknowledgements. This paper is based on the invited talk presented on 2nd International Conference on Computational Engineering (ICCE 2011), Darmstadt, 4-6 October 2011. This research is financed by the Polish Ministry of Higher Education and Science, project NN519 579338 and partially by AGH grant No.11.11.120.777. The Author thanks Professor A.Z. Dudek from Division of Hematology, Oncology, and Transplantation at the University of Minnesota, dr. Shea Miller and Margaret Balcerzak from Agriculture and Agri-Food Canada (AAFC), Ottawa for discussions about mediacal and biological aspects of this work, respectively. Thanks are also to Mr Przemysław Głowacki and dr Rafał Wcisło for their contribution to this paper.

References

1. Abraham, F., Broughton, J., Bernstein, N., Kaxiras, E.: Spanning the length scales in dynamic simulation. Computers in Physics 12(6) (1998)
2. Beazley, D.M., Lomdahl, P.S., Gronbech-Jansen, N., Giles, R., Tomayo, P.: Parallel algorithms for short range molecular dynamics. In: Annual Reviews of Computational Physics, vol. III, pp. 119–175. World Scientific (1996)
3. Bellomo, N., Angelis de, E., Preziosi, L.: Multiscale modeling and mathematical problems related to tumor evolution and medical therapy. J. Theor. Med. 5(2), 111–136 (2003)
4. Brown, N.A., Urban, M., Meene van de, A.M.L., Hammond-Kosack, K.E.: The infection biology of fusarium graminearum: Defining the pathways of spikelet to spikelet colonisation in wheat ears. Fungal Biology 114(7), 555–571 (2010)

5. Castorina, P., Carco, D., Guiot, C., Deisboeck, T.: Tumor growth instability and its implications for chemotherapy. Cancer Res. 69(21) (2009)
6. Chaplain, M.A.J.: Mathematical modelling of angiogenesis. J. Neuro.-Oncol. 50, 37–51 (2000)
7. Chopard, B., Droz, M.: Cellular Automata Modeling of Physical Systems. Cambridge University Press (1998)
8. Berge van den, L.A., Selten, F.M., Wiegerinck, W., Duane, G.S.: A multi-model ensemble method that combines imperfect models through learning. Earth Syst. Dynam. 2, 161–177 (2011)
9. Dzwinel, W.: Virtual particles and search for global minimum. Future Generation Computer Systems 12, 371–389 (1997)
10. Dzwinel, W., Alda, W., Kitowski, J., Yuen, D.A.: Using discrete particles as a natural solver in simulating multiple-scale phenomena. Molecular Simulation 20(6), 361–384 (2000)
11. Dzwinel, W., Alda, W., Pogoda, M., Yuen, D.A.: Turbulent mixing in the microscale. Physica D 137, 157–171 (2000)
12. Dzwinel, W., Alda, W., Yuen, D.A.: Cross-scale numerical simulations using discrete-particle models. Molecular Simulation 22, 397–418 (1999)
13. Dzwinel, W., Yuen, D.A.: Dissipative particle dynamics of the thin-film evolution in mesoscale. Molecular Simulation 22, 369–395 (1999)
14. Dzwinel, W., Yuen, D.A., Boryczko, K.: Bridging diverse physical scales with the discrete-particle paradigm in modeling colloidal dynamics with mesoscopic features. Chemical Engineering Sci. 61, 2169–2185 (2006)
15. Espanol, P.: Fluid particle model. Phys. Rev. E 57, 2930–2948 (1998)
16. Espanol, P., Serrano, M.: Dynamical regimes in dpd. Physical Review E 59(6), 6340–6347 (1999)
17. Flekkoy, E.G., Coveney, P.V.: Foundations of dissipative particle dynamics. Physics Review Letters 83, 1775–1778 (1999)
18. Folkman, J.: Tumor angiogenesis, therapeutic implications. N. Engl. J. Med. 285, 1182–1186 (1971)
19. Fuchslin, R.M., Eriksson, A., Fellermann, H., Ziock, H.J.: Coarse-graining and scaling in dissipative particle dynamics. J. Chem. Phys. 130 (2009)
20. German, T.C., Kadau, K.: Trillion-atom molecular dynamics becomes a reality. International Journal of Modern Physics C 19(9), 1315–1319 (2008)
21. Helbing, D., Farkas, I.J., Vicsek, T.: Simulating dynamical features of escape panic. Nature 407, 487–490 (2000)
22. Hoekstra, A.G., Lorenz, E., Falcone, J.-L., Chopard, B.: Towards a Complex Automata Framework for Multi-scale Modeling: Formalism and the Scale Separation Map. In: Shi, Y., van Albada, G.D., Dongarra, J., Sloot, P.M.A. (eds.) ICCS 2007. LNCS, vol. 4487, pp. 922–930. Springer, Heidelberg (2007)
23. Hoogerbrugge, P.J., Koelman, J.: Simulating microscopic hydrodynamic phenomena with dissipative particle dynamics. Europhysics Letters 19(3), 155–160 (1992)
24. Israeli, N., Goldenfeld, N.: Coarse-graining of cellular automata, emergence, and the predictability of complex systems. Phys. Rev. E 73(2) (2006)
25. Jemal, A., Siegel, R., Jiaquan, X., Ward, E.: Cancer statistics 2010. CA Cancer J. Clin. 60, 277–300 (2010)
26. Libersky, L.D., Petschek, A.G., Carney, T.C., Hipp, J.R., Allahdadi, F.A.: High strain lagrangian hydrodynamics. Journal of Computational Physics 109(1), 67–73 (1993)
27. Lowengrub, J.S., Frieboes, H.B., Jin, F., Chuang, Y.L., Li, X., Macklin, P., Wise, S.M., Cristini, V.: Nonlinear modelling of cancer: bridging the gap between cells and tumours. Nonlinearity 23 (2010)

28. Mantzaris, N., Webb, S., Othmer, H.G.: Mathematical modeling of tumor-induced angiogenesis. J. Math. Biol. 49(2), 1416–1432 (2004)
29. Marsh, C., Backx, G., Ernst, M.H.: Static and dynamic properties of dissipative particle dynamics. Physical Review E 56 (1997)
30. Miller, S.S., Chabota, D.M.P., Ouellet, T., Harrisa, L.J., Fedak, G.: Use of a fusarium graminearum strain transformed with green fluorescent protein to study infection in wheat (triticum aestivum). Canadian Journal of Plant Pathology 26(4), 453–463 (2004)
31. Nakano, A., Bachlechner, M., Campbell, T., Kalia, R., Omeltchenko, A., Tsuruta, K., Vashishta, P., Ogata, S., Ebbsjo, I., Madhukar, A.: Atomistic simulation of nanostructured materials. IEEE Computational Science and Engineering 5(4), 68–78 (1998)
32. Nakano, A., Bachlechner, M.E., Kalia, R.K., Lidorikis, E., Vashishta, P.: Multiscale simulation of nanosystems. Computing in Science and Engineering 3(4) (2001)
33. Nigiel, G.: Agent-based Models. Sage Publications, London (2007)
34. Oron, A., Davis, S.H., Bankoff, S.G.: Long-scale evolution of thin films. Rev. of Modern Phys. 69(3), 931–980 (1997)
35. Pelechano, N., Badler, N.I.: Improving the realism of agent movement for high density crowd simulation, `http://www.lsi.upc.edu/npelechano/MACES/MACES.htm`
36. Preziozi, L.: Cancer modelling and simulation. Chapman & Hall/CRC Mathematical Biology & Medicine (2003)
37. Raza, A., Franklin, M.J.: Pericytes and vessel maturation during tumor angiogenesis and metastasis. Am. J. Hematol. 85(8), 593–598 (2010)
38. Serrano, M., Espanol, P.: Thermodynamically consistent mesoscopic fluid particle model. Phys. Rev. E 64(4) (2001)
39. Sloot, P., Kroc, J.: Complex systems modeling by cellular automata. In: Encyclopedia of Artificial Intelligence, pp. 353–360. Informatio SCI, Harshey, New York (2009)
40. Vasilyev, O.V., Bowman, K.: Second-generation wavelet collocation method for the solution of partial differential equations. Journal of Computational Physics 165(2), 660–693 (2000)
41. Vasilyev, O.V., Zheng, X., Dzwinel, W., Dudek, A.Z., Yuen, D.A.: Collaborative research: Virtual melanoma — a predictive multiscale tool for optimal cancer therapy. NiH proposal (2011)
42. Wcisło, R., Dzwinel, W.: Particle Based Model of Tumor Progression Stimulated by the Process of Angiogenesis. In: Bubak, M., van Albada, G.D., Dongarra, J., Sloot, P.M.A. (eds.) ICCS 2008, Part II. LNCS, vol. 5102, pp. 177–186. Springer, Heidelberg (2008)
43. Wcisło, R., Dzwinel, W.: Particle Model of Tumor Growth and Its Parallel Implementation. In: Wyrzykowski, R., Dongarra, J., Karczewski, K., Wasniewski, J. (eds.) PPAM 2009. LNCS, vol. 6067, pp. 322–331. Springer, Heidelberg (2010)
44. Wcisło, R., Dzwinel, W., Yuen, D.A., Dudek, A.: A new model of tumor progression based on the concept of complex automata driven by particle dynamics. J. Mol. Mod. 15(12), 1517–1539 (2009)
45. Wcisło, R., Gosztyła, P., Dzwinel, W.: N-body parallel model of tumor proliferation. In: Proceedings of Summer Computer Simulation Conference, Ottawa, Canada, July 11-14, pp. 160–167 (2010)
46. Wolfram, S.: A New Kind of Science. Wolfram Media Incorporated (2002)

Graph Grammar Based Model for Three Dimensional Multi-physics Simulations

Maciej Paszyński, Anna Paszyńska, and Robert Schaefer

Abstract. This chapter presents a graph grammar model for the generation and refinements of three dimensional computational meshes as well as for the solution of multi-physics problems over them. It is assumed that the computational meshes consists in tetrahedral and prism elements, and the tetrahedral elements can be further broken into four new tetrahedral and two new pyramid elements, and the prism elements can be further broken into four new prism elements. The graph grammar expresses the generation and adaptation algorithms as a sequence of graph grammar productions. It is assumed that the computational mesh is represented as a graph, and the mesh generation is expressed as a sequence of execution of graph grammar productions, starting from an initial graph with a single node. The graph grammar based algorithm is then utilized to generate the three dimensional mesh representing the simplified model of the human head with internal ear. We concentrate on the solution of the linear elasticity coupled with acoustics (the problem of propagation of acoustic waves over the simplified model of the human head). In the following part of the chapter we introduce the graph grammar productions transforming the computational mesh into a sequence of element frontal matrices, being the input for the multi-frontal solver algorithm. Finally, the overview on the out-of-core solver algorithm is presented, and the numerical results with comparison to the state-of-the art MUMPS solver are described.

Maciej Paszyński · Robert Schaefer
AGH University of Science and Technology, Al. Mickiewicza 30, 30-059 Kraków, Poland
e-mail: {paszynsk, schaefer}@agh.edu.pl

Anna Paszyńska
Jagiellonian University, ul. Reymonta 4, 30-059 Kraków, Poland
e-mail: anna.paszynska@uj.edu.pl

A. Byrski et al. (Eds.): Advances in Intelligent Modelling and Simulation, SCI 416, pp. 299–324.
springerlink.com

1 Introduction

This chapter presents an application of graph grammar modeling for a three dimensional simulations of multi-physics problems by means of the finite element method (FEM). Such a simulation consists of three dimensional mesh generation followed by the execution of the multi-frontal direct solver. In addition, we include either h or p refinement procedure of the computational mesh in order to increase the accuracy of the solution. The mesh generation, mesh refinements and translation the mesh into a sequence of matrices for the multi-frontal solver algorithm execution are expressed by graph grammar productions (GGPs).

The mesh generation algorithm is expressed by a sequence of GGPs, where each GGP generates a new single finite element and connects the element to the face of an already existing element. The computational mesh is assumed to contain tetrahedral as well as prism elements, thus, we distinguish several GGPs for generating a new tetrahedral or prism element and adding the newly created element to a face of some other element that can be either tetrahedral or prism element. The mesh h refinement algorithm consists of breaking tetrahedral and prism elements. Each tetrahedral element is broken into four smaller tetrahedral elements and two pyramid elements. The prism element is broken into eight smaller tetrahedral elements.

The mesh generation and refinement is followed by the execution of the multi-frontal direct solver algorithm. The input for the multi-frontal solver is a sequence of element matrices with rows and columns associated with finite element vertices, edges, faces and interiors. We introduce GGPs for transferring the computational mesh into a sequence of element frontal matrices for the solver execution.

The representation of the topological structure of finite element meshes as a hierarchy of vertices, edges, faces and regions has been proposed by [1] to illustrate mesh generation and data storage. The first attempt to model mesh transformations by applying the graph grammar concept has been proposed by [8], for regular triangular two-dimensional meshes with h adaptation. This has been done using a quasi context sensitive graph grammar. However, the application of the quasi-context sensitive graph grammar for adaptive mesh transformation seems to be limited. The reason is that the mesh transformations, such as the enforcement of 1-irregularity rule or the minimum rule, are context dependent and cannot be modelled by the quasi-context sensitive graph grammar. In [18] a simple rewriting framework, based on topological chain rewriting, which models mesh refinements is described. This approach also allows only for modelling uniform refinements.

The graph grammar proposed in this chapter is the application of the Composite Programmable Graph Grammar (CP-GG) for modelling the mesh transformations. The CP-GG, which has been introduced in [5], is a generative tool for design process using graph transformations executed on the graph representation of the designed object [4, 10, 6].

The graph grammar presented in this chapter is the generalization of the two dimensional graph grammar derived for rectangular, triangular and mixed elements [15, 16, 14]. It is also a generalization of three dimensional graph grammar derived for tetrahedral and prism elements but without mesh refinements [7]. The multi-frontal solver algorithm is the state-of-the art way of solving linear systems of equations encounter during the finite element method simulations. The version of the algorithm presented here is based on the two dimensional algorithm discussed in [11, 12, 19] as well as three dimensional out-of-core algorithm presented in [13]. The acoustics of the human head considered in this chapter is modeled as linear elasticity coupled with acoustics, and the details of the formulations can be found in [2, 3].

In this chapter we present a novel application of the graph grammar for modeling generation and refinements of three dimensional meshes with tetrahedral, prism and pyramid elements, with distinguished stages for vertices, edges, faces and interior modeling, as well as the interfacing with the multi-frontal solver. Up to the knowledge of the authors this is the only one existing application of graph grammar for three dimensional multi-physics simulations.

The structure of the chapter is the following. We start with introducing GGPs expressing the process of mesh generation, in Section 1. There are GGPs for generation of tetrahedral and prism elements, as well as graph grammar productions for identification of common edges and faces, shared between adjacent elements. Next, in Section 2 we introduce GGPs responsible for partitioning of tetrahedral elements into new smaller tetrahedra and pyramid elements as well as prism elements into smaller ones. In the next Section 3 we present the application of the graph grammar for generating of three dimensional mesh which represents the simplified model of the human head. We also introduce the multi-physics problem of the acoustics coupled with linear elasticity for the simulation of the vibration of the human head as a response to incoming acoustic wave. Next Section 4 presents the GGPs generating element frontal matrices with rows and columns associated to mesh nodes. The matrices constitute the input for the multi-frontal direct solver algorithm. The idea of the algorithm is described in the next Section 5. The chapter ends Section 6 describing the application of the developed methodology for challenging multi-physics problem — the acoustics of the human head.

2 Graph Grammar Productions Expressing the Mesh Generation Algorithm

In this section we introduce a set of GGPs responsible for generating a mesh constructed with tetrahedral finite elements enriched with additional external layers of prism elements. The mesh is destined for multi-physics simulations of the linear elasticity coupled with acoustics. The details of the numerical formulations are postponed to the Section 6.

The process of mesh generation starts with generation of tetrahedral elements. The computational mesh is represented by a graph. The GGPs represent the mesh transformations performed during the mesh generation process. We start from an initial graph with a single node with **S** attribute. First GGP **(Pinittet)** presented in Fig. 1(a) replaces the graph with starting node by a new graph containing the vertex node **(v1)** of the first tetrahedral element being generated.

After generation of the first tetrahedral element, we generate additional tetrahedral elements. It is done by executing GGPs **(Paddtet1)** and **(Paddtet2)** summarized in Fig. 1(b) and 1(c). The first production adds a new tetrahedral element to a face of already generated tetrahedral element. The second production adds a new tetrahedral element to faces of two already generated elements.

Exemplary sequence of GGPs generating an eight finite element mesh is presented in Fig. 2. Note that, the index of graph vertex node represents number of tetrahedral elements where the vertex node belongs to.

The generation of vertex nodes for tetrahedral elements is followed by the generation of edge nodes. This is expressed by the GGP **(Paddtetedges)** presented in Fig. 3(a). The production just identifies edges of tetrahedral element and enrich them with edge nodes **(e)**. The edge nodes are necessary to generate since the FEM define polynomials over finite element edges for approximation of the solution.

There are two additional GGPs **(Paddtetedges1)** and **(Paddtetedges2)** presented in Fig. 3(b) and 3(c), responsible for generation of edge nodes for tetrahedral elements adjacent to tetrahedra with already generated edge nodes. Note that, the index at graph edge node represents number of tetrahedral elements where the edge node belongs to. The generation of edge nodes is followed by the generation of face nodes. This is expressed by GGP **(Paddtetface)** presented in Fig. 4(a).

Finally, the generation of face nodes is followed by the generation of interior nodes. This is expressed by GGP **(Paddtetinterior)** presented in Fig. 4(b). Here, the index at graph face node is supposed to represent the number of tetrahedral elements where the face node belongs to. Thus, we need to add one new GGP **(Pcountfaces)** presented in Fig. 4(c) responsible for identifying interior faces that belong to two elements.

We end up with three dimensional mesh that consists of tetrahedral finite elements. In this chapter, we consider the exemplary multi-physics application, the linear elasticity coupled with acoustics on the model of concentric balls, intended to simulate vibrations over the the simplified model of the human head as a response to incoming acoustic waves. The tetrahedral finite element mesh is intended to model the head. The boundary of the mesh consists of triangular faces.

The external layer representing the air surrounding the head is modeled by prism elements. That is way we introduce the GGPs responsible for adding prism elements on the triangular faces of tetrahedral elements. The first GGP **(Paddprism)** presented in Fig. 5(a) generates vertex nodes of a prism element attached to a triangular face of tetrahedral element. The following GGPs **(Paddprism1)** and **(Paddprism2)** presented in Fig. 5(b) and 5(c) generate vertex nodes of a prism element over a face of tetrahedral elements already generated and attach the prism element to faces of one or two prism elements already generated.

The generation of vertex nodes of prism elements is followed by the generation of the edge nodes. The GGP **(Pprismedges)** presented in Fig. 5(d) generates edge nodes of a prism element. The following GGPs **(Pprismedges1)** and **(Pprismedges2)** presented in Fig. 6(a) and 6(b) generate edge nodes of a prism element adjacent to one or two prism elements with edge nodes already generated.

After the generation of edge nodes, we can generate face nodes for all faces of prism elements. The GGP **(Pprismface)** generating a single face node for a single face of a prism element is presented in Fig. 6(c). Finally, having all face nodes of a prism element generated, we can generate its interior node, as it is summarized by GGP **(Pprisminterior)** presented in Fig. 6(d).

All presented GGPs allow us to generate the computational mesh with the interior built with tetrahedral elements and with additional layers of prism elements. One possible application of such meshes are the multi-physics simulations of the acoustics of human head, but the graph grammar is obviously not limited to such simulations.

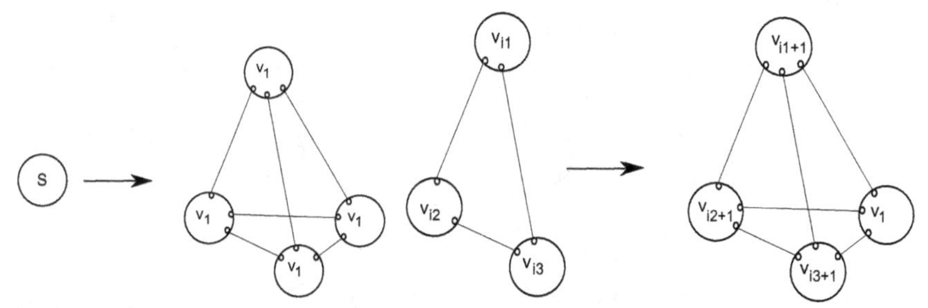

(a) GGP **(Pinittet)** generating vertices of a single tetrahedron.

(b) GGP **(Paddtet1)** adding vertices of a single tetrahedron neighboring a face of tetrahedra already generated.

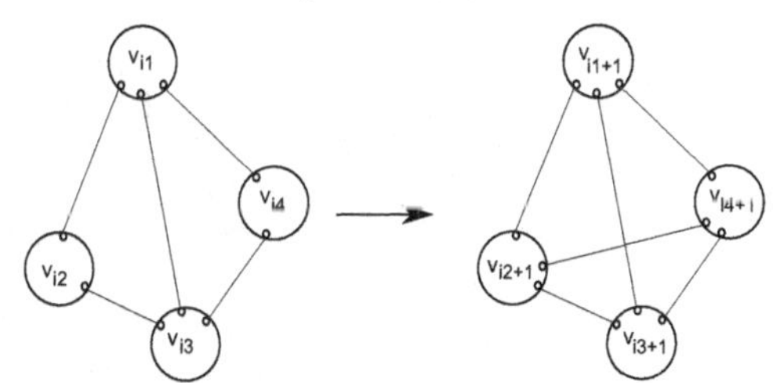

(c) GGP **(Paddtet2)** adding vertices of a single tetrahedron neighboring faces of two tetrahedra already generated.

Fig. 1 GGPs **(Pinittet)**, **(Paddtet1)** and **(Paddtet2)**

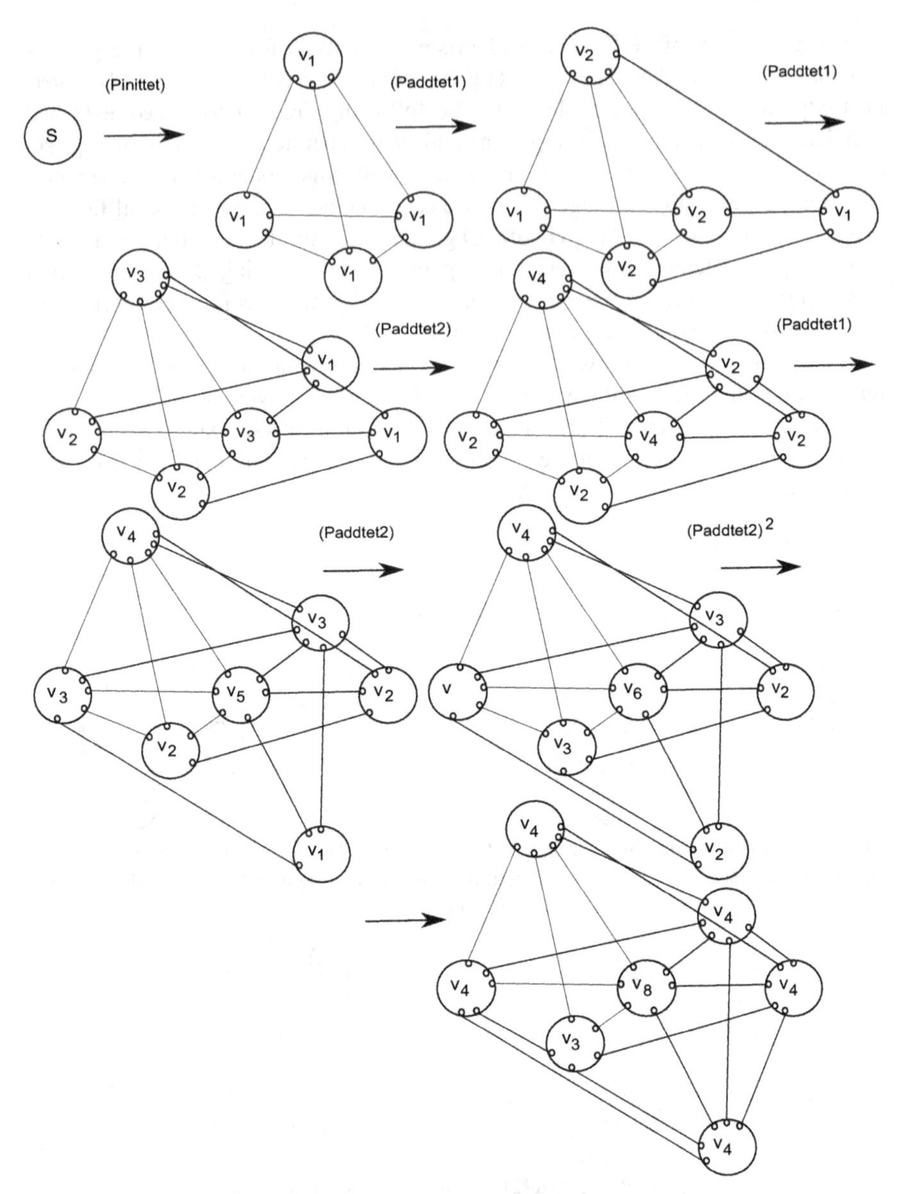

Fig. 2 Sequence of GGPs **(Pinittet)–(Paddtet1)–(Paddtet1)–(Paddtet2)–(Paddtet1)–(Paddtet2)–(Paddtet2)–(Paddtet2)** generating exemplary eight finite element mesh.

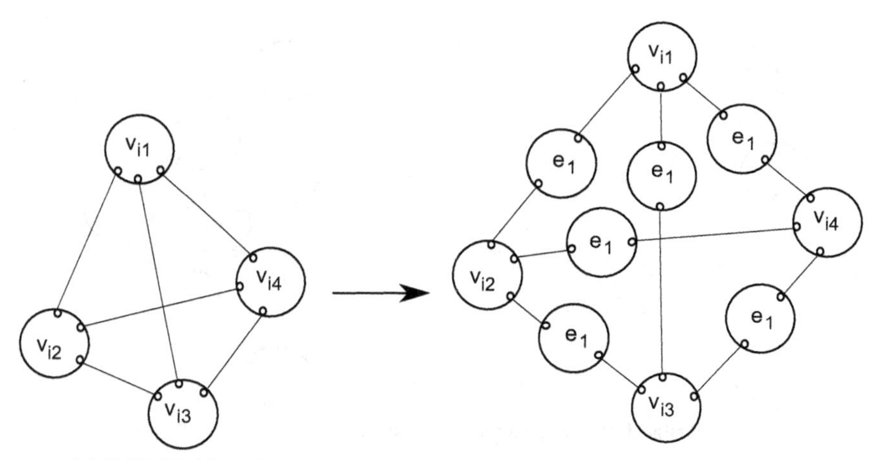

(a) GGP **(Paddtetedges)** generating edge nodes of a single face of a tetrahedra.

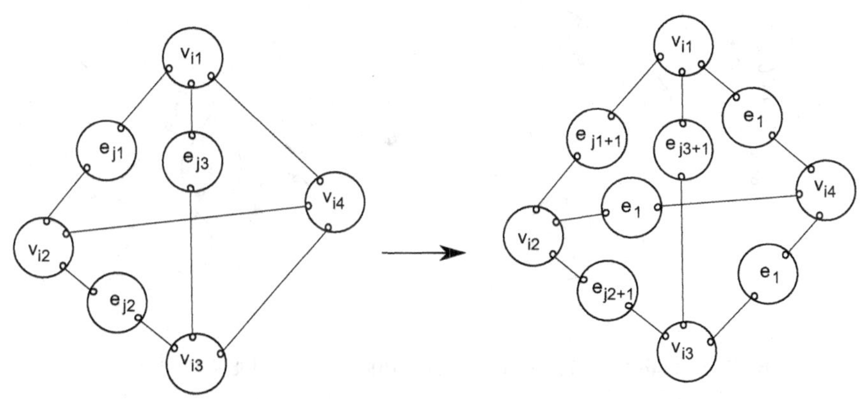

(b) GGP **(Paddtetedges1)** generating edge nodes of a single tetrahedron neighboring a face of a tetrahedron with edges already generated.

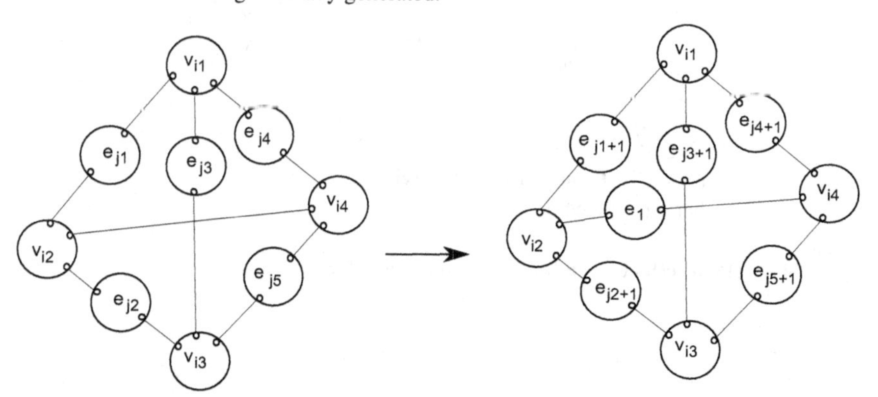

(c) GGP **(Paddtetedges2)** generating edge nodes of a single tetrahedron neighboring faces of two tetrahedra with edges already generated.

Fig. 3 GGPs (**Paddtetedges**), (**Paddtetedges1**) and (**Paddtetedges2**)

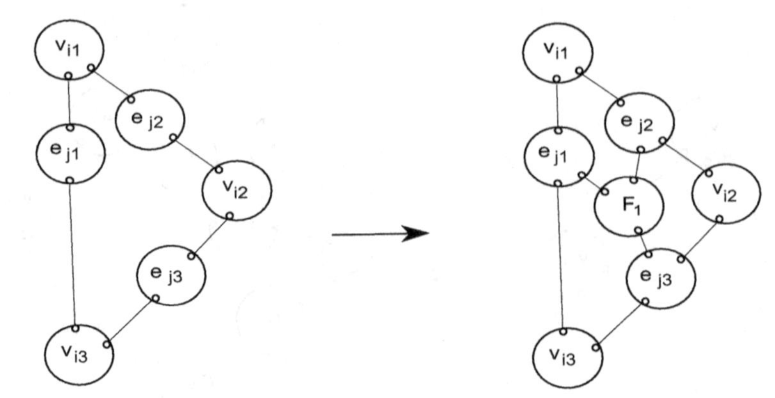

(a) GGP **(Paddtetface)** generating a face node of a single tetrahedron.

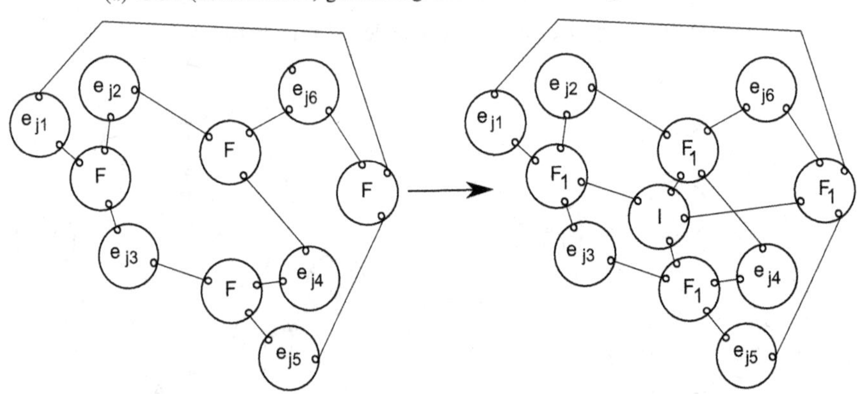

(b) GGP **(Paddtetint)** generating an interior node of a single tetrahedron.

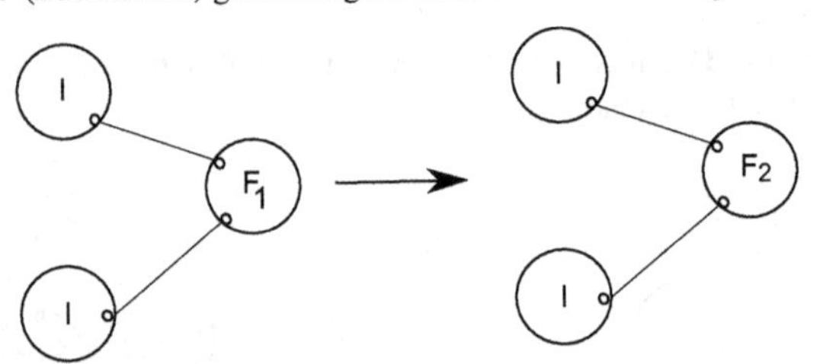

(c) GGP **(Pcountfaces)** identifying faces of two neighboring tetrahedra.

Fig. 4 GGPs **(Paddtetface)**, **(Paddtetint)** and **(Pcountfaces)**

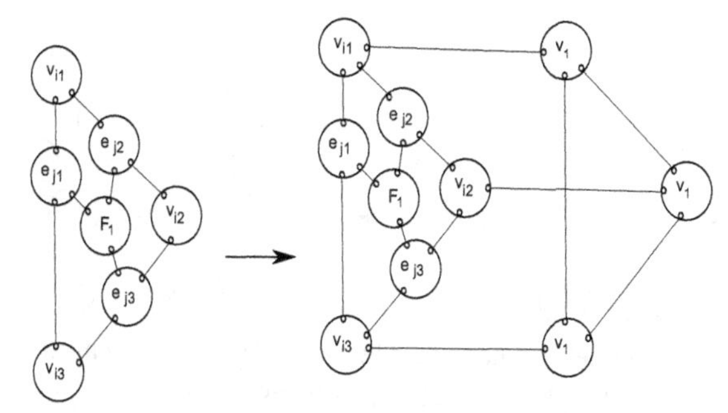

(a) GGP **(Paddprism)** adding prism element on a face of triangular element.

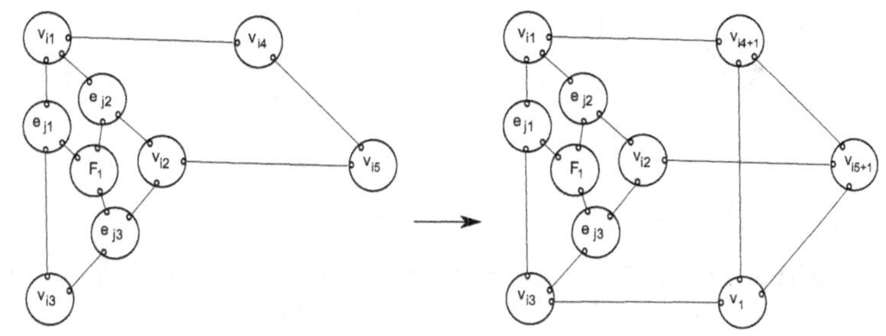

(b) GGP **(Paddprism1)** adding prism element on a face of triangular element, and connecting the prism element with a face of adjacent prism element already generated.

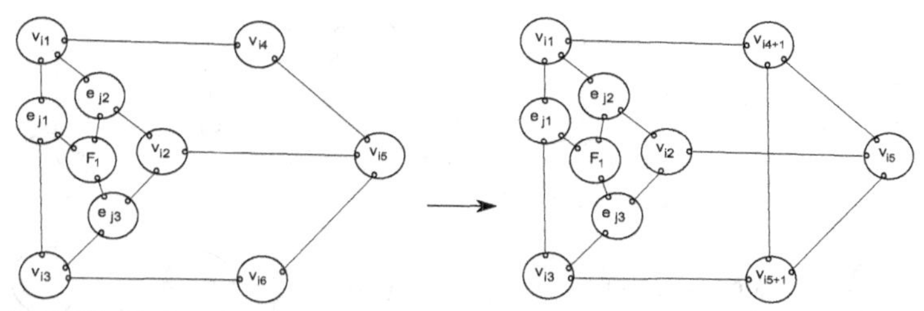

(c) GGP **(Paddprism2)** adding prism element on a face of triangular element, and connecting the prism element with faces of two adjacent prism elements already generated.

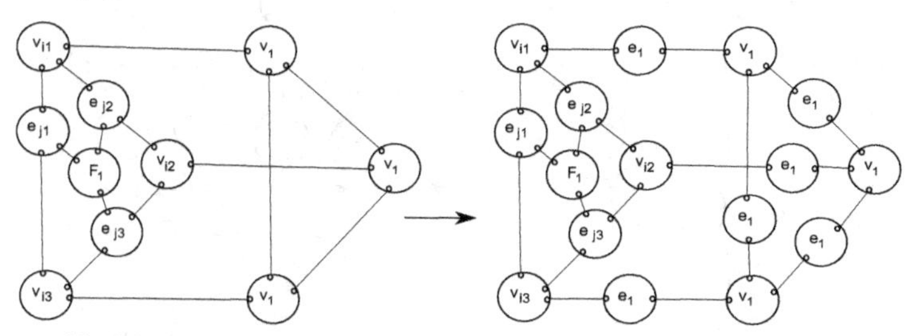

(d) GGP **(Pprismedges)** generating edge nodes for a prism element.

Fig. 5 GGPs **(Paddprism)**, **(Paddprism1)**, **(Paddprism2)** and **(Pprismedges)**

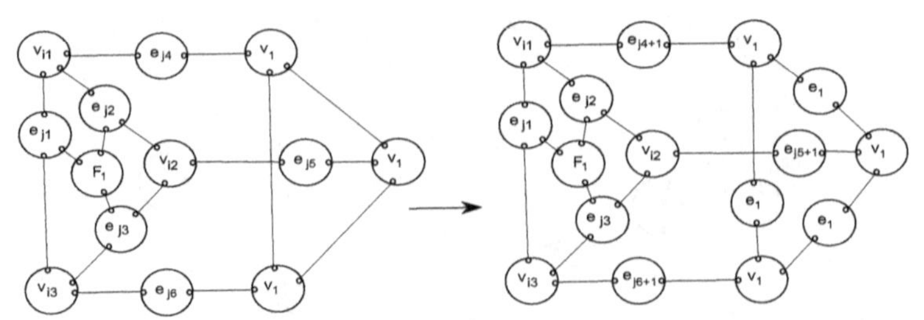

(a) GGP (**Pprismedges1**) generating edge nodes for a prism element adjacent to a prism element with edge nodes already generated.

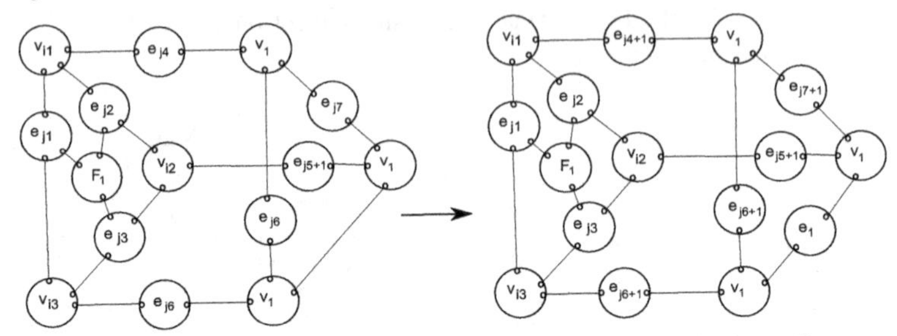

(b) GGP (**Pprismedges2**) generating edge nodes for a prism element adjacent to two prism elements with edge nodes already generated.

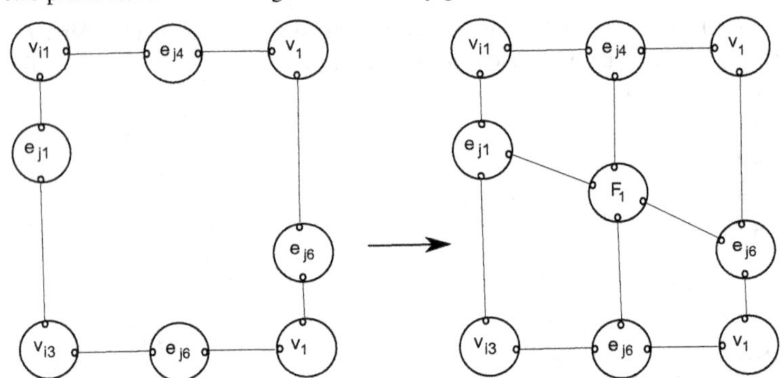

(c) GGP (**Pprismface**) generating face node for a prism element.

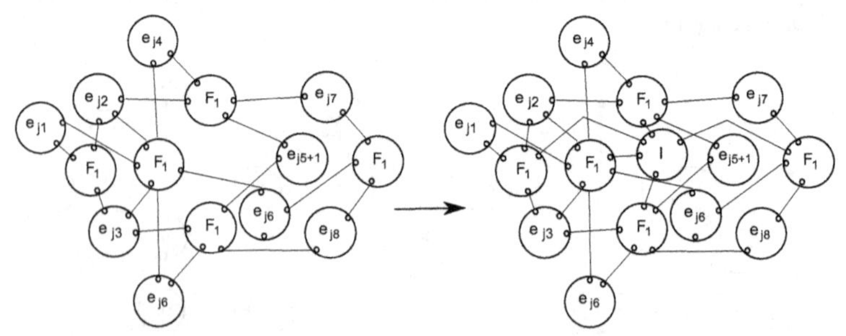

(d) GGP (**Pprisminterior**) generating interior node for a prism element.

Fig. 6 GGPs (**Pprismedges1**), (**Pprismedges2**), (**Pprismface**) and (**Pprisminterior**)

3 Graph Grammar Productions Expressing the Mesh Refinements Algorithm

The computational meshes may suffer from low accuracy of the FEM solution. In such a case, the accuracy may be improved by refining the mesh by breaking the original "big" finite elements into smaller ones, so the FEM solution may be performed with smaller and more accurate mesh. We distinguish the refinement of a tetrahedral element and the refinement of a prism element.

We allow for the following refinements, summarized in Fig. 7. A tetrahedral element is broken into four smaller tetrahedral elements and two pyramids. The pyramid elements are further broken into two tetrahedral elements, on the other hand, a prism element is broken into four new prisms.

We distinguish refinements of element edges, faces and interiors. For example, when we break a tetrahedral element, we first break all six edges of the tetrahedral element. Each edge is broken into two smaller edges. In the next step, we break all four triangular faces of the tetrahedral element into four new triangular faces. Finally, we break the interior of the tetrahedron, into four new tetrahedral and two new pyramid elements. This is consistent with the mesh refinement philosophy introduced by Leszek Demkowicz [2]. Such a strategy allows to store mesh refinements as trees of nodes growing from nodes of the initial mesh. However, this approach requires some smart algorithms for extractions of finite element data.

The refinement of an edge is expressed by the GPP (**Pbreaktetedge**) presented in Fig. 8(a). When we break an edge, one new son vertex node and two new son edge nodes are created. The newly created nodes are pinned into the edge node of the farther edge.

The refinement of a face is expressed by the GGP (**Pbreaktetface**) presented in Fig. 8(b). When we break a face, three new face nodes and three new edge nodes are created. The newly created nodes are connected as son nodes to the face node of the farther face.

The refinement of an interior is expressed by the GGPs (**Pbreaktetinterior**) and (**Pbreakprisminterior**) presented in Fig. 8(c) and 8(d). For example, when we break an interior of a tetrahedral element, four new interior nodes of tetrahedral elements, two new interior nodes of pyramid elements and five new face nodes separating the interior nodes are created. The newly created nodes are son nodes of the interior node of the parent tetrahedron.

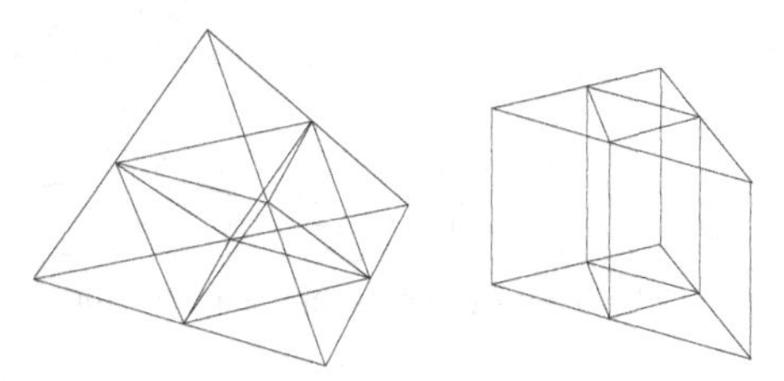

Fig. 7 Refinement of tetrahedral and prism elements.

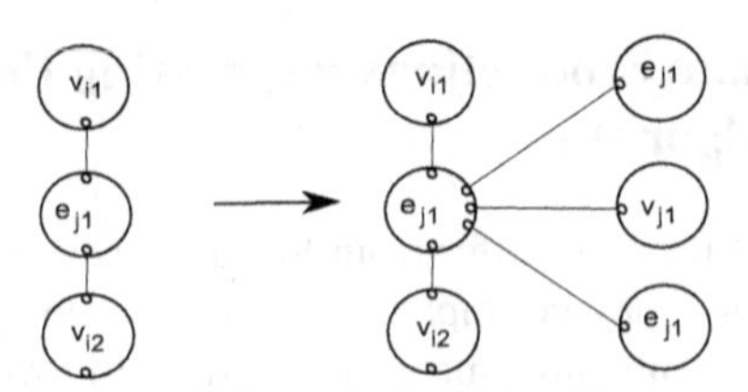

(a) GGP **(Pbreakedge)** breaking an edge of a tetrahedron.

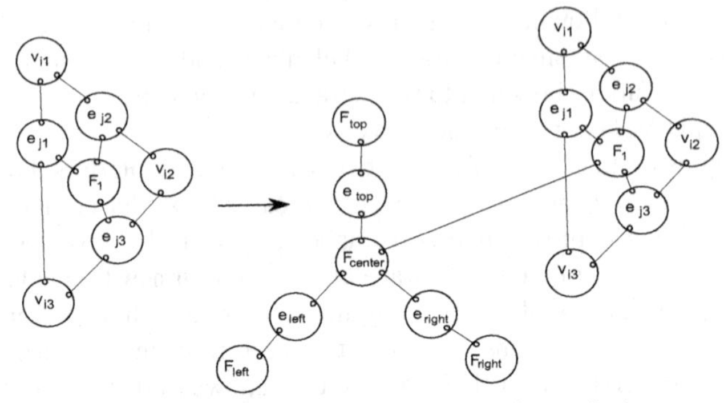

(b) GGP **(Pbreakeface)** breaking a face of a tetrahedron.

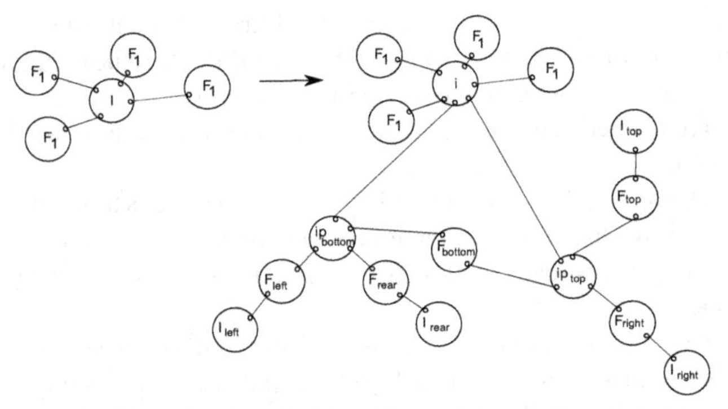

(c) GGP **(Pbreaktetinterior)** breaking an interior of a tetrahedron.

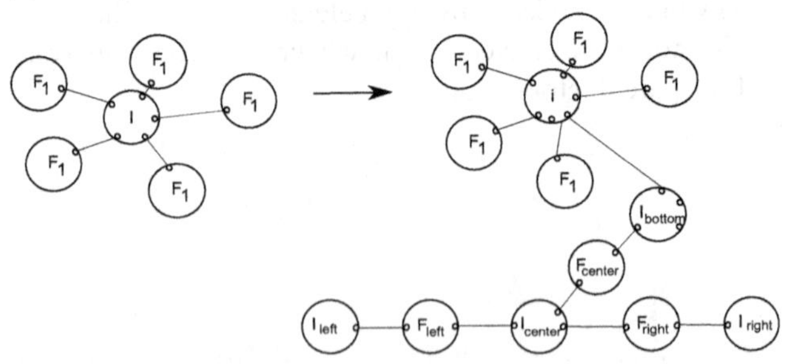

(d) GGP **(Pbreakprisminterior)** breaking an interior of a prism.

Fig. 8 GGPs **(Pbreakedge)**, **(Pbreakeface)**, **(Pbreaktetinterior)** and **(Pbreakprisminterior)**

4 Mesh Generation and Problem Formulation

The presented GGPs are utilized to generate computational mesh for the model problem of linear elasticity coupled with acoustics [3]. The domain Ω in which the problem is defined, presented in Fig. 9, is the interior of a ball including a simplified model of the human head, and it is split into an acoustic part Ω_a, and an elastic part Ω_e.

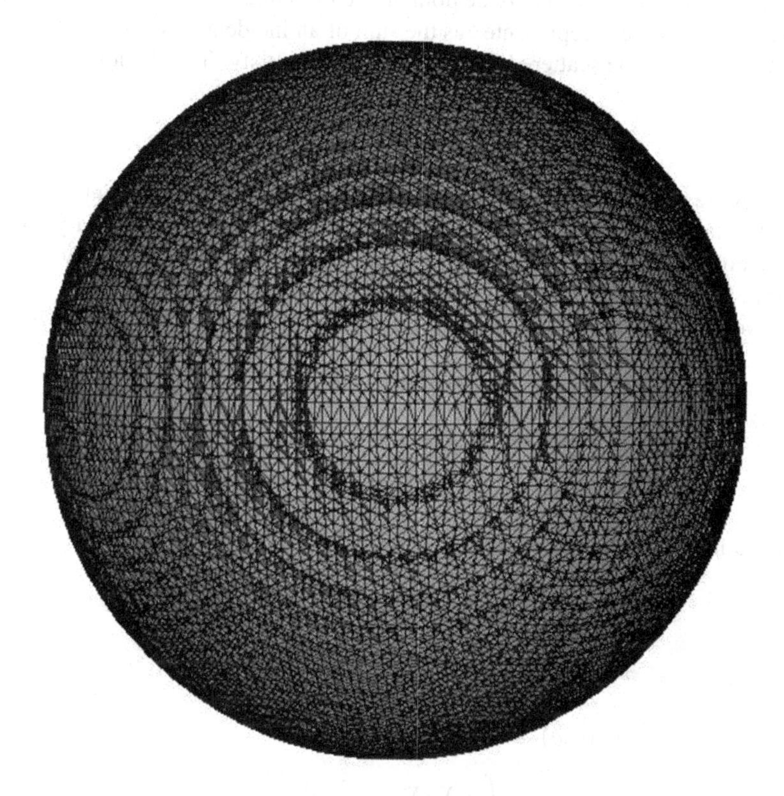

Fig. 9 Computational mesh.

The acoustic part, Ω_a, includes:

- air surrounding the human head, bounded by the head surface and a truncating sphere; this part of the domain include portions of air ducts leading to the middle ear through mouth and nose openings;
- cochlea,
- an additional layer of air bounded by the truncating sphere and the outer sphere terminating the computational domain, where the equations of acoustics are replaced with the corresponding *Perfectly Matched Layer* (PML) modification.

The elastic part, Ω_e, includes:

- skull,
- tissue.

The term *tissue* refers here all parts of the head that are not occupied by the skull (bone) and the cochlea. This includes the thin layer of the skin and the entire interior of the head with the brain. In the current stage of the project, it is assumed that the elastic constants for the whole tissue domain are the same.

The acoustic wave is represented as the sum of an incident wave p^{inc} and a scattered wave p. Only the scattered wave is assumed to satisfy the radiation (Sommerfeld) condition,

$$\frac{\partial p}{\partial r} + ikp \in L^2(\boldsymbol{R}^3) \tag{1}$$

with the wave number $k = \omega/c$. The boundary conditions include the interface between the elastic and acoustic subdomains and the outer, Dirichlet boundary for the acoustic domain. Material interfaces between the skull and tissue, and between the air and PML air do not require any special treatment.

The weak (variational) final formulation of the problem has the following form

$$\begin{cases} u \in \tilde{u}_D + V,\ p \in \tilde{p}_D + V, \\ b_{ee}(u,v) + b_{ae}(p,v) = l_e(v), \quad \forall v \in V \\ b_{ea}(u,q) + b_{aa}(p,q) = l_a(q), \quad \forall q \in V \end{cases} \tag{2}$$

with the bilinear and linear forms defined as follows:

$$\begin{aligned} b_{ee}(u,v) &= \int_{\Omega_e} \left(E_{ijkl}u_{k,l}v_{i,j} - \rho_s\omega^2 u_i v_i\right) dx \\ b_{ae}(p,v) &= \int_{\Gamma_I} p v_n \, dS \\ b_{ea}(u,q) &= -\omega^2 \rho_f \int_{\Gamma_I} u_n q \, dS \\ b_{aa}(p,q) &= \int_{\Omega_a} \left(\nabla p \nabla q - k^2 pq\right) dx \\ l_e(v) &= -\int_{\Gamma_I} p^{inc} v_n \, dS \\ l_a(q) &= -\int_{\Gamma_I} \frac{\partial p^{inc}}{\partial n} q \, dS \end{aligned} \tag{3}$$

where V and V are the spaces of the test functions,

$$\begin{aligned} V &= \{v \in H^1(\Omega_e) : v = \mathbf{0} \text{ on } \Gamma_{De}\} \\ V &= \{q \in H^1(\Omega_a) : q = 0 \text{ on } \Gamma_{Da}\} \end{aligned} \tag{4}$$

Here, ρ_f is the density of the fluid; ρ_s is the density of the solid; E_{ijkl} is the tensor of elasticities; ω is the circular frequency; c denotes the sound speed; $k = \omega/c$ is the acoustic wave number, and p^{inc} is the incident wave impinging from the top. For more details regarding the problem formulation we refer to [3]. Coupled problem (2) is symmetric if and only if diagonal forms b_{ee} and b_{aa} are symmetric and,

$$b_{ae}(p,u) = b_{ea}(u,p)\,.$$

Thus, in order to enable the symmetry of the formulation, it is necessary to rescale the problem by, for instance, dividing the equation (2) by factor $-\omega^2\rho_f$.

$$\begin{aligned}
b_{ee}(u,v) &= \int_{\Omega_e} \left(E_{ijkl}u_{k,l}v_{i,j} - \rho_s\omega^2 u_i v_i\right) dx \\
b_{ae}(p,v) &= \int_{\Gamma_I} p v_n \, dS \\
b_{ea}(u,q) &= -\int_{\Gamma_I} u_n q \, dS \\
b_{aa}(p,q) &= \frac{1}{\omega^2\rho_f} \int_{\Omega_a} \left(\nabla p \nabla q - k^2 pq\right) dx \\
l_e(v) &= -\int_{\Gamma_I} p^{inc} v_n \, dS \\
l_a(q) &= -\frac{1}{\omega^2\rho_f} \int_{\Gamma_I} \frac{\partial p^{inc}}{\partial n} q \, dS
\end{aligned} \tag{5}$$

The symmetry of the problem is essential, among other reasons, from the point of view of using a direct solver. Notice that the outer normal unit vector n is referred always *locally*, i.e. in the formula for the coupling bilinear form b_{ae} involving elasticity test functions v, versor n points outside of the elastic domain, whereas in the formula for the coupling bilinear form b_{ea} involving acoustic test functions q, versor n points outside of the acoustic domain. The normal components v_n and u_n present in the coupling terms are thus opposite to each other, and the formulation is indeed symmetric. The symmetry of the problem allows us to reduce the direct solver memory usage by half as well as allows us to give up the global pivoting, which simplifies the computations.

In the PML part of the acoustical domain, the bilinear form b_{aa} is modified as follows:

$$b_{aa}(p,q) =$$
$$\int_{\Omega_{a,PML}} \left(\frac{z^2}{z' r^2} \frac{\partial p}{\partial r} \frac{\partial q}{\partial r} + \frac{z'}{r^2} \frac{\partial p}{\partial \psi} \frac{\partial q}{\partial \psi} + \frac{z'}{r^2 \sin^2\psi} \frac{\partial p}{\partial \theta} \frac{\partial q}{\partial \theta} \right) r^2 \sin\psi \, dr\, d\psi\, d\theta \tag{6}$$

Here r, ψ, θ denote the standard spherical coordinates and $z = z(r)$ is the PML stretching factor defined as follows:

$$z(r) = \left(1 - \frac{i}{k}\left[\frac{r-a}{b-a}\right]^{\alpha}\right) r \tag{7}$$

Here, a is the radius of the truncating sphere; b is the external radius of the computational domain ($b-a$ is thus the thickness of the PML layer); i denotes the imaginary unit; k is the acoustical wave number, and r is the radial coordinate. In computations, all derivatives with respect to spherical coordinates are expressed in terms of the standard derivatives with respect to Cartesian coordinates. In all reported computations, parameter $\alpha = 5$. For a detailed discussion on derivation of PML modifications and the effects of higher-order discretisations one may refer to [9].

The analysis domain is divided into four parts as shown in Fig. 10(a): cochlea, skull, tissue and air. The skull is represented by a spherical shell, and the cochlea (see Fig. 10(b)) is located inside the ear canal represented with a cylindrical cavity. Except cochlea and skull, the remaining part of region within the outer bounding sphere is classified as air. The cochlea is connected with the skull by growing manually an additional bone structure in between the skull and the cochlea, i.e. a number of air elements in the cavity is manually reclassified as the skull elements. The head is excited with the plane wave. The purpose of this example is to investigate the proportion of the energy transferred to the cochlea directly through air and indirectly through the bone.

Except for the PML domain, both acoustic and elastic domains are generated with the tetrahedra elements by the following sequence of GGPs: **(Pinittet)–(Paddtet1)**^{k}**–(Paddtet2)**^{l}**–(Paddtetedges)**^{m}**–(Paddtetedges1)**^{n} **–(Paddtetedges2)**^{o} **–(Paddtetface)**^{p}**–(Paddtetint)**^{r} **–(Pcountfaces)**^{s}. We utilise third-order polynomials over the computational mesh.

The triangular mesh on the (approximate) truncating sphere is extended in the radial direction to form two layers of prismatic elements. This is done by executing the following sequence of GGPs: **{(Paddprism)**^{a}**–(Paddprism1)**^{b}**–(Paddprism2)**^{c}**–(Paddprismedges)**^{d} **–(Paddprismedges1)**^{e} **–(Paddprismedges2)**^{f}**–(Pprismface)**^{g}**–(Paddprisminterior)**^{h} **}**^{2}. In order to approximate well the PML induced layer, higher order polynomials in the radial direction are used, $p = 4$ in the first layer, and $p = 2$ in the second layer. This is in accordance with our experience of resolving PML induced boundary layers with hp-adaptive elements (see [9] for examples).

5 Graph Grammar Productions Interfacing with the Multi-frontal Direct Solver

The multi-frontal direct solver solution of the weak form of the variational formulation (5) requires to transform each finite element into an element local matrix (see [2]). This is done by executing the GGPs presented in Fig. 11 and 12(a). The matrix consists of sub-matrices corresponding to the number of nodes within an element, and the size of each sub-matrix corresponds to the polynomial order of approximation assigned to a node. The matrices are ordered in such a way that first goes

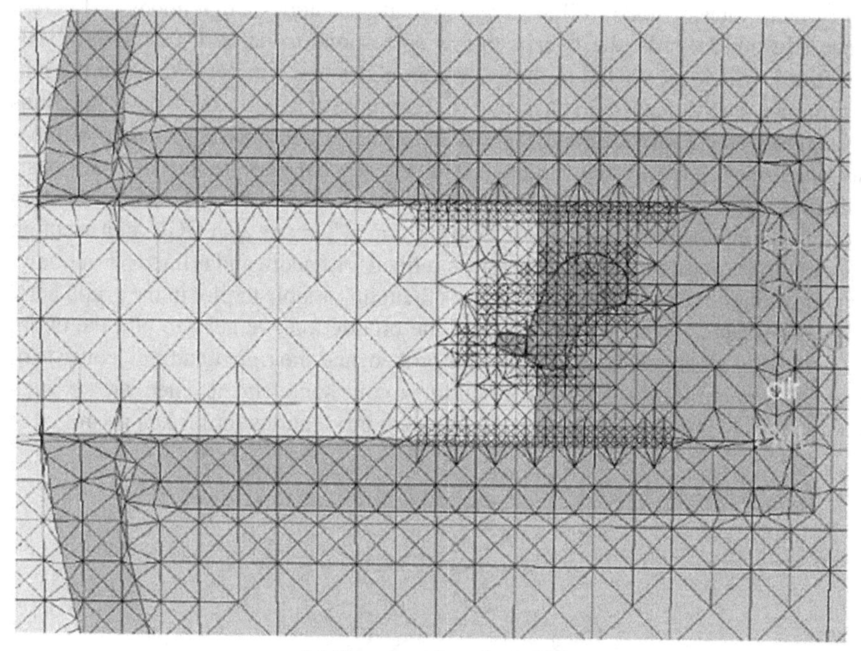

(a) Cross-section of a mesh.

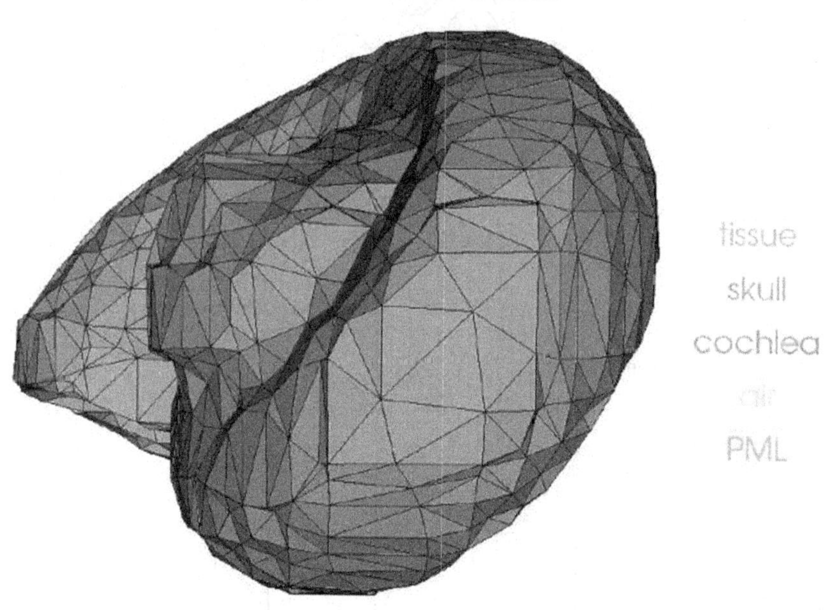

(b) Part of the mesh representing the cochlea.

Fig. 10 Cross-section of a mesh and part of the mesh representing the cochlea.

an element interior node, followed by element face nodes, edge nodes and vertex nodes. In other words, the matrix entries correspond to pairs of interacting basis functions, which are related to finite element nodes and vertices. We order the matrix in such a way, that we first list rows (or columns) of a matrix with basis functions related to element interior node, then we list rows (or columns) of a matrix with basis functions related to face nodes, and finally we list rows (columns) with basis functions related to element vertices. This corresponds to the order of elimination of rows during the multi-frontal solver algorithm. There are also GGPs constructing element local matrices for elements from refined tetrahedra and refined prism, however the complexity of these GGPs makes them impossible to plot in the graph here. This is because, on the left hand side of the production, we need to plot the entire element with all edges, faces and interior broken, and the right hand side consists of just one element local matrix corresponding to one son element. There are six such productions for broken tetrahedra and five such productions for broken prisms.

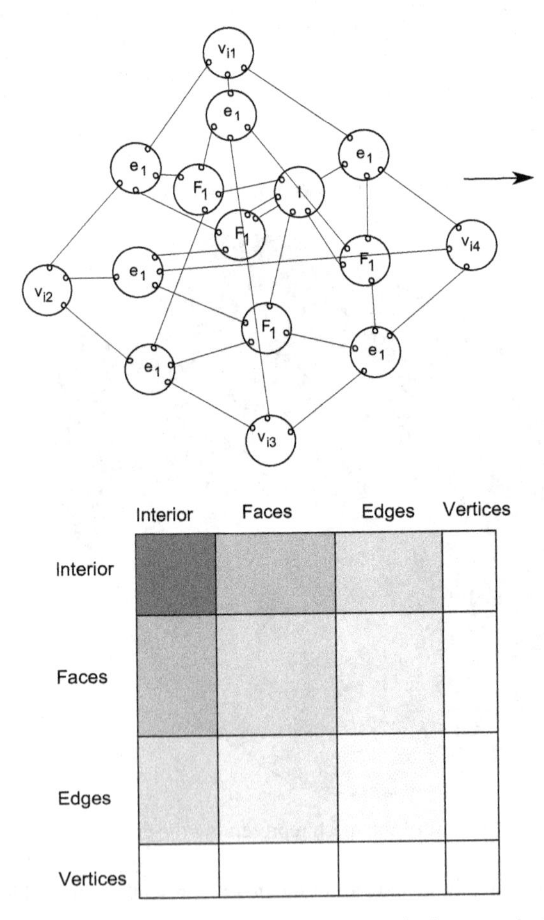

Fig. 11 GGP (**Ptet2matrix**) generating element local matrix for a tetrahedral element.

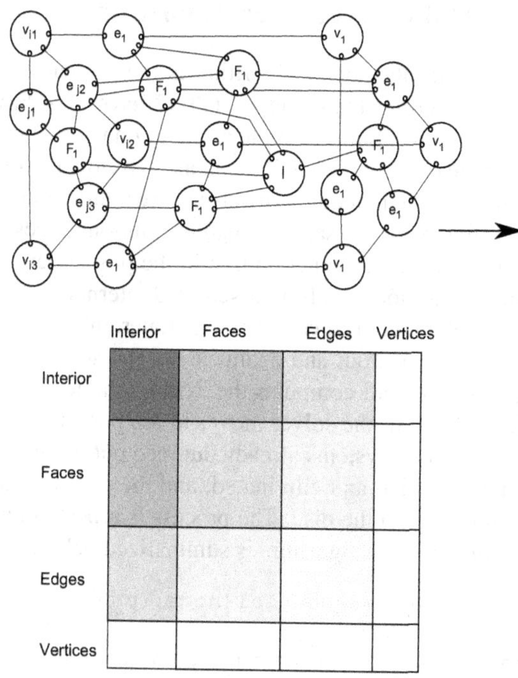

(a) GGP (**Pprism2matrix**) generating element local matrix for a prism element.

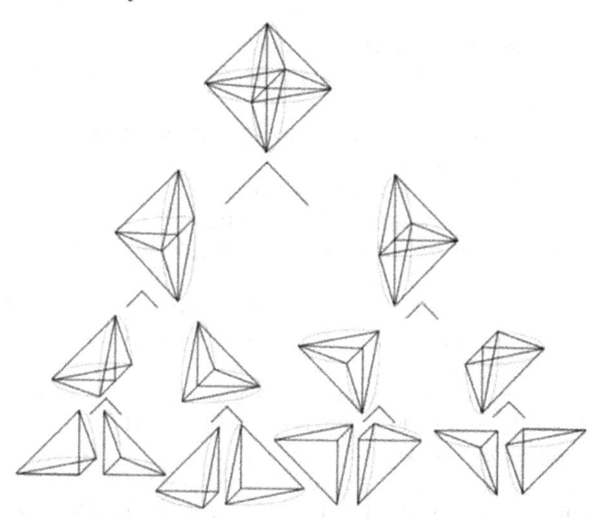

(b) The construction of the elimination tree for a simple eight elements sub-domain.

Fig. 12 GGP (**Pprism2matrix**) and elimination tree.

6 The Multi-frontal Direct Solver Algorithm

The input for the multi-frontal solver algorithm is the sequence of element local matrices generated by executing the sequence: **(Ptet2matrix)**^{m}**–(Pprism2matrix)**^{n}. The input for the solver algorithm is also the elimination tree, which is obtained by recursive partition of the computational domain, as it is shown in Fig. 12(b).

We utilize sequential version of our own out-of-core multi-frontal solver [13] that is intended to reduce the memory usage during the solution process. The sequential out-of-core solver browses the elimination tree in the post-order, starting from the left bottom leaf. In the leaf node the fully assembled internal nodes are eliminated, leaving the Schur complement of interface nodes. Later, the Schur complement for the left bottom leaf is dumped out and deallocated. The solver moves to the right neighbor of the left bottom leaf, computes the Schur complement at the node, and dumps it out to the disc. Next, the solver moves to the parent node, and constructs the new system based on two systems already dumped out. In the new system fully assembled nodes are identified and eliminated, and the newly created Schur complement is also dumped out to the disc. The process is repeated until we reach the root of the elimination tree. The algorithm is summarized below:

```
1 function out_of_core_sequential(node, proc, iret)
2 if node is a leaf then
3   generate local system assigned to node
4   find internal nodes at node
5   eliminate internal nodes at node
6 else
7   loop through son_nodes of node
8   if proc is assigned to son_node then
9     iret =0
10    call out_of_core_sequential(son_node,proc,iret)
11    if iret==1 then return
12      compute schur1 complement at son_node
13      dumpout the system from son_node
14      deallocate the system at son_node
15    endif
16    if proc is 1st proc assigned to 2nd son of node then
17      BUFFER = schur1; deallocate schur1;
18      iret =1; return
19    else if proc is 1st proc assigned to node then
20      dumpout schur1; deallocate(schur1)
21      proc_org = proc
22      iret =1; proc =1st proc assigned to 2nd son of node
23      call out_of_core_sequential(son_node,proc,iret)
24      iret =0; proc = proc_org
25      schur2 = BUFFER; deallocate(BUFFER)
26    endif
27  endif
28  end loop
29  if proc is 1st proc assigned to node then
```

```
30      dumpin previously dumpout schur1
31      merge schur1,2 into new system at node
32      deallocate(schur1,schur2)
33      find nodes to eliminate at node
34    endif
35    compute schur complement at node
36    dumpout the system at node
37 endif
```

and it is called as follows:

```
iret=0; proc=0; node=root;
call out_of_core_sequential(node, proc, iret)
```

7 Numerical Results

We conclude the presentation with some numerical results obtained for the propagation of the acoustic waves with the simplified model of the human head with cochlea. The cochlea, presented in Fig. 10(b) has been connected with the skull by growing manually an additional bone elements between the skull and the cochlea, as illustrated in Fig. 10(a). In this model, the domain consists of four concentric spheres with a hole. The most inner sphere is filled with an elastic material with data corresponding to human brain. The first layer is also elastic with constants corresponding to human skull. The second layer corresponds to air, and the last one to the PML air. The material data corresponding to the considered layers are presented in Table 1. The Young modulus E and Poisson ration ν correspond to the elastic domain, the density ρ corresponds for both elastic and acoustics domains, the speed of compressional waves c_p is defined for both domains and the speed of shear waves c_s is defined for elastic domain.

Table 1 Material contrasts

Material	$E[MPa]$	ν	$\rho[kg/m^3]$	$c_p[m/s]$	$c_s[m/s]$
skull (bone)	6500	0.22	1412	2293	1374
cochlea (water)			1000	1500	
air			1.2	344	

The incident wave is assumed to be in the form of a plane wave impinging from the top, $p^{inc} = p_0 e^{ikex}$, $e = (0,0,-1)$, $p_0 = 1[Pa]$. The test problem is being solved with frequency $f = 200Hz$. The precise geometry data are as follows: brain $r < 0.1m$, skull $0.1m < r < 0.125m$, air $0.125m < r < 0.2m$ and PML air $0.2m < r < 0.3m$.

We have solved two computational problems. In the first problem we assumed that the interior of the human head is filled with air (thus the material constants from the third row of Table 1 have been used); in the second problem, we assumed that the interior of the human head is filled with water (thus the material constants from the second row of Table 1 have been used). The results are presented in Fig. 13 and 14(a), with zoom at the cochlea part presented in Fig. 14(b). The solution to the computational problem is the complex valued time-harmonic wave. The results presented in Fig. 13, 14(a) and 14(b) presents the real part of the solution (denoted by $Re(p)$) which corresponds to the wave length. The imaginary part, not presented here, corresponds to the amplitude of the wave. The two numerical problems presented here are utilized as the proof of concept, showing that our graph grammar

Table 2 Parameters of the mesh with uniform p=4.

Parameter	Value
Number of finite elements	19,288
Number of degrees of freedom	559,141
Number of non-zero entries	97,486,955
Maximum memory usage out-of-core solver	6.014 GB
Maximum memory usage MUMPS	7.346 GB
Execution time sequential MUMPS	6349 s
Execution time sequential out-of-core solver	38367 s (6X slower than MUMPS)

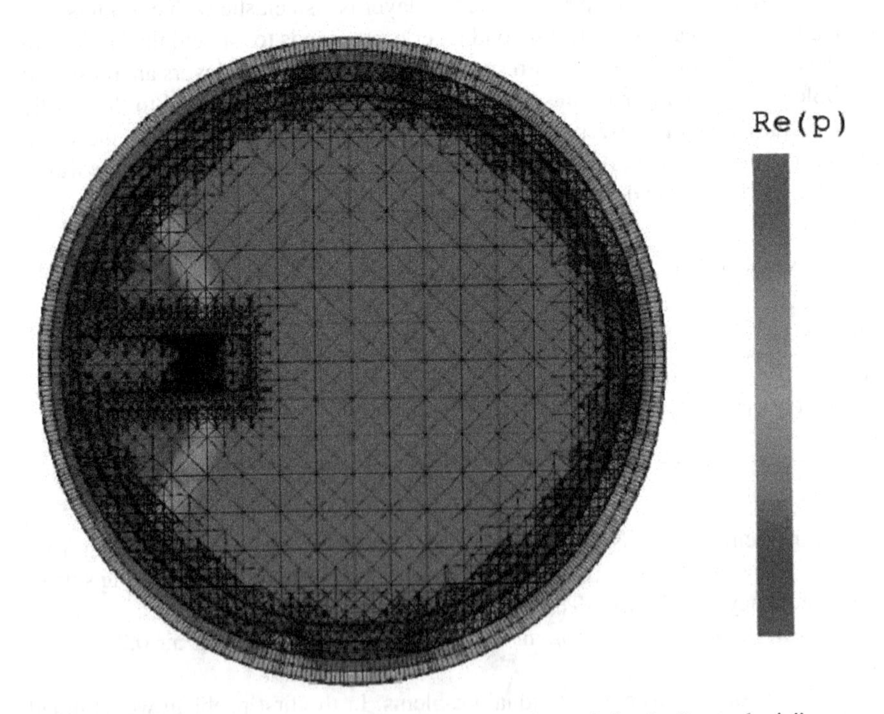

Fig. 13 Real part of the solution at the cross-section of a mesh for the "water-brain".

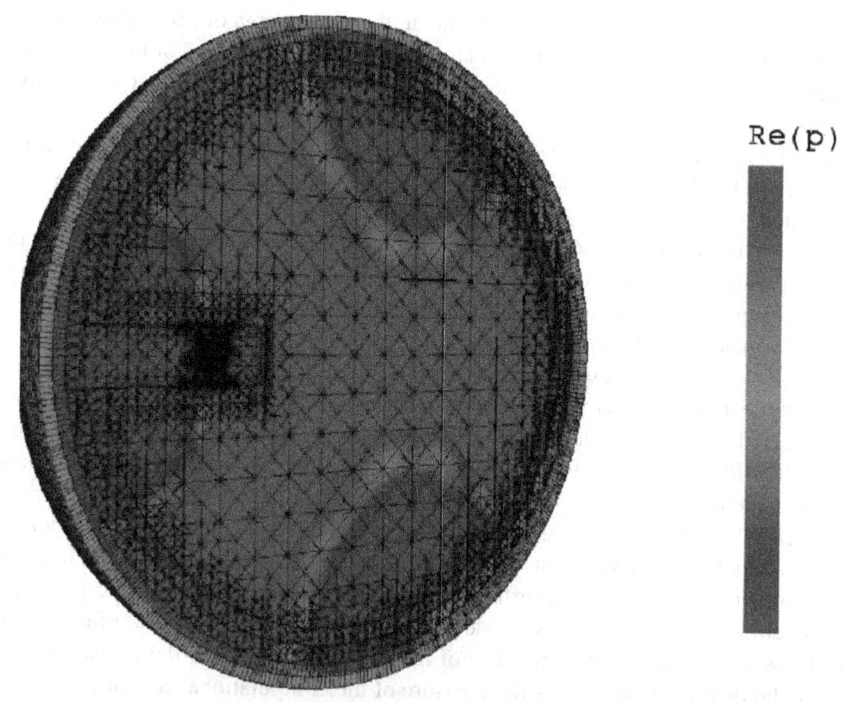

(a) Real part of the solution at the cross-section of a mesh for the "air-brain".

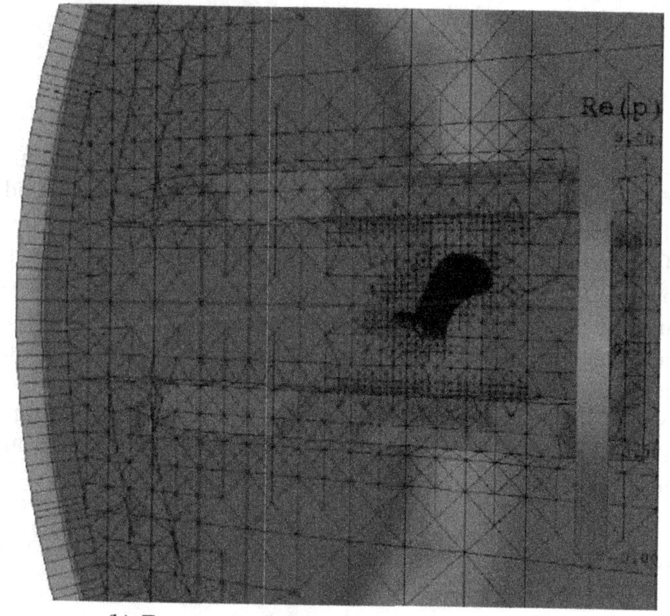

(b) Zoom towards the cochlea.

Fig. 14 Part of the solution and zoom towards the cochlea.

based computational code is able to simulate the phenomena of propagation of the acoustic wave in the three dimensional model of the human head. For more detailed simulations and the discussion on the correctness of the numerical results we refer to [3].

To visualize our models, meshes and solutions, we have implemented a simple interface to the Visualization Toolkit (VTK) [17], a collection of C++ classes that implement a wide range of visualization algorithms. The central data structure for this project is the vtkUnstructuredGrid, which represents volumetric data as a collection of points with corresponding scalar values (in this case, the real and imaginary part of the pressure), connected by cells of arbitrary type and dimension (i.e. lines, triangles, quads, tetrahedra, prisms, etc.).

We also compare the execution time and memory usage of the sequential version of our multi-frontal solver with the state-of-the art MUMPS solver. The comparison is presented in Table 2. The execution time of our solver is about six times slower than MUMPS solver. The first row represents the total number of tetrahedral and prism elements forming the simplified model of the human head. The second row represents the number of degrees of freedom over the mesh, equal to the number of unknown coefficients over them entire mesh (usually one coefficient over each mesh node for acoustic part, three coefficients over each mesh node for elastic part and four coefficients over each mesh node over the acoustic/elastic part interface). The third row represents the total number of non-zero entries within the global matrix before the factorization. This is the measure of the computational cost of the direct solver algorithm. However, our solver uses less memory, which may be critical when solving large computational problems.

8 Conclusions

In this chapter we presented a graph grammar model for generation and adaptation of three dimensional computational meshes with tetrahedral and prism elements. The graph grammar based algorithm was used for generation of the simplified model of the human head: the concentric spheres with the tissue, the skull, and the surrounding air. We focused then on the solution of the problem of acoustic of the human head, modeled as linear elasticity coupled with acoustics. We introduced the graph grammar productions transforming the mesh into a sequence of matrices, the input for the multi-frontal solver algorithm. Finally, we introduced the direct solver algorithm intended to minimize the memory usage in comparison to the state of the art MUMPS solver. We concluded the chapter with some numerical results as the proof of concept of the introduced methodology.

Acknowledgements. The work reported in this paper was supported by the Polish Ministry of Science and Higher Education grant no. NN 519 405737.

References

1. Beal, M.W., Shephard, M.S.: A General Topology-Based Mesh Data Structure. International Journal for Numerical Methods in Engineering 40, 1573–1596 (1997)
2. Demkowicz, L., Kurtz, J., Pardo, D., Paszyński, M., Rachowicz, W., Zdunek, A.: Computing with hp-Adaptive Finite Element Method. In: Frontiers: Three Dimensional Elliptic and Maxwell Problems, vol. II. Chapmann & Hall / CRC Applied Mathematics & Nonlinear Science (2007)
3. Demkowicz, L., Gatto, P., Kurtz, J., Paszyński, M., Rachowicz, W., Bleszyński, E., Bleszyński, M., Hamilton, M., Champlin, C., Pardo, D.: Modeling of bone conduction of sound in the human head using hp-finite elements: code desing and verification. Computer Methods in Applied Mechanics and Engineering 200(21-22), 1757–1773 (2011)
4. Grabska, E.: Theoretical Concepts of Graphical Modeling. Part One: Realization of CP-Graphs. Machine Graphics and Vision 2(1), 3–38 (1993)
5. Grabska, E.: Theoretical Concepts of Graphical Modeling. Part Two: CP-Graph Grammars and Languages. Machine Graphics and Vision 2(2), 149–178 (1993)
6. Grabska, E., Hliniak, G.: Structural Aspects of CP-Graph Languages. Schedae Informaticae 5, 81–100 (1993)
7. Paszyński, M., Pardo, D., Paszyńska, A.: Parallel multi-frontal solver for p adaptive finite element modeling of multi-physics computational problems. Journal of Computational Science 1, 48–54 (2010)
8. Flasiński, M., Schaefer, R.: Quasi context sensitive graph grammars as a formal model of FE mesh generation. Computer-Assisted Mechanics and Engineering Science 3, 191–203 (1996)
9. Michler, C., Demkowicz, L., Kurtz, J., Pardo, D.: Improving the performance of perfectly matched layers ny means of hp-adaptivity. ICES-Report 06-17, The University of Texas at Austin (2006)
10. Olkusnik, R.: Edytor graficzny do projektowania gramatyk grafowych. Msc Thesis, Jagiellonian University, Krakow (1996)
11. Paszyński, M., Pardo, D., Torres-Verdin, C., Demkowicz, L., Calo, V.: A Parallel Direct Solver for Self-Adaptive hp Finite Element Method. Journal of Parall and Distributed Computing 70, 270–281 (2010)
12. Paszyński, M., Schaefer, R.: Graph grammar driven partial differential eqautions solver. Concurrency and Computations: Practise and Experience 22(9), 1063–1097 (2010)
13. Paszyński, M., Pardo, D., Paszyńska, A., Demkowicz, L.: Out-of-core multi-frontal solver for multi-physics hp adaptive problems. Procedia Computer Science 4, 1788–1797 (2011)
14. Paszyńska, A., Paszyński, M., Grabska, E.: Graph Transformations for Modeling *hp*-Adaptive Finite Element Method with Mixed Triangular and Rectangular Elements. In: Allen, G., Nabrzyski, J., Seidel, E., van Albada, G.D., Dongarra, J., Sloot, P.M.A. (eds.) ICCS 2009. LNCS, vol. 5545, pp. 875–884. Springer, Heidelberg (2009)
15. Paszyńska, A., Paszyński, M., Grabska, E.: Graph Transformations for Modeling *hp*-Adaptive Finite Element Method with Triangular Elements. In: Bubak, M., van Albada, G.D., Dongarra, J., Sloot, P.M.A. (eds.) ICCS 2008, Part III. LNCS, vol. 5103, pp. 604–613. Springer, Heidelberg (2008)
16. Paszyński, M., Paszyńska, A.: Graph Transformations for Modeling Parallel *hp*-Adaptive Finite Element Method. In: Wyrzykowski, R., Dongarra, J., Karczewski, K., Wasniewski, J. (eds.) PPAM 2007. LNCS, vol. 4967, pp. 1313–1322. Springer, Heidelberg (2008)

17. Schroeder, W., Martin, K., Lorensen, B.: The Visualization Toolkit An Object-Oriented Approach To 3D Graphics, 3rd edn. Kitware, Inc., http://www.kitware.com
18. Spicher, A., Michel, O., Giavitto, J.-L.: Declarative Mesh Subdivision Using Topological Rewriting in MGS. In: Ehrig, H., Rensink, A., Rozenberg, G., Schürr, A. (eds.) ICGT 2010. LNCS, vol. 6372, pp. 298–313. Springer, Heidelberg (2010)
19. Szymczak, A., Paszyński, M.: Graph grammar based Petri net controlled direct solver algorithm. Computer Science 11, 65–79 (2010)

Review of Graph Invariants for Quantitative Analysis of Structure Dynamics

Wojciech Czech and Witold Dzwinel

Abstract. In this work we review graph invariants used for quantitative analysis of evolving graphs. Focusing on graph datasets derived from structural pattern recognition and complex networks fields, we demonstrate how to capture relevant topological features of networks. In an experimental setup, we study structural properties of graphs representing rotating 3D objects and show how they are related to characteritics of undelying images. We present how evolving strucure of Autonomous Systems (ASs) network is reflected by non-trivial changes in scalar graph descriptors. We also inspect characteristics of growing tumor vascular networks, obtained from a simulation. Additionally, the overview of currently used graph invariants with several possible groupings is provided.

1 Introduction

Structural data appear in many domains of contemporary science. This is because graphs constitute abstraction layer convenient in modeling binary relations between elements of any type. Heterogeneity of those relations can be represented using edge weights. Additionally, they allow for integration of large amounts of data into one high-level structure capturing system as a whole. This is particularly useful in high-throughput experiments producing high volume data that cannot be easily tackled without previous synthesis. In this context, graphs simplify system-level analysis of large ensembles.

1.1 Sources of Structured Datasets

In the last decade, theory of complex networks emerged as an interdisciplinary research area providing deep insight into topology and dynamics of networks and

Wojciech Czech · Witold Dzwinel
AGH University of Science and Technology, Al. Mickiewicza 30, 30-059 Kraków, Poland
e-mail: {czech,dzwinel}@agh.edu.pl

A. Byrski et al. (Eds.): Advances in Intelligent Modelling and Simulation, SCI 416, pp. 325–343.
springerlink.com

a common viewpoint for their analysis [16]. The structured datasets derived from various disciplines were analyzed by means of this theory. Just to briefly mention social networks modeling relations and associations between individuals [4, 5], cellular networks integrating expression data for genes, proteins and metabolites [27] or protein folding networks encoding transitions between three-dimensional tertiary protein configurations [31, 22].

The study of real-world networks revealed that they commonly exhibit inhomogeneous structure reflected by power-law degree distributions [6], for which average values and standard deviations are not good representatives of underlying data. The network following *scale-free* degree distribution has heterogenic topology with a few densely connected *hub* nodes and many low-degree nodes. As a consequence such a network is resistant to random attacks and vulnerable to targeted attacks, in which *hubs* are eliminated with a high probability. In the model of evolving network, proposed as a generating mechanism for *scale-free* networks, newly-created vertices connect to existing vertices with a probability proportional to their degree. This rule is known as *preferential attachment* [2]. As the *scale-free* structure of a network potentially reveals underlying dynamics, the significant research efforts was put on finding power-laws in real-world data. Vast amount of networks was claimed to be *scale-free* including cellular networks, World Wide Web (WWW) and social networks.

The experiments on chains of correspondence conducted by Stanley Milgram initiated long-standing debate about shortest path structure of social-networks and gave rise to *six degrees of separation* idea [2]. Since then, many real-world networks was shown to possess *small-world* property, that is small average shortest path scaling logarithmically with a graph size. As many types of graphs, including random ones, share this characteristic, the more strict definition of *small-world* networks requires additionally high *clustering coefficients*. The *small-world* property influences dynamics on a graph as it provides fast transport through short links, what is a crucial factor for such processes on networks as rumor propagation or disease spread.

Graph-based representations and graph learning are also the core of structural pattern recognition field [10, 33]. Here, the graph comparison is a task of particular importance, as measuring graph similarity or dissimilarity allows for application of statistical pattern recognition techniques for objects originally located in the space of graphs which does not possess inbuilt metric. Graphs allow for capturing structure of image or shape elements in a manner which is invariant under rotation, scaling, translation and changes in viewpoint. Encoding relations between scene primitives is particularly benefitial in content-based image retrieval and analysis. Therefore graph-based representations are gaining considerable attention in computer vision and image processing communities [1]. Transformation of pixel matrix into graph requires selection of objects mapped to node primitives and binary relations between them to form edges. This can be achieved in several ways, e.g., using skeletonization of binary images or treating image segments as nodes and connecting neighboring segments.

1.2 Graph Comparsion

In order to perform quantitative analysis of graph structure and dynamics, one needs to have efficient method of graph comparison in hand. The practical approach to this task uses graph invariants to construct feature vector and embed graph into the metric space. This is frequently referred as *explicit graph embedding*, being particularly useful in inexact matching of noisy structural patterns. The question how to quantitatively capture relevant structural properties of graphs provoked the development of many graph measures such as *clustering coefficient*, *efficiency* and *betweenness centrality* [11]. Nevertheless, with a continuous rise of structured data volume, quantitative graph analysis using graph descriptors becomes complex and computationally challenging task. This is observed both in structural pattern recognition and complex networks areas, where increasing size of analyzed graphs makes currently used algorithms impractical. Moreover, the abundance of scalar graph descriptors and frequent correlations between typical topological measures [26] make the selection of relevant features a difficult task.

The great part of network research is centered on assessing structure of static graphs. However, in this work we focus on graph evolution in time presenting how dynamics of a graph can be tracked using topological descriptors. To this end, we study three sample structured datasets, namely image graph representations of rotating objects (Dataset I), the set of 733 daily instances of autonomous systems graphs obtained from Stanford Large Networks Dataset Collection [28, 29] (Dataset II) and tumor vascular networks generated by simulation (Dataset III). We also review state-of-art graph descriptors providing some groupings and recall definitions of arbitrarily selected invariants. At the end we provide conclusions with remarks on utility of descriptors and experimental results.

2 Graph Invariants

The question how to capture quantitatively relevant structural properties of graphs provoked development of various graph descriptors. Frequently, we call them invariants, to emphasize that invariance under graph isomorphism is their key property.

2.1 Types of Graph Invariants

Diverse origins of graph descriptors result in large quantity and several ways of grouping. We review the most common groups of graph invariants and list sample descriptors from each group.

2.1.1 Invariant Target

The first distinction of graph descriptors is derived from the graph primitive they describe.

a. Graph descriptors describe graph as a whole. Typically they reflect a single topological property and cannot be used to reconstruct graph univocally. Examples are as follows: *density, diameter, radius, average path length, efficiency, average degree, graph clustering coefficient, Laplacian matrix spectrum, Estrada Index, fractal dimension* [11, 5, 15, 40].

b. Vertex descriptors are used to indicate vertex importance called centrality or assess other vertex-related topological feature. Frequently, they are computed by applying an iterative procedure on a graph resulting in steady state and scalars assigned to its vertices. Random walks on graphs are typical examples for such an iterative procedure. The representatives of this class of descriptors are *degree, clustering coefficient, betweenness centrality, random walks betweenness centrality, vertex distance, eccentricity, closeness, Page Rank, communicability betweenness* [37, 35, 18].

c. Edge descriptors reflect importance of an edge in a certain dynamical processes on a graph or are values of two argument function on edge ends vertex descriptors. Examples include *edge connectivity, range of edge, edge betweenness* [20].

d. Pair descriptors are assigned to a pair of graph vertices or edges. The typical examples here include vertex-vertex similarity or dissimilarity measures such as *shortest-path length, longest-path length, f-communicability* and *commute time* [39, 17].

Last three types of descriptors can be treated in common as **element descriptors**. Given single graph, the distribution of selected element descriptor may be used to compute permutation invariant graph descriptor.

2.1.2 Domain Descriptors

Graph descriptors have been commonly developed in the context of their application in a specific domain. The main sources of graph topological measures are listed below. The classification is imprecise as some descriptors may have been introduced in one domain and then adopted in a different one.

a. Theory of complex networks
A great part of systems described as a complex network revealed common structural and dynamical properties such as *scale-freeness, small worldliness, modularity, assortativity* and *disassortativity*. Those characteristics are captured using graph descriptors like *clustering coefficient, average path length, average nearest neighbors degree, efficiency, transitivity, motif Z-score* [6].

b. Sociology
Graphs are used in sociology to encode relations between members of a population. The vertex of social network usually models an individual (or a group of individuals) and the basic concern of network analysis is to evaluate the relative importance of a vertex within the network. This has lead to the development of several vertex descriptors called centrality measures: *betweenness centrality*

[19], *random walks betweenness centrality* [35], *closeness* [42] or *eigenvector centrality* [36].

c. Chemistry

A number of graph descriptors were derived from the topological studies of molecules. The examples are *Wiener index* [46], *normalized Wiener index*, *Platt index* [38], *Gordon-Scantleburry Index* [24]. Generally, they reflect branching of the molecue and are correlated with its van der Waals surface.

d. Biology

Biological systems can be described at the cellular level as a large network integrating expression data for genes, proteins and metabolites. Some of graph descriptors were specifically designed for investigating these kinds of graphs, just to mention generalized *walk-based centrality measures* by Estrada [15].

e. Structural pattern recognition

Representing patterns as graphs posed a new problem of measuring similarity between objects non-vectorial in nature. This gave rise to the development of new graph matching algorithms, including ones based on graph to feature vector transformation. Such transformation can be performed with the use of graph descriptors of any type, however, algebraic graph descriptors proved their superior robustness in many cases [47, 49].

2.1.3 Algebraic Descriptors

Graphs can be represented using matrices of several types. The group of well-known graph matrices contains *adjacency matrix*, *incidence matrix*, *Laplace matrix*, *normalized Laplace matrix* [23, 9] and *heat kernel* [49, 48]. The study of graph matrices originating in algebraic graph theory has led to valuable findings about connections between algebraic and structural properties of matrices and corresponding graphs. Particularly, the analysis of spectral decomposition of Laplace matrices brought interesting results. The set of eigenvalues and eigenvectors contain full information about graph structure and can be used to embed graph [30, 47, 49].

This type of graph feature generation is slightly different from gathering well-known scalar descriptors, as we have plenty of available numbers (e.g. eigenvalues); but only a small part of them has established connection with a graph structure. The descriptors created using such blind procedure exhibit their advantage in the field of structural pattern recognition.

2.1.4 Local and Global Descriptors

This distinction is directly linked with element descriptors. Local descriptors are computed on the basis of elements close to a given element (*k-level neighbors*). The global ones quantify a property of a selected element from the whole graph perspective. The examples from each family are presented below.

Local descriptors : *degree, vertex clustering coefficient, local efficiency*
Global descriptors : *Page Rank, betweenness centrality, vertex eccentricity*

2.1.5 Statistical Descriptors

The distributions of element descriptor values bring significant portion of information about graph structure. For that reason statistical moments or Shannon entropy can be used to generate valuable graph descriptors. The examples of statistical descriptors are *graph clustering coefficient*, *characteristic path length* and *average degree*. Statistical descriptors are useful for comparison of different size graphs.

2.2 Sample Invariants

In this section we recall definitions of five graph invariants, which will be used in the experimental section to track dynamics of network structures. Let us denote the number of graph vertices as n, length of the shortest path between vertex u and v, i.e. distance as $d(u,v)$ and a set of graph vertices as $V(G)$.

2.2.1 Efficiency and Local Efficiency

Efficiency (E) [6] is the harmonic mean of geodesic lengths over all couples of nodes. The normalization factor $n(n-1)$, proportional to the number of node pairs, ensures that it lies within the range $[0,1]$.

$$E(G) = \frac{1}{n(n-1)} \sum_{u,v \in V(G), u \neq v} \frac{1}{d(u,v)} \tag{1}$$

Efficiency measures the traffic capacity of a network and reflects its parallel-type transfer ability. In turn, local efficiency [6] captures clustering of a network understood as presence of triangles. It is defined as follows

$$E_{loc}(v) = E(G_v), \tag{2}$$

where G_v is a subgraph constructed from neighbors of v and $E(\ldots)$ is the graph efficiency. The local efficiency for a whole graph is computed as the average value of vertex local efficiency, as given below

$$E_{loc}(G) = \frac{1}{n} \sum_{v \in V(G)} E_{loc}(v). \tag{3}$$

2.2.2 Estrada Index

The Estrada Index (EE) is a spectral descriptor, defined as a trace of the exponential adjacency matrix. It quantifies the content of subgraphs in the graph [15] taking closed walks of lengths 2 to ∞ into account, encoded on the diagonal of powered adjacency matrix. It is formulated as below

$$EE(G) = \sum_{i=1}^{n} \exp(\lambda_i) = tr(e^{\mathbf{A}}), \tag{4}$$

where λ_i is i-th eigenvalue of the adjacency matrix, $\mathbf{A}$, and $e^{\mathbf{A}}$ is defined as follows

$$e^{\mathbf{A}} = \mathbf{I} + \mathbf{A} + \frac{\mathbf{A}^2}{2!} + \frac{\mathbf{A}^3}{3!} + \ldots \tag{5}$$

2.2.3 Heat-Kernel Matrix Invariants

Denoting adjacency matrix of a graph as $\mathbf{A}$ and diagonal matrix of vertex degrees as $\mathbf{D}$, we obtain Laplace matrix $\mathbf{L}$ as $\mathbf{L} = \mathbf{A} - \mathbf{D}$. This matrix is well-suited to graph analysis as it possesses only nonnegative eigenvalues which helps to order eigenvectors and exploit them in graph features extraction. The heat equation, associated with Laplace matrix describes diffusion process on a graph.

$$\frac{\partial \mathbf{h_t}}{\partial t} = -\mathbf{L}\mathbf{h_t} \tag{6}$$

$$\mathbf{h_t} = \exp(-\mathbf{L}t) = \sum_{i=1}^{n} \exp(-\lambda_i t)\phi_i \phi_i^T, \tag{7}$$

Heat kernel - the fundamental solution of this equation (see Equation 6 and Equation 7, ϕ_i denotes eigenvector associated with an eigenvalue λ_i) allows for construction of valuable graph descriptors that additionally can be scaled by the time parameter [49]. This enables to navigate between local properties (low values of t) and global properties (high values of t) of a graph.

Graph characteristics generated on the basis of heat kernel include zeta-function derivative, which is related to the number of spanning trees in the graph and degrees of its vertices [49]. It is defined as follows

$$\zeta'(0) = \ln\left(\prod_{\lambda_i \neq 0} \lambda_i^{-1}\right). \tag{8}$$

2.2.4 Distance k-Graphs Invariants

For a graph $G = (V(G), E(G))$ we define distance k-graph G_k as a graph with vertex set $V(G_k) = V(G)$ and edge set $E(G_k)$ so that $\{u,v\} \in E(G_k)$ iff $d(u,v) = k$. If $d(u,v) = k$, the vertices u and v are called k-neighbors. From this definition it follows that $G_1 = G$ and for $k > diam(G)$, G_k is an empty graph. As a set of distance k-graphs is ordered, it constitutes a good basis for extracting features invariant under graph isomorphism. With growing value of k, the information about connectivity of G encoded in G_k graphs moves from local to global. By enumerating scalar invariants of G_k graphs we can obtain different graph feature vectors, depending on selected basic invariant. Particularly, we can use $E(G)$ or $EE(G)$ descriptors defined earlier. Distance k-graphs degree distributions are exploited in construction of 2D graph invariants called B-matrices [12], which are rich source of information about graph structure.

3 Graph Datasets

In this section we briefly present two types of structured data, which will be used further in this text for the demonstation of graph analysis using topological descriptors.

3.1 *Dataset I: Graph Representations of Images*

Extracting feature points using corner detector and applying Delaunay triangulation yields purely structural image representation. This transformation is frequently used in image recognition tasks owing to its good correspondence with a structure of underlying image [30]. The schematic view of corners-based graph generation is depicted in Fig. 1. Typically, Harris corner detector [25] is employed to obtain feature points. A corner reflects image point where intensities change rapidly in two perpendicular directions (crossing edges).

In order to add connectivity information to discrete feature set, 2D Voronoi cells are generated around each corner. Next, dual graph is constructed, in which edges join corners with adjacent Voronoi cells. Delaunay graphs possess several properties that make them well-suited for encoding image structure [32]. For instance, no point appears inside circumcircle of a triangle, therefore any additional noise point affects only local structure, in the region indicated by neighboring circumcircles. In turn, partial occlusions leave remaining part of a graph similar to the original one.

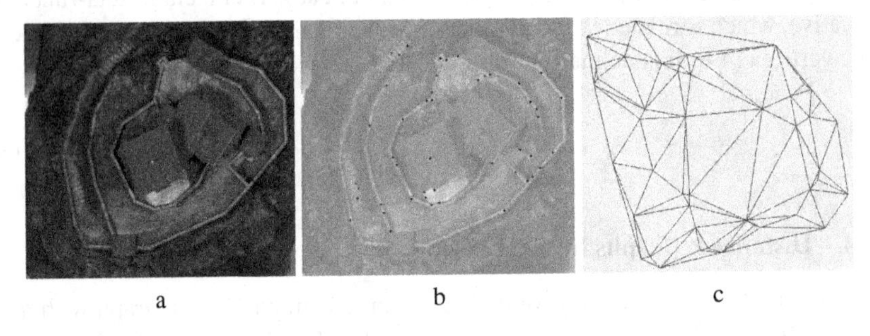

Fig. 1 Image to graph transformation using Delaunay triangulation of corners: a. original image, b. applying Harris corner detector [25], c. Delaunay triangulation.

We exploit graph representations of images from *Columbia object image database* (COIL) [34] to show how properities of rotating objects can be captured with the use of graph invariants.

3.2 *Dataset II: Technological Networks*

The set of 733 complex networks representing daily snapshots of communication between Autonomous Systems via Border Gateway Protocol is available in

Stanford Large Networks Dataset Collection [28]. Here, the vertices model Autonomous Systems and edges join those AS-es, which have exchanged information during a day. The data was collected from November 8, 1997 to January 2, 2000. The number of nodes and edges changes with time and spans the range 103-6473 (4183) for vertices and 239-12572 for edges (7783) (average values in parentheses). In the work of Leskovec et al. [29], the authors demonstrated that AS networks exhibit interesting dynamics, which cannot be explained using *preferential attachment* or *small-world* models. Particularly, the empirical study on evolution of average degree revealed that the number of edges grows faster than linearly with the number of nodes, yielding power-law relation. Moreover, it was shown that effective diameter is shrinking with time, and it is opposite to predictions derived from *small-world* models, which demonstrate diameters growing logarithmically or slower with a number of vertices.

In this work we compute distance matrices for the AS networks using GPU-enabled all-pair shortest-paths algorithm and present how distance-based graph descriptors change with time, reflecting complex, non-monotonic evolution of a graph structure.

3.3 Dataset III: Tumor Vascular Networks from Angiogenesis Simulation

Vascular networks deliver oxygen and nutrients to tissues, therefore modeling their formation and behavior plays significant role in understanding processes taking place in organs. Normal vasculature at capillary level resembles regular, planar mesh. This is because it evolves to provide approximately uniform spatial distribution of chemicals in surrounding tissue. After changing scale from micrometers to millimeters, we can observe hierarchically organized arterio-venous vessels forming tree-like structures with many three-way junctions. This type of vessels provide long-range transport. The situation changes when vasculature in presence of tumor is examined. Here, the observed structures are heterogeneous, non-hierarchical and fragmented into zones of different microvascular density [45]. Contrary to normal vasculature, tumor blood vessels exhibit fractal geometry with many irregularities of different sizes [3].

In recent years many efforts were made to elucidate complex dynamics of tumor vasculature growth and investigate how its structure influences tumor resistance to drugs [44, 41]. Quantitative analysis of vascular networks topology has considerable impact on advances of anti-angiogenesis therapy in cancer. Nevertheless, digitalized vascular networks are expensive and troublesome to obtain. Images from angiography or confocal microscopy can be segmented and after skeletonization transformed to adjacency matrices. However, such a two-dimensional approach is definitely not perfect since vasculature is most likely a complex 3D structure. Two-dimensional real-world vascular networks are also obtained by special tumor inoculation into laboratory mice [21] but the cost of such a procedure is prohibitively high. Therefore, *in-silico* experiments which can deliver vast amount of simulation data, play prominent role in providing new insights into tumor growth mechanisms. Graph comparison algorithms are applied here to validate the results of simulation with a real-world data [43].

In recent years many competing models of tumor-induced angiogenesis were created. Among them, hybrid 3D model which combines particle dynamics, cellular automata and continuum approach [44] appears to be the most advanced one. In turn, the model presented in work [41] takes into consideration newly discovered angiogenic factors and includes remodeling of existing vasculature. It employs graph of cellular automata as a substrate for tumor and vasculature growth. In this chapter we also demonstrate utilization of graph descriptors for keeping track of angiogenesis simulation results.

4 Examples

In this section we first present two experiments on real-world structured datasets derived from pattern recognition and complex networks domains. Then, we demonstrate another experiment based on angiogenesis simulation dataset.

4.1 Dataset I: COIL Images

We extracted three groups of rotating toy cars from COIL database (see Fig. 2). Each group contains 70 samples (rotation with 5° step, starting at 0° ending at 355°). The images were transformed into Dealaunay graphs, as described in Section 3.1. The resulting graphs cannot be easily distinguished on the basis of vertex or edge count - the respective histograms overlap heavily (see Fig. 3).

The group or graph descriptors including efficiency, Estrada Index and zeta-function derivative was computed for the structured dataset. We explore how those invariants reflect underlying dynamics of rotating objects. The values of selected descriptors for angles changing from 0° to 355° are depicted in Fig. 4. Following observations regarding correspondence of extracted topological features with underlying images can be made on the basis of those charts.

- For angles 0°, 180° and 350°, the values of descriptors are relatively close, as corresponding photos represent structurally similar objects observed from similar perspective (see Fig. 4(a),(e)).
- The values for subsequent angles are close to each other except for local efficiency invariant, which exhibits significant fluctuations (Fig. 4(d)). This is because local descriptors reflect more specific features being vulnerable to small changes of viewing angle.
- For angles near to 90° and 270°, we observe rapid changes of descriptor values reflecting much different look of an object and greater zoom (see Fig. 2, second row).
- For angles 25° to 150°, object 15 is well-separated from two others, this probably reflects lack of rear window in this car (Fig. 2, second row).

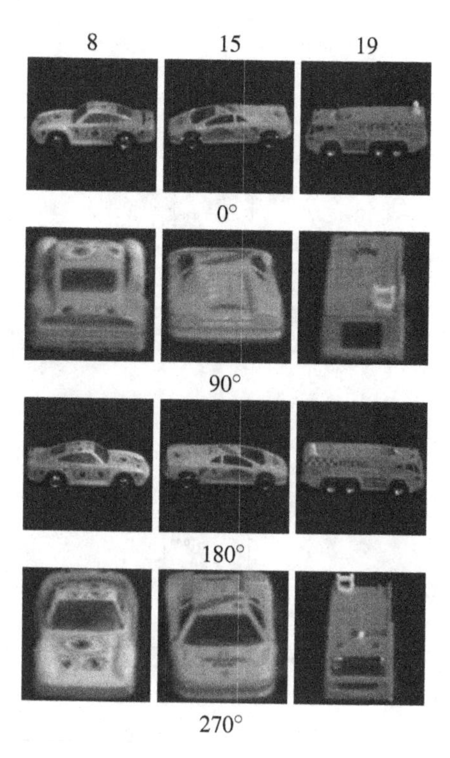

Fig. 2 The samples from three groups of rotating objects (toy cars, COIL database). Objects from group number 8 (COIL database numeration) in first column, group 15 - middle column and group 19 - right column. Subsequent rows present cars rotated by 0°, 90°, 180° and 270° accordingly.

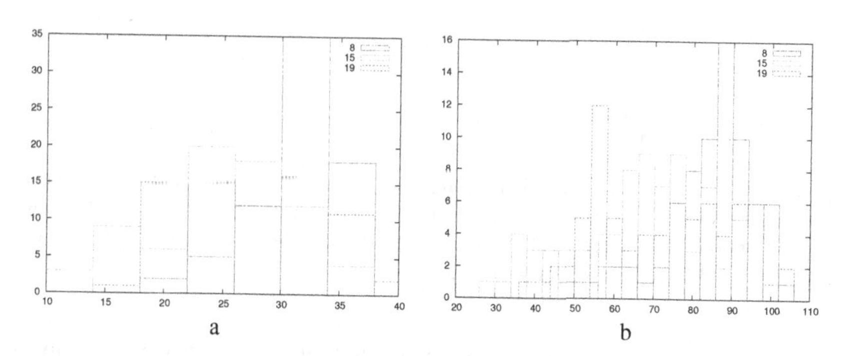

Fig. 3 Histograms of vertex count (a) and edge count (b) for Delaunay graphs representing rotating 3D objects.

- The global graph descriptors such as efficiency (Fig. 4(a)), Estrada Index (Fig. 4(e)) and zeta-function derivative (Fig. 4(f)) follow similar pattern of changes, with two last ones providing better separation of objects.

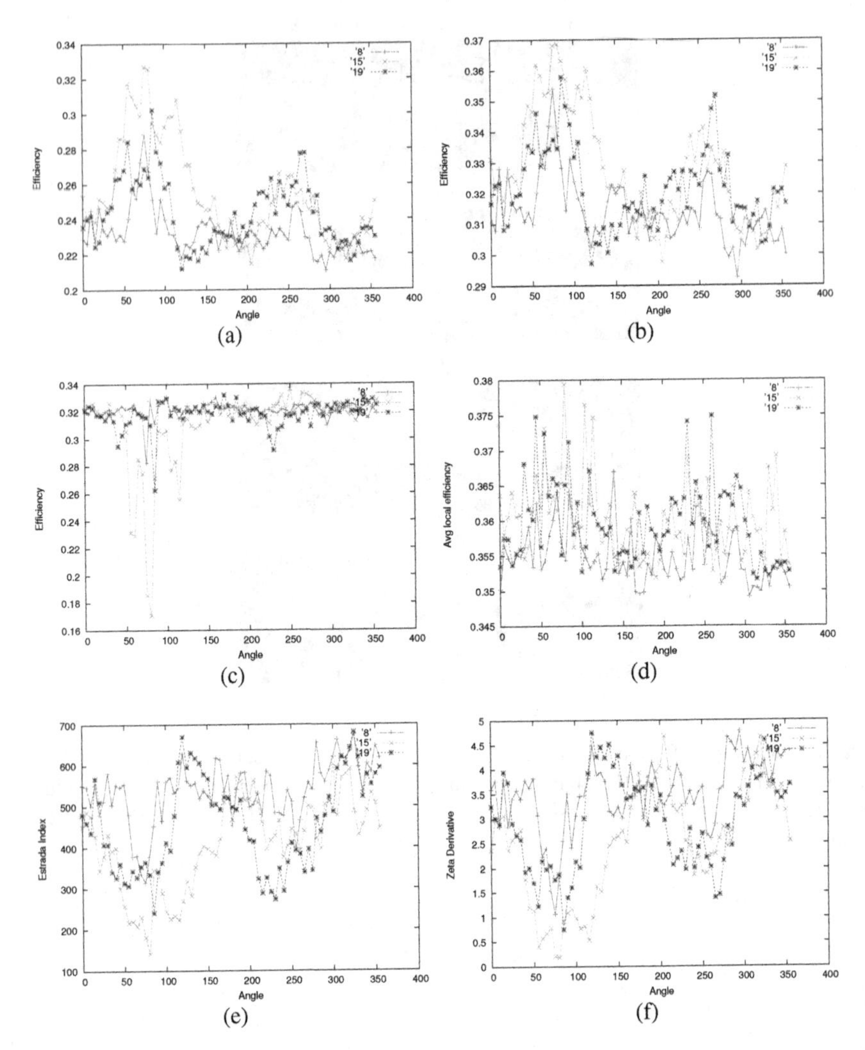

Fig. 4 The values of graph invariants computed for different angles and three groups of objects from COIL database (a) efficiency, (b) efficiency of distance 2-graph, (c) efficiency of distance 3-graph, (d) average local efficiency, (e) Estrada Index, (f) zeta-function derivative. Labels 8, 15 and 19 represent groups of objects.

- Subsequent G_k graphs bring more specific information about distance structure of underlying patterns, with G_3 giving priority to different rear-view of cars (angles 50°-100°, see Fig. 4(c))
- For angles reflecting head-view of cars, the object labeled 19 is more easily distinguishable from the two others. Indeed, for viewing angles near to 270°, the ladder visible on the fire truck makes the object 19 much different from two sport cars (see Fig. 2, row 4).

4.2 Dataset II: Autonomous Systems Networks

We computed vertex-vertex distance matrices for all 733 graphs representing communication between Autonomous Systems (see Section 3.2). As average size of AS graphs is major (4183), we took advantage of Compute Unified Device Architecture (CUDA) implementation of R-Kleene all-pair shortest-paths algorithm [7, 14] available in *Graph Investigator* application [13]. This allows for analysis of large graphs in an interactive way. Many graph invariants rely on distance information (see Table 1), therefore, the approach of employing Graphics Processing Unit (GPU) to descriptor generation considerably improves efficiency of graph mining.

Particularly, we focus on two graph descriptors: efficiency and average path length. The changes in values of these two invariants are depicted in Fig. 5. As the number of edges grows super-linearily with the number of vertices (see Fig. 5(e),(f)), the graph becomes more dense in time. General view of efficiency and average path length evolution (see Fig. 5(a),(c)) shows relatively stable behavior of those measures with slowly decreasing average path length and approximately constant efficiency. However, this is due to two abnormal events, appearing around sample 300 and 700, in which more connections between AS-es were established. After excluding these two groups of outliers we revealed more complex pattern of structural changes (see Fig. 5(b),(d)). The average shortest path decreases slowly and non-monotonically, what confirms shrinking diameter observation described in [29]. In turn, efficiency exhibits small relative changes, so that it remains approximately constant (≈ 0.145) during network growth. The traffic capacity of AS network appears to be preserved during its evolution.

4.3 Dataset III: Tracking Angiogenesis Simulation

The experiment presented in this section is based on simulation data obtained from cellular automata model of tumor-induced angiogenesis [41] (see also Section 3.3). We analyze four sets of evolving vascular networks obtained for different parameters of the model, investigating how topological properties of generated networks change during simulation and how those characteristics reflect functions of simulated growth factors.

The model consists of two interacting parts: transportation network of blood vessels and consuming environment (tissue). [41]. The vascular network changes in response to distribution of Tumour Angiogenesis Factors (TAFs), which are produced by starving tumor cells to provoke vessels growth [8]. The generated network delivers oxygen and nutrients to the tissue forming a gradient which affects tumor cells. The tissue is modeled by a mesh of cellular automata while the transportation network by a graph of cellular automata built over the CA mesh. After exceeding certain concentration of TAF, the vessels form sprouts that develop towards tumor tissue. The vessels become mature with time and more mature vessels provide better transport capability.

We focus on three types of parameters that control branching: TAF threshold T_c, baseline branching probability P_b, and level threshold multipliers which allow

Table 1 Graph descriptors which can be computed on the basis of information on shortest-paths in a graph (v denotes a vertex of a graph G).

Descriptor	Remarks
Vertex distance	Sum of distances between v and all other vertices from the graph G, $d_v = \sum_{w \in V(G)} d(v,w)$
Eccentricity	Maximum distance between the vertex v and any of the remaining graph vertices, $e_v = \max_{w \in V(G)} d(v,w)$
Vertex B Index	$b_v = \frac{k_v}{d_v}$, where d_v is vertex distance of vertex v
Closeness	Mean distance to any other vertex, $c_v = \frac{1}{n-1} \sum_{u \in V(G), u \neq v} d(u,v)$
Distance k-graphs invariants	see Section 2.2.4
Efficiency	Harmonic mean of geodesic lengths over all pairs of vertices, $E(G) = \frac{1}{n(n-1)} \sum_{u,v \in V(G), u \neq v} \frac{1}{d(u,v)}$
Diameter of a graph	Maximum distance between any two vertices in the graph, $diam(G) = \max_{u,v \in V(G)} d(u,v)$
Radius of a graph	$r(G) = \min_{v \in V(G)} e_v$, where e_v is eccentricity of the vertex v
Wiener Index	Sum of distances between every pair of vertices in a given graph, $W(G) = \frac{1}{2} \sum_{u \in V(G)} \sum_{v \in V(G)} d(u,v)$
Average path length	$Apl(G) = \frac{2}{n(n-1)} \sum_{u,v \in V(G)} d(u,v)$

for increasing branching probabilities for higher TAF concentrations. Provided that TAF threshold in certain locations is exceeded, the relevant vascular cell can start branching with certain probability P_b. This probability can be adjusted using two pairs of parameters: threshold T_1, multiplier m_1 and threshold T_2, multiplier m_2. If $T_c > T_1$ then P_b is multiplied by m_1 and if $T_c > T_2$ then branching probability is set to $m_2 P_b$.

We run 4 simulations with different parameters (see Table 2). Next, for each simulation step the selected graph descriptors were computed to capture changing topology of growing networks. As presented in Fig. 6(f), after 40 steps, the sizes of networks start to grow exponentially with different rates driven by branching probabilities. The fastest growth is observed for the Set 2, characterized by highest branching probability for immature, young cells. Comparing it to the slowest growth rate for Set 4, which has similar TAF threshold for immature vessels T_{ci} and greater multiplier for threshold 0.8, it seems that P_{bi} parameter affects growing rate to the greatest extent.

We observe that efficiency of vascular network starts converging to about 0.03 (Fig. 6(a)) with slight differences visible after zooming-in (Fig. 6(b)). After passing initial simulation steps, the efficiency starts to decrease, what reflects path-like growth of vascular networks towards tumor, with minor branching rate. Then,

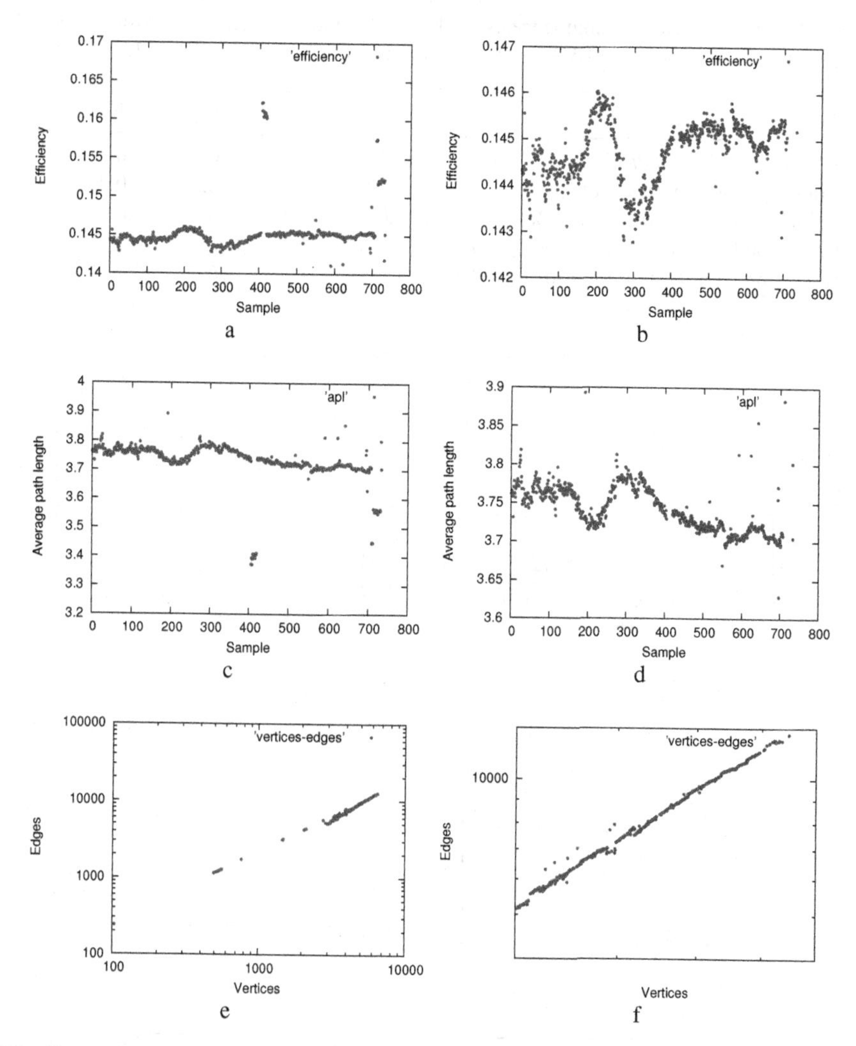

Fig. 5 The values of invariants for evolving Autonomous Systems network: (a) efficiency, (b) zoomed-in efficiency without outliers, (c) average path length, (d) zoomed-in average path length without outliers, (e) log-log plot of edges count vs. vertices count showing densification law, (f) log-log plot of edges count vs. vertices count restricted to graphs with a number of vertices ranging from 3000 to 7000.

following the 90th step, the value of invariant stabilizes because of increased branching rate implied by large number of mature cells. The charts showing evolution of average degree (Fig. 6(c),(d)) for the Set 1 and Set 3 have similar shape. It seems that TAF threshold for immature vessels is the dominant discrimination factor, as it possesses similar value for the Set 2 and Set 4 (0.4 and 0.5 respectively) and different value for the Set 1 and the Set 3 (0.1). Relatively high values of T_{ci} for the Set 2

Table 2 The parameters of angiogenesis model for 4 simulation runs (only parameters that vary over sets are presented). TAF concentration is a real value between 0 and 1.

	set 1	set 2	set 3	set 4
TAF threshold (*immature*) T_{ci}	0.1	0.4	0.1	0.5
Branch probability (*mature*) P_{bm}	0.02	0.01	0.01	0.01
Branch probability (*immature*) P_{bi}	0.02	0.08	0.02	0.02
Branch probability multiplier for $T_c > 0.6$	1	2	2	2
Branch probability multiplier for $T_c > 0.8$	1	4	6	6

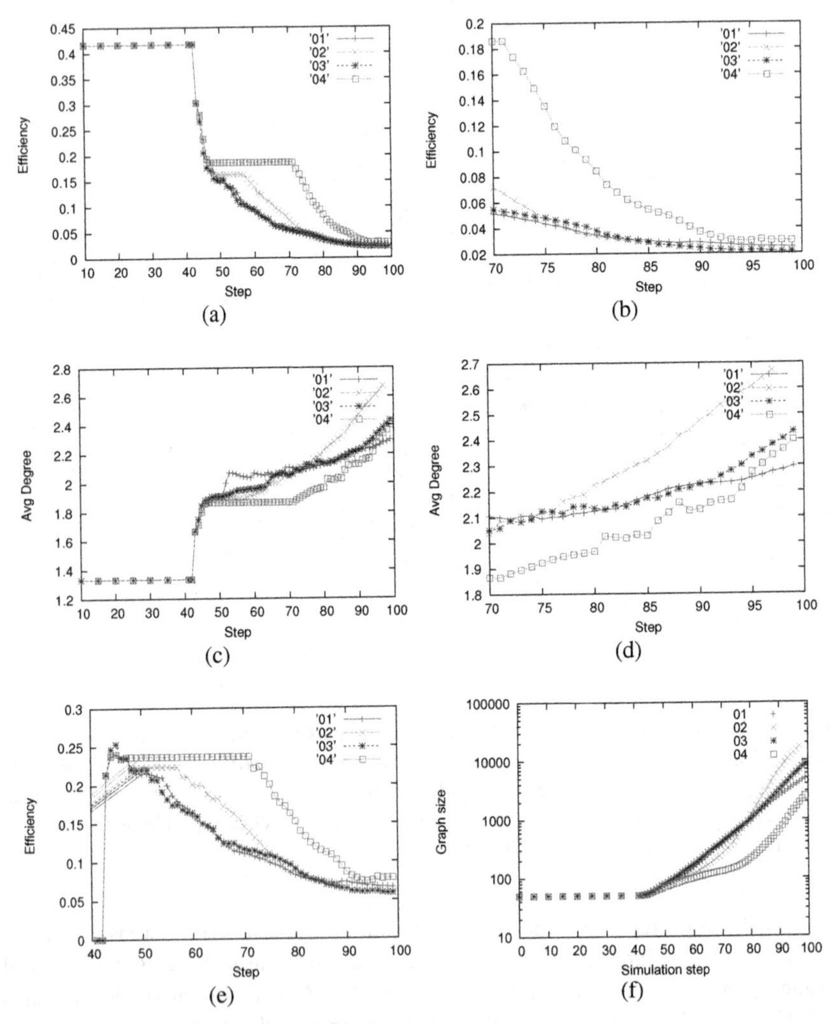

Fig. 6 The values of invariants for growing tumor vascular network obtained from simulation: (a) efficiency, (b) zoomed-in efficiency, (c) average degree, (d) zoomed-in average degree, (e) efficiency of distance 3-graph, (f) number of vertices.

and Set 4 prevent young immature cells (dominating at the beginning of simulation) from forming branches, therefore we can observe constant values of descriptors on the charts. Nevertheless, a few sprouts created from mature cells can develop towards tumor, extending size of vascular network. After passing through *mature* age, the number of *mature* vascular cells increases rapidly, resulting in a high branching rate at the end of the simulation.

5 Conclusions

The variety of available graph descriptors allows for capturing relevant features of structural patterns and investigating network properties over time. The invariants can be aggregated into a feature vector embedding a graph into metric space and making the graph comparison a straightforward task. This enables us to perform graph mining using methods of statistical pattern recognition. Nevertheless, the selection of relevant set of descriptors is a challenging, domain-dependent task, which can be approached empirically or more automatically with the use of feature selection algorithms.

Graph invariants can be grouped according to several criteria such as descriptor target, domain of origin, locality vs. globality of encoded features and a method of computation (algebraic or statistical). The first distinction, by descriptor target, is related to applicability of descriptors for certain tasks. Element descriptors can be used for community detection and finding correspondences between graph primitives, while global ones are applied in higher-level comparison of whole structures. Domain of invariant origin has only historical meaning as general graph invariants usually have wide range of applications and are not binned to a single field. In turn, global descriptors providing an overall view of structure can be compared to local, detailed invariants which take into account a number of specific constraints and short-range anomalies.

The studies performed on dynamics of graphs representing 3D images, AS complex networks and tumor vascular networks have shown that properly selected graph descriptors reflect high-level properties of encoded objects and bring insights into undelying growth patterns.

Acknowledgements. This work is financed by the Polish Ministry of Science and Higher Education, Project No. N N519 579338.

References

1. Iapr technical commitee 15 (January 31, 2012), `http://www.greyc.ensicaen.fr/iapr-tc15/`
2. Albert, R., Barabási, A.L.: Statistical mechanics of complex networks. Reviews of modern physics 74(1), 47 (2002)
3. Baish, J.W., Jain, R.K.: Fractals and cancer. Cancer Research 60(14), 3683 (2000)

4. Barabási, A.L., Oltvai, Z.N.: Network biology: understanding the cell's functional organization. Nature Reviews Genetics 5(2), 101–113 (2004)
5. Barrat, A., Barthlemy, M., Vespignani, A.: Dynamical processes on complex networks. Cambridge University Press (2008)
6. Boccaletti, S., Latora, V., Moreno, Y., Chavez, M., Hwang, D.U.: Complex networks: Structure and dynamics. Physics Reports 424(4-5), 175–308 (2006)
7. Buluç, A., Gilbert, J.R., Budak, C.: Solving path problems on the gpu. Parallel Computing 36(5-6), 241–253 (2010)
8. Carmeliet, P., Jain, R.K.: Angiogenesis in cancer and other diseases. NATURE-LONDON, 249–257 (2000)
9. Chung, F.R.K.: Spectral Graph Theory. CBMS Regional Conference Series in Mathematics, vol. 92, p. 3, 8. American Mathematical Society (1997)
10. Conte, D., Foggia, P., Sansone, C., Vento, M.: Thirty years of graph matching in pattern recognition. International Journal of Pattern Recognition and Artificial Intelligence 18(3), 265–298 (2004)
11. Costa, L.F., Rodrigues, F.A., Travieso, G., Boas, P.: Characterization of complex networks: A survey of measurements. Advances in Physics 56(1), 167–242 (2007)
12. Czech, W.: Graph descriptors from b-matrix representation. In: Jiang, X., Ferrer, M., Torsello, A. (eds.) GbRPR 2011. LNCS, vol. 6658, pp. 12–21. Springer, Heidelberg (2011)
13. Czech, W., Goryczka, S., Arodz, T., Dzwinel, W., Dudek, A.: Exploring complex networks with graph investigator research application. Computing and Informatics 30(2) (2011)
14. Czech, W., Yuen, D.A.: Efficient graph comparison and visualization using gpu. In: Proceedings of the 14th IEEE International Conference on Computational Science and Engineering (CSE 2011), pp. 561–566 (2011), doi:10.1109/CSE.2011.223
15. Estrada, E.: Generalized walks-based centrality measures for complex biological networks. Journal of Theoretical Biology 263(4), 556–565 (2010)
16. Estrada, E., Fox, M., Higham, D., Oppo, G.: Network Science. Springer (2010)
17. Estrada, E., Higham, D.: Network properties revealed through matrix functions. SIAM Review 52(4), 696–714 (2010)
18. Estrada, E., Higham, D., Hatano, N.: Communicability betweenness in complex networks. Physica A: Statistical Mechanics and its Applications 388(5), 764–774 (2009)
19. Freeman, L.: A set of measures of centrality based on betweenness. Sociometry, 35–41 (1977)
20. Freeman, L.: Centrality in social networks conceptual clarification. Social Networks 1(3), 215–239 (1979)
21. Gazit, Y., Berk, D., Leunig, M., Baxter, L., Jain, R.: Scale-invariant behavior and vascular network formation in normal and tumor tissue. Physical Review Letters 75(12), 2428–2431 (1995)
22. Gfeller, D., Rios, P.D.L., Caflisch, A., Rao, F.: Complex network analysis of free-energy landscapes. Proceedings of the National Academy of Sciences 104(6), 1817 (2007)
23. Godsil, C., Royle, G.: Algebraic graph theory. Springer, New York (2001)
24. Gordon, M., Scantlebury, G.: Non-random polycondensation: Statistical theory of the substitution effect. Transactions of the Faraday Society 60, 604–621 (1964)
25. Harris, C., Stephens, M.: A combined corner and edge detector. In: Alvey Vision Conference, Manchester, UK, vol. 15, p. 50 (1988)
26. Jamakovic, A., Uhlig, S.: On the relationships between topological measures in real-world networks. Networks and Heterogeneous Media 3(2), 345 (2008)
27. Képès, F.: Biological networks. World Scientific Pub. Co. Inc. (2007)

28. Leskovec, J.: Stanford large network dataset collection, http://snap.stanford.edu/data/
29. Leskovec, J., Kleinberg, J., Faloutsos, C.: Graph evolution: Densification and shrinking diameters. ACM Transactions on Knowledge Discovery from Data (TKDD) 1(1), 2–es (2007)
30. Luo, B., Wilson, R., Hancock, E.: Spectral embedding of graphs. Pattern Recognition 36(10), 2213–2230 (2003)
31. Luo, Y., Lin, H., Huang, M., Liaw, T.: Conformation-networks of two-dimensional lattice homopolymers. Physics Letters A 359(3), 211–217 (2006)
32. Marchette, D.: Random graphs for statistical pattern recognition. Wiley-IEEE (2004)
33. Marfil, R., Escolano, F., Bandera, A.: Graph-Based Representations in Pattern Recognition and Computational Intelligence. In: Cabestany, J., Sandoval, F., Prieto, A., Corchado, J.M. (eds.) IWANN 2009. LNCS, vol. 5517, pp. 399–406. Springer, Heidelberg (2009)
34. Nene, S., Nayar, S., Murase, H.: Columbia object image library (coil-20). Dept. Comput. Sci., Columbia Univ., New York (1996), http://www.cs.columbia.edu/CAVE/coil-20.html
35. Newman, M.: A measure of betweenness centrality based on random walks. Social networks 27(1), 39–54 (2005)
36. Newman, M.: The mathematics of networks. The New Palgrave Encyclopedia of Economics (2007)
37. Page, L., Brin, S., Motwani, R., Winograd, T.: The pagerank citation ranking: Bringing order to the web (1999)
38. Platt, J.: Influence of neighbor bonds on additive bond properties in paraffins. The Journal of Chemical Physics 15, 419 (1947)
39. Qiu, H., Hancock, E.: Clustering and embedding using commute times. IEEE Transactions on Pattern Analysis and Machine Intelligence 29(11), 1873–1890 (2007)
40. Song, C., Havlin, S., Makse, H.: Self-similarity of complex networks. Nature 433 (2005)
41. Topa, P.: Dynamically reorganising vascular networks modelled using cellular automata approach. Cellular Automata, 494–499 (2010)
42. Wasserman, S., Faust, K.: Social network analysis: Methods and applications. Cambridge Univ. Pr. (1994)
43. Wcisło, R., Dzwinel, W., Gosztyla, P., Yuen, D.A., Czech, W.: Interactive visualization tool for planning cancer treatment. Tech. rep., University of Minnesota Supercomputing Institute Research Report UMSI 2011/7, CB number 2011-4 (2011)
44. Wcisło, R., Dzwinel, W., Yuen, D., Dudek, A.: A 3-d model of tumor progression based on complex automata driven by particle dynamics. Journal of Molecular Modeling 15(12), 1517–1539 (2009)
45. Welter, M., Rieger, H.: Physical determinants of vascular network remodeling during tumor growth. Eur. Phys. J. E 33, 149–163 (2010)
46. Wiener, H.: Structural determination of paraffin boiling points. Journal of the American Chemical Society 69(1), 17–20 (1947)
47. Wilson, R., Hancock, E., Luo, B.: Pattern vectors from algebraic graph theory. IEEE Transactions on Pattern Analysis and Machine Intelligence 27(7), 1112–1124 (2005)
48. Xiao, B., Hancock, E., Wilson, R.: A generative model for graph matching and embedding. Computer Vision and Image Understanding 113(7), 777–789 (2009)
49. Xiao, B., Hancock, E., Wilson, R.: Graph characteristics from the heat kernel trace. Pattern Recognition 42(11), 2589–2606 (2009)

A Consolidated Study Regarding the Formation of the Aero-Inlet Vortex

Wei Hua Ho

Abstract. Under certain flow conditions, when an air inlet is aspirated in close proximity to a solid surface, an inlet vortex will form between the inlet and the surface under certain flow conditions. This phenomenon can manifest itself during the operation of aircraft engines either when the aircraft is on the runway prior to take-off or during engine ground run, or when the engine is in a test cell during post maintenance tests. The vortex can pitch debris into the intake causing foreign object damage (**F.O.D.**) to the engine blades or result in compressor stall. The take-off problem can be partially solved by keeping the runway clear of debris and scheduling the throttle appropriately whenever possible. However throttle scheduling will not be appropriate during engine tests both on the ground and in a test cell. The characteristics of the vortex depends on a number of geometric and flow conditions such as the position of the engine relative to the surface, intake flow capture ratio and upstream flow. To eliminate these vortices at the design stage of aircraft configuration or new test cell, it is essential to be able to predict the onset of the vortex or at least understand the factors affecting their formation. In addition, it is also very important to understand the characteristics of such a vortex to be able to determine the potential damage if complete prevention is not possible.

This paper seeks to provide an understanding of this flow phenomenon by collating and analysing previous investigations and implemented solutions in both aircraft ground operations as well as during engine tests in a jet engine test cell. It will contain information from computational as well as experimental studies with emphasis on computational methods. The paper will present computational and experimental results from various sources with regards to the threshold conditions at which such vortices are formed in addition to their characteristics. To permit interested readers sufficient information to perform similar calculations, it will also contain details of the computational methods and parameters, such as the required computational

Wei Hua Ho
Department of Mechanical and Industrial Engineering University of South Africa,
Private Bag X6, Florida 1710, South Africa
e-mail: howh@unisa.ac.za

A. Byrski et al. (Eds.): Advances in Intelligent Modelling and Simulation, SCI 416, pp. 345–364.
springerlink.com

domain as well as the solution schemes. The computational simulations were performed on commercial computational fluid dynamics (CFD) package Fluent 13.0 which utilises the finite volume method to solve the governing Navier-Stokes equations.

1 Introduction

Aircraft engines operating at high power in close proximity to solid surfaces, either the ground or the walls of a jet engine test cell **(JETC)**, can, under certain conditions, cause the formation of a surface-to-intake vortex. This can either happen during the operation of the engine on an aircraft at take-off or during engine ground runs or in a test cell during post maintenance engine tests. The structure of the vortex is very similar to the vortex seen in the draining of a basin or bath tub, where the streamlines spiral into the suction inlet (or outlet) with a decreasing radius of gyration as the vortex approaches the engine inlet (or basin outlet). One end of the vortex is anchored to the nearby solid surface, in the case of the aircraft engine, or to a fluid-fluid interface in the basin, bath tub or similar scenarios.

When such a vortex is formed and ingested into the engine, it could adversely affect the engine performance and operation. In more serious situations, it can lead to catastrophic physical damage to the engine when the vortex lifts debris from the surface into the intake resulting in foreign object damage **(F.O.D.)**. The vortex can also cause mass flow and pressure fluctuations as well as strong changes in the direction of the flow resulting in compressor stall if ingested into the engine core. In addition, ingestion of such vortices can also increase engine wear due to additional blade loads and vibrations. Even if the vortex is ingested into the bypass duct of a turbofan engine, it can still cause instability in the low pressure spool. This is typical of high bypass engine with vortices entering from the edge of the fan base [5].

Kline [19] stated three conditions for the formation of a vortex: non-zero ambient vorticity, a stagnation point on the solid surface, and an updraught occuring above the stagnation point to the inlet. These are necessary but not sufficient conditions. The first condition states that the ambient flow must have a certain vorticity, else two counter-rotating and sometimes alternating vortices [33, 37] will be formed instead of a single core vortex. In such circumstances, the vorticity is induced by the boundary layer [1]. If the irrotational flow is a headwind, the two vortices will originate from the solid surface [4]. If the irrotational flow is a crosswind, one of the vortex (the trailing vortex) will be from the side [2]. Jermy and Ho [17] showed, using computational methods, that an ambient flow with a velocity gradient less than 0.001 /s results in the same counter-rotating vortices. This indicates that 0.001 /s is the minimum shear required to trigger a single vortex. Murphy [22] argued against this minimum shear value using experimental investigation and previously reported investigations, stating that vortices have been known to form under quiescent conditions. However this does not contradict the findings of Jermy and Ho [17] as the minimum shear reported by Jermy and Ho was for a vortex of consistent rotation as opposed to the unsteady alternating counter-rotating vortices.

The second condition indicates that a stagnation point on the solid surface (or fluid interface) is necessary for the vortex to form. This condition manifests itself in three ways and no vortex can form if:

1. The solid surface is too far from the engine inlet.
2. The suction of the engine is too weak.
3. The flow velocity near the solid surface is too high.

If any of the above three conditions are present, then the capture streamtube does not reach the ground plane and a vortex cannot form.

The third condition is similar to the second condition and indicates the presence of a suction source and this source causes and updraught above the stagnation point. The stagnation point and updraught forms the vortex core that is anchored to the solid surface.

In all instances of the previous experiments and simulations, a table vortex core or stagnation point is always vertically below the suction engine or suction tube. Fig. 1 and 2 below illustrates this. Fig. 1 is generated from a CFD simulation using Fluent, which uses the finite volume method to solve the governing Navier-Stokes equations.

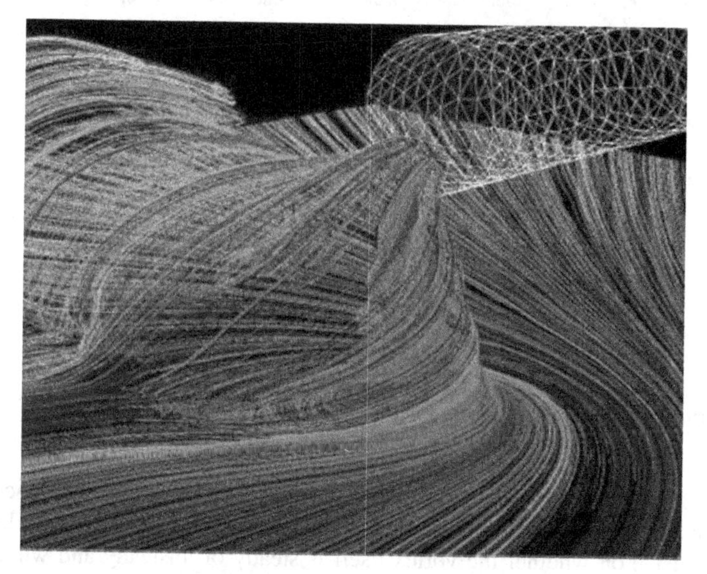

Fig. 1 Stagnation point with vertical updraught below suction engine (CFD Simulation)

The main parameters involved in the formation of such a vortex consist of an ambient (or upstream) velocity $\mathbf{V}_0$, engine (or suction) inlet velocity $\mathbf{V}_i$, height of engine (or suction) inlet from the solid surface **H**, diameter of engine (or suction) inlet $\mathbf{D}_i$ and ambient (or upstream) velocity gradient (or vorticity) **W**. These parameters are illustrated in Fig. 3 below. They are applicable both to the case of an engine on a plane on a runway and to an engine in a JETC.

Fig. 2 Stagnation point with vertical updraught below suction engine (Actual Occurance) - Photo by Peter Thomas

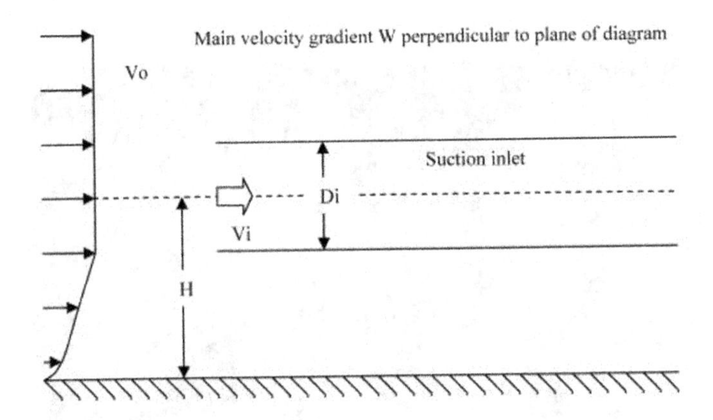

Fig. 3 Principal parameters in a aero-inlet vortex system

When such vortices were observed in experiments and in actual occurances, they are always unsteady in nature with the core seen to precess. However findings are not conclusive on whether the vortex itself is steady or unsteady and whether the unsteadiness observed are due to the unsteadiness of the upstream conditions. It must be noted that Glenny reported very minor disturbances to the surroundings such as opening a door and the experimenter moving around the vicinity could upset a vortex [9]. This makes maintaining an absolute steady surrounding conditions in any experiment or in actual occurances highly difficult to achieve.

2 Vortex Formation Threshold

As mentioned in the previous section (**Introduction**), there are a number of conditions that affect if a vortex will form. Essentially they are the five parameters shown in Fig. 3:

1. Ambient (or upstream) velocity $\mathbf{V}_0$
2. Engine (or suction) inlet velocity $\mathbf{V}_i$
3. Distance between engine (or suction) inlet **H**
4. Engine (or suction) inlet diameter $\mathbf{D}_i$
5. Velocity Gradient **W**

There has been many experiments conducted to identify and quantify the threshold at which a vortex forms and/or dissipates in a scenario representing an engine on an aircraft on a runway. Besides a non-zero ambient vorticity single cored vortices require the distance from inlet to free surface to be lower than a certain threshold, and the velocity of the air between the inlet and the solid surface to be lower than a certain velocity, also known as the "blow-away velocity". Many investigators [9, 20, 26, 27, 28] represent their findings in plots of $\frac{V_i}{V_0}$ against $\frac{H}{D_i}$. Their findings show that a single linear upward sloping boundary exists to separate the vortex forming and non-vortex forming regimes. Their findings are collated and shown in Fig. 4 below.

Computational simulation by Jermy and Ho [17, 13] also exhibit the same trends. A particular set of their simulation results is shown in Fig. 5 below. Both results show a single linear upward sloping threshold boundary separating the vortex and non-vortex forming flow regimes. This indicates that for a certain engine (or suction) inlet, increasing the distance between the inlet and the solid surface will require a higher $\frac{V_i}{V_0}$ ratio (i.e. a higher engine setting at prevailing ambient condition).

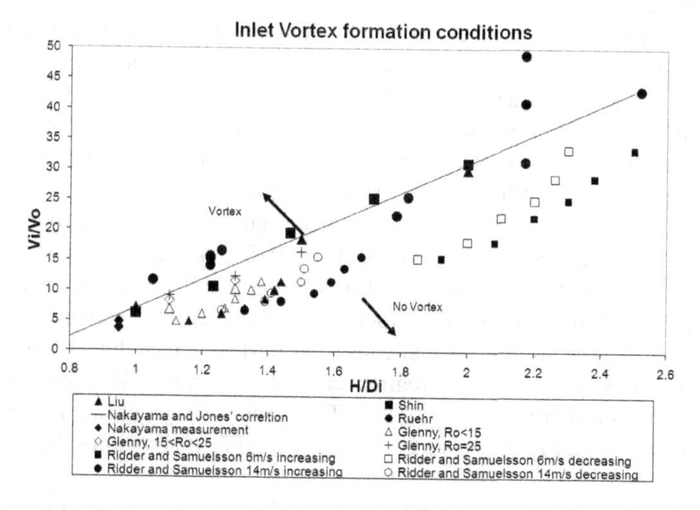

Fig. 4 Experimental data showing the boundary between the vortex forming and non-vortex forming flow regimes

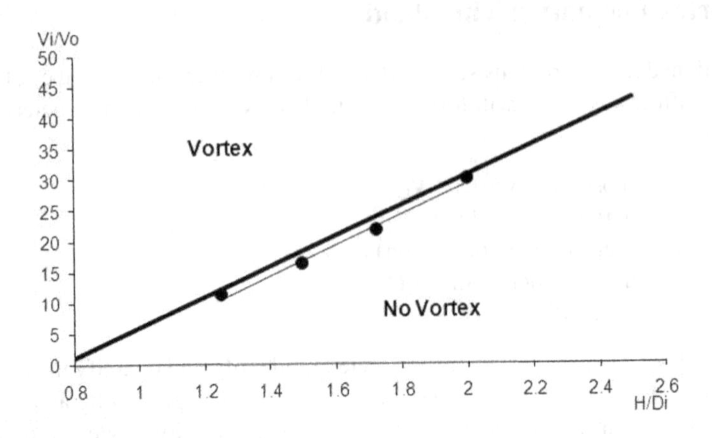

Fig. 5 Computational data showing the boundary between the vortex forming and non-vortex forming flow regimes

Conversely the opposite is also true. The same is also true if we replace the engine (or suction) inlet with a smaller one and vice-versa.

In a test cell, there is a flow of excess air beyond that required by the engine, driven by entrainment by the exhaust plume. This flow passes between the engine and the internal walls of the cell. It is quantified by the cell bypass ratio (CBR):

$$CBR = \frac{\dot{m}_{cell} - \dot{m}_{engine}}{\dot{m}_{engine}} \tag{1}$$

The CBR is distinct from the engine bypass ratio, which is the ratio of the fan to core flow rate of the engine.

The cell bypass ratio can be seen as the equivalent of the blow-away velocity parameter in the runway scenarios. Thus in a test cell, no vortices will form when the CBR_{actual} rises above a certain $\mathrm{CBR}_{blow-away}$.

A common rule of thumb used in test cell design is that a cell must have a bypass ratio of more than 80% to avoid vortex formation. Typically, cells are designed with CBRs up to 100%, and in some cases exceeding 200%.

Ho and Jermy used the same CFD methodologies as the above to investigate and quantify the vortex formation threshold in a JETC-like scenario [13]. Details of these are presented in the section 2.2. It was found that the threshold is no longer **ONE** single linear line but **TWO** lines separating three regions. The region between the lines denote flow conditions that causes an unsteady, unstable vortex to form. The vortex formed at this stage is irregular, elongated in shape and the core is located away from the bottom of the suction inlet. The other two regions are the same as in the runway scenario. Fig. 6 shows the model while Fig. 7 illustrates the three flow regimes.

Both experimental and computational methods have been used to quantify the thresholds. These will be discussed in the following sections.

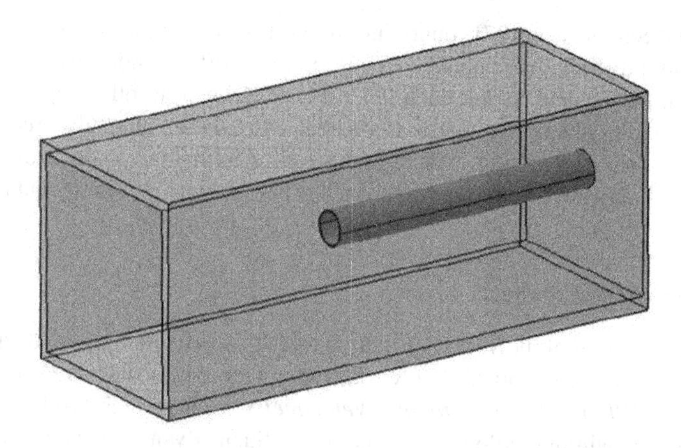

Fig. 6 CFD Flow Model (JETC Scenario)

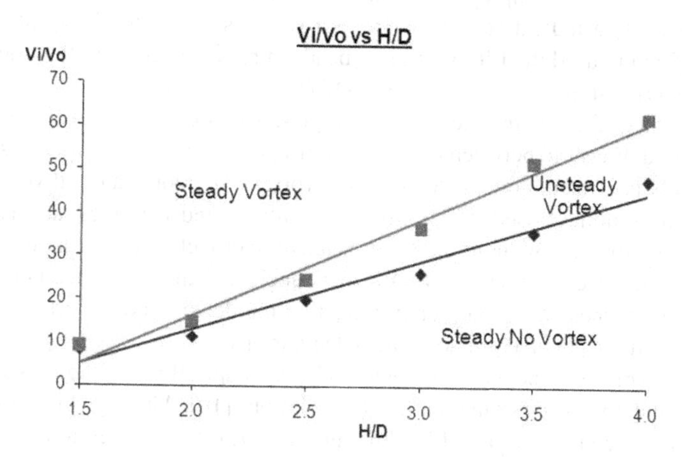

Fig. 7 Computational data showing the boundary between the different vortex regimes (JETC Scenario)

2.1 Experimental Techniques

This section will provide some details on the methodologies, both experimental and computational, employed by the various researchers to investigate and quantify the threshold location.

The main areas of importance when designing an investigation set-up are the following:

1. Vortex detection techniques
2. Vorticity generation
3. Control of ambient conditions

Ridder and Samuelsson [27] found that the vortex system has a "memory" effect. A different threshold exists between a vortex to no vortex condition and vice versa. Other than Ridder and Samuelsson [27], there has been no other reported studies on this. Computational investigations are been carried out but results are being processed at the time of writing. Although this indicates that thresholds from previous experiments may not be entirely accurate, it is not known how significant this effect has.

Vortex detection techniques

Since the detection of the vortex formation threshold does not require details of the vortex core, advanced optical flow visualisation techniques such as *Particle Image Velocimetry (PIV)* and *Laser Doppler Velocimetry (LDV)* which produces velocity measurements within the flow field is not used. Such advanced techniques are only used when vortex characteristics and detailed velocity and pressure measurements are desired. Generally simple flow visualisation techniques such as the use of smoke or techniques that indicate the presence of a vortex such as the lifting of particles are used. Glenny used the lifting of glass beads in his experiments [9] while Ridder and Samuelsson used smoke visualisation [27].

Due to the difference in detection techniques, it is possible to have variations in the measured threshold between the different investigations. Jeong and Hussain [16] conclusively pointed out the inadequacies of vortex detection via intuitive measures (pressure minimum, closed or spiralling streamlines and pathlines, and isovorticity surface). However without advance measurement techniques, it is not possible to calculate the second invariant of ∇u. It is unclear if the difference between the different techniques has a significant impact on the detection of a vortex in lab experiments and the adverse impact in actual situations.

Ho and Jermy used the intuitive criteria (closed or spiralling streamlines and pathlines) for detecting a vortex in their CFD simulations [10]. Although they could have calculate the second invariant of ∇u, the intuitive criteria was used to be consistent with the the other experimental results.

Vorticity Generation

The most common form of vorticity generation is placing the inlet at an angle to the upstream flow simulating crosswind scenario. Glenny [9], Liu et al. [20], Shin et al. [30] and Nakayama and Jones [26] all used this method. The other methods include the use of screens by Motycka et al. [21] and the use of triangular obstacles by Ridder and Samuelsson [27].

Control of Ambient Conditions

Simple experimental studies merely conducted the experiments in ambient environment under normal laboratory conditions such as Glenny [9], and Colehour and Farquhar [7]. Slightly more complex setups by by Liu et al. [20] and Nakayama and Jones [26] conduct the experiments within a wind tunnel. Generally the setups that uses a wind tunnel will include measurements of the vortex characteristics such as vortex strength etc. These will be covered in later sections.

2.2 Computational Techniques

Computational studies are a little more complex as considerations need to be made for the size and boundaries of the ambient air. To date, the only computational studies that aim only to quantify the vortex formation threshold are the ones performed by Jermy and Ho [17]. There are numerous other computational studies on the subject but they seek to calculate more detailed information about the vortex core and its immediate vicinity. Such studies typically require more stringent computational requirements than those performed by Ho and Jermy in order to get accurate velocity and pressure maps of the flow region. Jermy and Ho [17] showed that less stringent computational requirements are sufficient to accurately quantify the threshold. The computations were performed on commercial CFD package Fluent which uses the finite volume method to solve the governing Navier-Stokes equations:

$$\rho \frac{D\mathbf{V}}{Dt} = -\nabla p + \rho g + \mu \nabla^2 \mathbf{V} \tag{2}$$

The flow model used in their simulations are shown in Fig. 8 below. Other information regarding the mesh density, turbulence model and discretisation scheme used is presented in the proceeding sections.

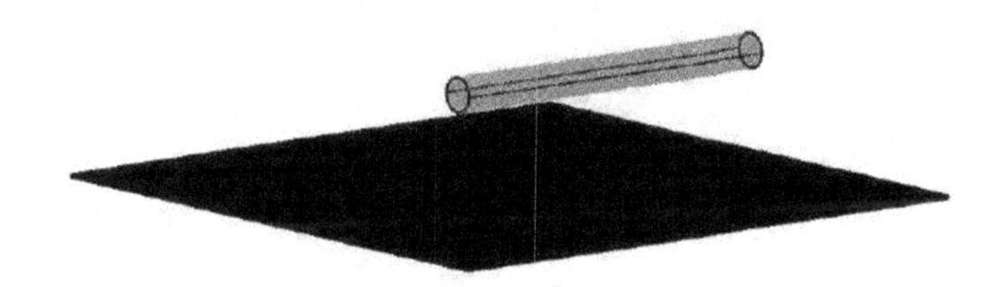

Fig. 8 CFD Flow Model (Runway Scenario) [10]

The final mesh had about 300 000 cells. Increasing the mesh density within the flow region will yield more a more accurate calculation of the pressures within the vortex core but it does not alter the location of the vortex core or the threshold significantly. The same is true for the different turbulence models and discretisation schemes. Figs. 9–11 illustrates this. More details of these convergence exercises can be found in the paper [17]. Thus if a particular mesh density, turbulence model and discretisation scheme is sufficient to show a vortex in a clear-cut scenario, it will be sufficient for the determination of the threshold. These figures seek to illustrate the effects of the differing conditions and not to quantify the threshold. Hence they are generated from simulations in regions where it was certain that a vortex will be formed.

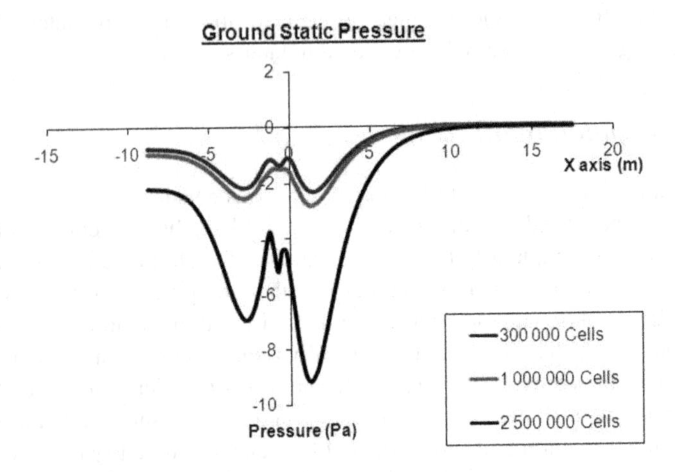

Fig. 9 Vortex core location with increase mesh density (Pressure Plot)

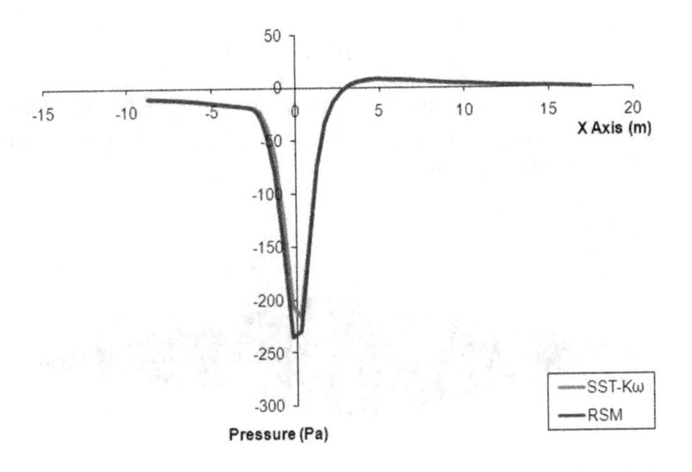

Fig. 10 Vortex core location for different turbulence model (Pressure Plot)

Jermy and Ho [17] also found that a rectangular flow domain around the engine (or suction) inlet with the following criterion is sufficient to model the ambient environment:

- Upstream, 5 x suction inlet diameter
- Downstream, 10 x suction inlet diameter
- Sides, Ĺ8 x suction inlet diameter
- Ceiling, Ĺ7 x suction inlet diameter

Ho and Jermy [17] showed that an incompressible solver is sufficient to accurately determine the vortex formation threshold and a compressible solver is not necessary [10]. This is applicable to both the runway and the JETC scenario.

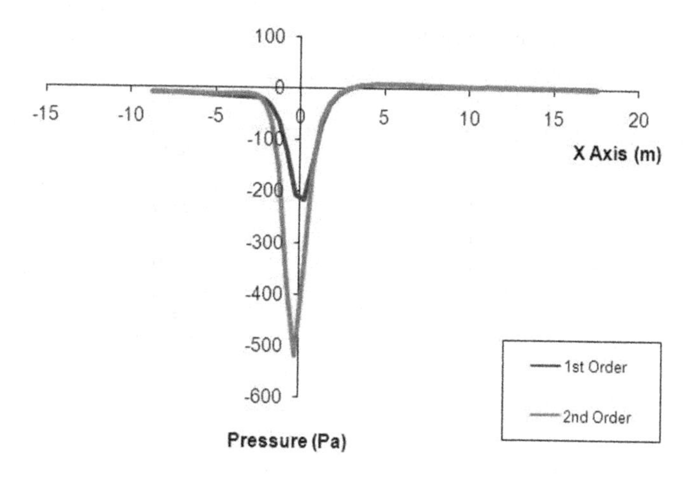

Fig. 11 Vortex core location for different discretisation scheme (Pressure Plot)

3 Factors That Shift the Threshold

A number of geometric and flow parameters affect the location of the threshold increasing or decreasing the range of conditions that permits the formation of a vortex thus making it more or less likely for a vortex to form.

It was found that each of the five principal parameters shown in fig 3 have an effect on the location of the threshold. The relationship of each parameter was found to be consistent between the runway and JETC scenario.

Ambient (or Upstream) Velocity Gradient (or Vorticity)

The type of vortex of concerned in this paper is the type that concentrates ambient vorticity, hence it is intuitive that an increase in the source of vorticity will increase the probability of vortex formation *cetaris paribus*. Experimental studies by Glenny [9] and computational calculations by Ho and Jermy [17, 11, 14, 15] reported this trend and their results are shown in table 1 and Figs. 12–13. Glenny used the dimensionless Rossby (**Ro**) number to indicate vorticity.

$$Ro = \frac{V_i}{W \cdot D} \tag{3}$$

where: V_i=engine inlet velocity; W=velocity gradient;, D=engine inlet diameter. An increase in **Ro** is equivalent to a decrease in velocity gradient.

Reynolds Number

Reynolds number is not a parameter that was investigated intensely in the experimental setups probably because the controlling of the upstream velocity is not a trivial task. In a CFD simulation, it is a significantly easier undertaking. Although

Table 1 Glenny's experimental results

$\frac{V_i}{V_0}$	$\frac{H}{D_i}$	Ro
6.84	1.1	8.0
8.33	1.1	18.0
9.09	1.1	25.0
10.4	1.3	12.0
11.6	1.3	17.0
12.2	1.3	25.0
13.2	1.5	25.0

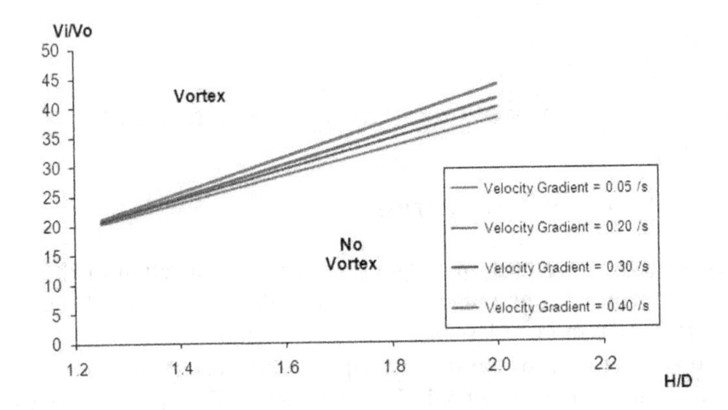

Fig. 12 Effect of different velocity gradient on vortex formation threshold (runway scenario)

Reynolds number is not properly investigated in any one single experiment, Ridder and Samuelsson's measurements were conducted in a significantly lower Re (8000–26 000) compared to the other experiements (>25 000) illustrated in Fig. 4 and lie lower than the rest of the results. The values of Re were calculated using the ambient velocity and not the engine (or suction) inlet velocity.

Computational studies by Ho and Jermy [14, 10] showed similar trends i.e. a decrease in Re lowered the threshold thus increasing the probability of a vortex forming. This results are illustrated in Figs. 14 and 15 below.

Engine (or Suction) Inlet Diameter

Suction inlet diameter is the representation of engine diameter in real test cells. It is a very important investigation parameter as it investigates the usability of existing cells for new engines, which are likely to be larger. It is also of interest to builders and designers of new test cells, as it provides information on how much larger these test cells should be in order to retain the same safety standards. Investigating the effects of a change in suction inlet diameter in experimental setups is not a trivial task due to the difficulty in isolating the effects of changes in Reynolds number

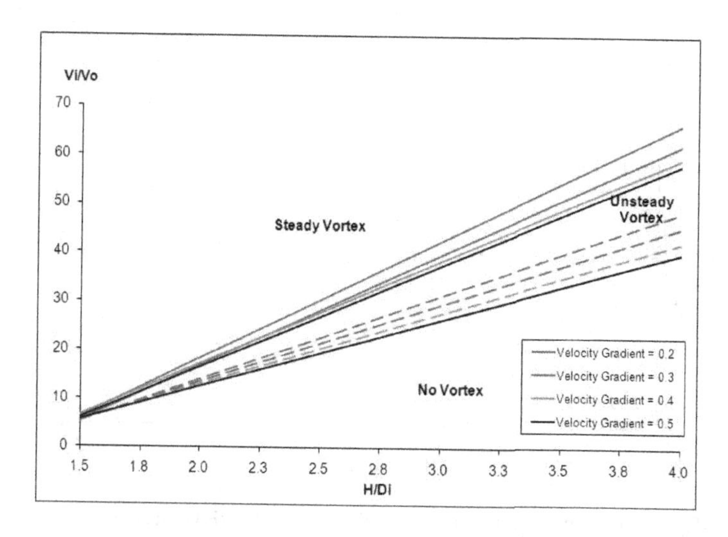

Fig. 13 Effect of different velocity gradient on vortex formation threshold (JETC scenario)

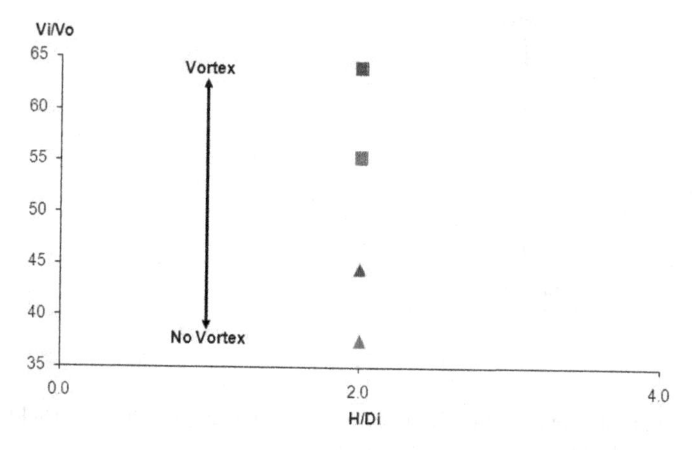

Fig. 14 Effect of different Re on vortex formation threshold (runway scenario) - red indicate lower Re and blue indicate higher Re

and suction inlet size and has not been reported. Similarly, the task is considerably simplified using computational methods.

Ho and Jermy [14, 10] shows that increasing the engine (or suction) inlet diameter *cetaris paribus* lowers the threshold thus increasing the probability of a vortex forming. These results are illustrated in Figs. 16 and 17 below.

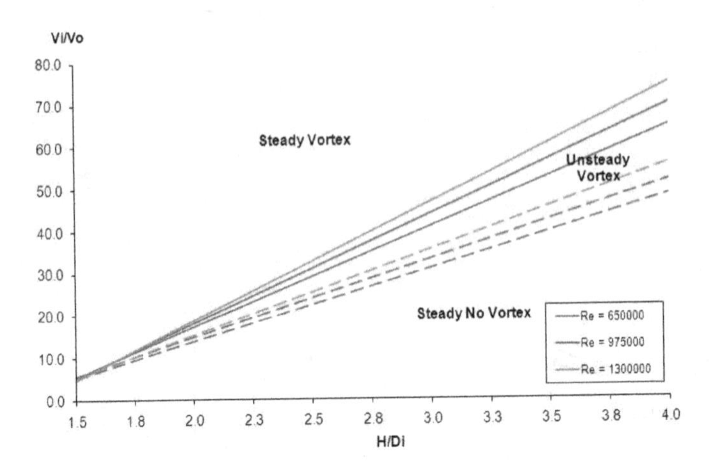

Fig. 15 Effect of different Re on vortex formation threshold (JETC scenario)

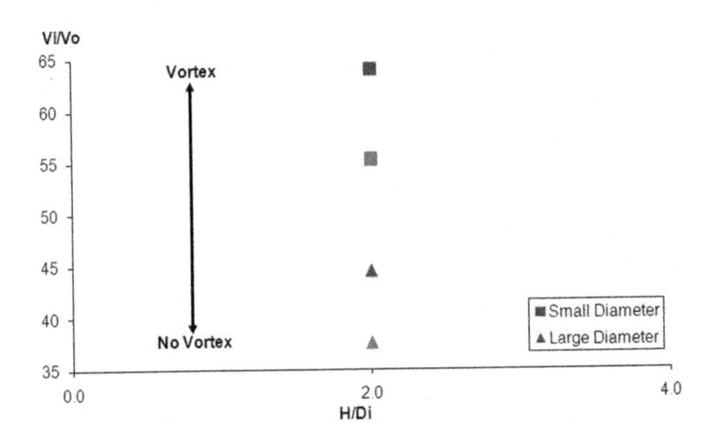

Fig. 16 Effect of different engine (or suction) inlet on vortex formation threshold (runway scenario) - the different colours indicate different Re.

Ground Boundary Layer

An increase in ground boundary layer at the ground plane in a runway scenario lowers the threshold thus increasing the probability of a vortex forming. Fig. 18 illustrates this result.

An increase in the height of the boundary layer in the approaching flow indicates a lower near ground velocity for a particular average velocity. This would increase the probability of a vortex forming because the flow in the vicinity of the vortex core on the solid surface (if it is formed) is lowered. This is the principle behind some of the vortex prevention techniques currently used and is presented in section 5.

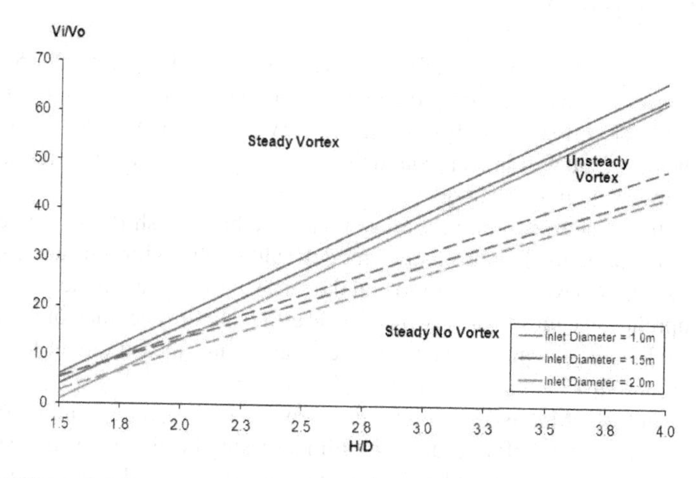

Fig. 17 Effect of different engine (or suction) inlet on vortex formation threshold (JETC scenario)

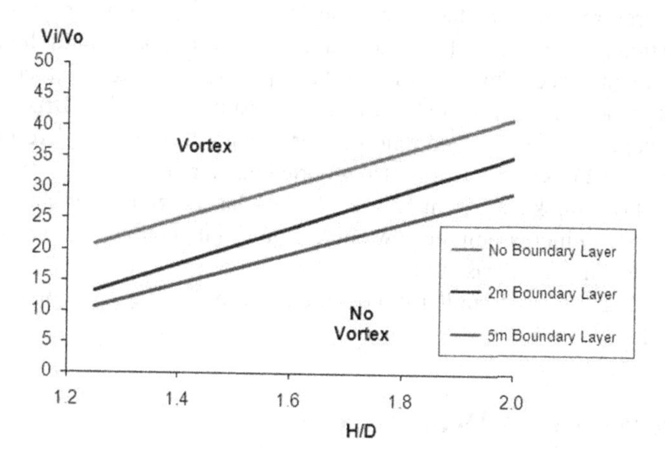

Fig. 18 Effect of different boundary layers on vortex formation threshold (runway scenario)

Turbulence Intensity

Ho and Jermy [12] investigated the effect of increasing the upstream turbulence intensity in a JETC using computational methods and reported that increasing turbulence intensity lowered the threshold thus increasing the probability of a vortex forming. It was reported that the threshold changed by up to 5%.

4 Vortex Strength

Both computational [29, 31, 36, 2, 35, 37] and experimental [30, 29, 23, 25, 24] investigations have been used to investigate the strength when such a vortex is formed. Determination of vortex strength $\Gamma*$ requires very accurate measurement or calculation of the flow field within and around the vortex, thus a much higher requirement than those described in the section 2.

Computational studies require the use of a very high mesh density model and the use of LES or equivalent turbulence models, although Yadlin [36] used k-ε and *SST* models. There have limited reports on the mesh density used in the simulations, thus a comprehensive mesh convergence test is critical to the accuracy of the results. Experimental studies require advanced visualisation and measurement techniques such as *PIV* and *laser doppler techniques (LDV)*.

Murphy and MacManus [23] reported vortex strength increases monotonically up to a local maximum before reducing with increasing headwind. They explained this to be the result of a balance between V_i and V_0. At low wind speeds the capture stream-tube size is large but the boundary layer vorticity levels are low. As V_0 increases vorticity within the ground boundary layer increases resulting in a stronger vortex and therefore higher fan face distortion. However with further increases in V_0 the capture stream-tube size will also decrease at a rate proportional to the inverse of the free-stream speed. This seem to be the explanation between the contradiction in results reported by Shin et al. [30] and the results reported by Brix et al [4]. In the same paper, it was reported that a lower ground clearance ($\frac{H}{D_i}$) results in a stronger vortex (0.25 vs 0.40). It was also reported that a reduction in the approaching boundary layer thickness, from $\frac{\delta*}{D_i}$ = 0.11–0.03, has no notable effect. Lastly it was reported that an increase in the yaw angle (equivalent to an increase in ambient vorticity) resulted in an increase in $\Gamma*$.

Barata et al. [2] reported similar trends in effects for $\frac{V_i}{V_0}$ and $\frac{H}{D_i}$ using computational methods.

5 Vortex Prevention Methods

The best and most obvious method of preventing the formation of such vortices is to design the engine, JETC and other accompanying system in the no vortex region far away from the threshold. However this may require drastic re-design and/or retrofitting of existing system. This may not be possible due to many reasons, cost often being the most important factor. Many solutions have been proposed and/or adopted as stop-gap measures. Most of these solutions work by increasing the local velocity flow field around the vortex core as compared to altering the entire flow field. Two of the most common ones are presented.

Building a ramp at the location of the vortex core in a JETC

Clark et al. [5, 6] reported the use of a ramp at the location of the core of an observed vortex at the V2500 test cell at the Christchurch Engine Centre. The strategic installation of a ramp increased local velocity by reducing the cross-sectional area of the test chamber of the JETC. This increased the local velocity beyond the "blow-away" velocity thus preventing the vortex from forming. After the installation of one such ramp at the said JETC, not vortices have been reported but it does not conclusively proof the effectiveness of the device. To date, there has not been any reported controlled experiment to investigate the effectiveness of the device.

The blow-away jet

This is one of the first inlet-vortex prevention system to be put into commercial service was the blow-away jet used on the DC-8 aircraft in the late 1950s and early 1960s [18]. This system creates a continuous single jet of air using bleed air from the engine. This jet is exhausted near the inlet lip and is directed toward the vortex stagnation line. The introduced air jet impinging on the vortex weaken the vortex which tend to disappear.

The effectiveness of the system depended on accurately directing the air jet at the vortex stagnation line. The location of the vortex stagnation line is influenced by different external environmental factors like the wind velocity and engine setting. One other disadvantage of the system is that the jet also impinged debris on the ground increasing the chances of ***F.O.D.***.

To address these limitations, there have been various other improvements based on similar principles such as the ones by Smith et al. [34], Bigelis et al. [3], Funk et al. [8] and Shmilovich et al. [32].

6 Conclusion

This paper details the various investigations and results for a engine (or suction) inlet-ground vortex system. The threshold conditions at which a vortex forms and dissipates as well as the vortex strength are discussed together with parameters that affect them. These effects are compiled in tables 2 and 3

Table 2 Parameters that affect the vortex formation probability

Parameter	Vortex Formation Probability
Ambient (or upstream) Velocity Gradient	Direct
Reynolds number	Inverse
Engine (or Suction) Inlet Diameter	Direct
Ground Boundary Layer Height	Direct
Turbulence Intensity	Direct

Table 3 Parameters that affect the vortex strength

Parameter	Vortex Strength
Ambient (or upstream) average velocity	Direct followed by inverse
Ground clearance	Inverse
Yaw angle (Equivalent to velocity gradient)	Direct

Acknowledgements. The author wishes to thank Prof. Moses Strydom for his valuable comments on the manuscript.

References

1. Barata, J.M.M., Maneta, A.M., Silva, A.: Numerical Study of a Gas Turbine Engine Ground Vortex. In: 45th AIAA/ASME/SAE/ASEE Joint Propulsion Conference & Exhibit, Denver, Colorado (August 2009)
2. Barata, J.M.M., Manquinho, P., Silva, A.: A Comparison of Different Gas Turbine Engines Ground Vortex Flows. In: 46th AIAA/ASME/SAE/ASEE Joint Propulsion Conference & Exhibit, AIAA, Nashville (July 2010)
3. Bigelis, C., Colehour, J., Davidson, G., Farqhar, B., Helberg, A.: Vortex preventing apparatus for aircraft jet engines. US Patent 3599429 (1969)
4. Brix, S., Neuwerth, G., Jacob, D.: The Inlet-Vortex System of Jet Engines Operating Near the Ground. In: 18th AIAA Applied Aerodynamics Conference, pp. 75–85. AIAA, Denver (2000)
5. Clark, T., P, J.W., Jermy, M., Harris, G., Gilmore, J., Agmen, K.: Suppression of vortex ingestion in the Christchurch Engine Centre V2500 cell Christchurch Engine Centre V2500 test cell (2005)
6. Clark, T.A., Peszko, M.W., Roberts, J.H., Muller, G.L., Nikkanen, J.P.: Gas turbine engine test cell. US Patent 5293775 (1994)
7. Colehour, J.L., Farquhar, B.W.: Inlet vortex. Journal of Aircraft 8(1), 39–43 (1971), http://doi.aiaa.org/10.2514/3.44224, doi:10.2514/3.44224
8. Funk, R., Parekh, D., Smith, D., Dorris, J.: Inlet vortex alleviation. In: AIAA Applied Aerodynamics Conference. AIAA (June 2001)
9. Glenny, D.E.: Ingestion of Debris into Intakes by Vortex Action. Aeronautical Research Council Papers 1114 (1970)
10. Ho, W.H.: Investigation into the Vortex Formation Threshold and Infrasound Generation in a Jet Engine Test Cell. Ph.D. thesis, University of Canterbury (2009)
11. Ho, W.H., Dumbleton, H., Jermy, M.: Effect of Upstream Velocity Gradient on the Formation of Sink Vortices in a Jet Engine Test. In: Proceedings of the International MultiConference of Engineers and Computer Scientists (IMECS), pp. 1767–1772. IAENG, Hong Kong (2008)
12. Ho, W.H., Gilmore, J., Jermy, M.: Reduction of engine exhaust noise in a jet engine test cell. Noise Control Engineering Journal 59(2), 194–201 (2011)
13. Ho, W.H., Jermy, M.: Validated CFD simulations of vortex formation in jet engine test cells. In: 16th Australasian Fluid Mechanics Conference, Gold Coast, pp. 1102–1107 (December 2007)

14. Ho, W.H., Jermy, M.: Formation And Ingestion Of Vortices Into Jet Engines During Operation. In: IAENG Transactions on Engineering Technologies, Special Editions of the International Multi Conference of Engineers and Computer Scientists 2008, pp. 132–143. Springer (2008)
15. Ho, W.H., Jermy, M., Dumbleton, H.: Formation of Sink Vortices in a Jet Engine Test Cell. Engineering Letters 16(3), 406–411 (2008)
16. Jeong, J.H., Hussain, F.: On the identification of a vortex. Journal of Fluid Mechanics 285, 69–94 (1991)
17. Jermy, M., Ho, W.H.: Location of the vortex formation threshold at suction inlets near ground planes by computational fluid dynamics simulation. Proceedings of the Institution of Mechanical Engineers, Part G: Journal of Aerospace Engineering 222(3), 393–402 (2008), http://pig.sagepub.com/lookup/doi/10.1243/09544100JAERO265, doi:10.1243/09544100JAERO265
18. Johns, C.: The aircraft engine inlet vortex problem. In: AIAA's Aircraft Technology, Integration, and Operations (ATIO) Technical Forum. AIAA (October 2002)
19. Klein, H.: Small scale tests on jet engine pebble aspiration tests. Report SM-14885, Douglas Aircraft Company (1953)
20. Liu, W., Greitzer, E.M., Tan, C.S.: Surface Static Pressures in an Inlet Vortex Flow Field. Journal of Engineering for Gas Turbines and Power 107(2), 387 (1985), http://link.aip.org/link/JETPEZ/v107/i2/p387/s1&Agg=doi, doi:10.1115/1.3239738
21. Motycka, D., Walter, W., Muller, G.: An analytical and experimental study of inlet ground vortices. In: 9th Propulsion Conference. AIAA (November 1973)
22. Murphy, J.P.: Intake Ground Vortex Aerodynamics. Ph.D. thesis, Cranfield University (2008)
23. Murphy, J.P., MacManus, D.G.: Ground vortex aerodynamics under crosswind conditions. Experiments in Fluids 50(1), 109–124 (2010), http://www.springerlink.com/index/10.1007/s00348-010-0902-4, doi:10.1007/s00348-010-0902-4
24. Murphy, J.P., MacManus, D.G.: Inlet ground vortex aerodynamics under headwind conditions. Aerospace Science and Technology 15(3), 207–215 (2011), http://linkinghub.elsevier.com/retrieve/pii/S1270963810001604, doi:10.1016/j.ast.2010.12.005
25. Murphy, J.P., MacManus, D.G.: Intake Ground Vortex Prediction Methods. Journal of Aircraft 48(1), 23–33 (2011), http://doi.aiaa.org/10.2514/1.46221, doi:10.2514/1.46221
26. Nakayama, A., Jones, J.R.: Vortex Formation in Inlet Flow Near a Wall. In: 34th Aerospace Sciences Meeting & Exhibit (1996)
27. Ridder, S.O., Samuelsson, I.: An experimental study of strength and existence domain of groun-to-air inlet vortices by ground board static pressure measurements. Tech. rep., The Royal Institute of Technology Stockholm KTH AERO TN62 (1982)
28. Ruehr, W.: Technical Information Series Report R75AEG384, General Electric (1975)
29. Secareanu, A., Moroianu, D.: Experimental and Numerical Study of Ground Vortex Interaction in an Air-Intake. In: 43rd AIAA Aerospace Sciences Meeting and Exhibit, Reno, Nevada (January 2005)
30. Shin, H.W., Greitzer, E.M., Cheng, W.K., Tan, C.S., Shippee, C.L.: Circulation measurements and vortical structure in an inlet-vortex flow field. Journal of Fluid Mechanics 162, 463–487 (1986)

31. Shmilovich, A., Yadlin, Y.: Engine Ground Vortex Control. In: 24th Applied Aerodynamics Conference, Francisco, California (June 2006)
32. Shmilovich, A., Yadlin, Y., Smith, D., Clark, R.: Active system forwide area suppression of engine vortex. US Patent 6763651 (2002)
33. Siervi, F.D., Viguier, H.C., Greitzer, E.M., Tan, C.S.: Mechanisms of inlet-vortex formation. Journal of Fluid Mechanics 124, 173–207 (1982)
34. Smith, J.: Protective air curtain for aircraft engine inlet. US Patent 3527430 (1970)
35. Trapp, L.G., Girardi, R.D.M.: Crosswind Effects on Engine Inlets: The Inlet Vortex. Journal of Aircraft 47(2), 577–590 (2010), http://doi.aiaa.org/10.2514/1.45743, doi:10.2514/1.45743
36. Yadlin, Y., Shmilovich, A.: Simulation of Vortex Flows for Airplanes in Ground Operations. In: 44th AIAA Aerospace Sciences Meeting and Exhibit. AIAA, Reno (2006)
37. Zantopp, S., Macmanus, D.G., Murphy, J.P.: Computational and experimental study of intake ground vortices. The Aeronautical Journal 114(1162), 769–784 (2010)

Author Index